Primate Tourism

A Tool for Conservation?

Primate tourism is a growing phenomenon, with increasing pressure coming from several directions: the private sector, governments, and conservation agencies. At the same time, some primate sites are working to exclude or severely restrict tourism because of problems that have developed as a result. Indeed, tourism has proven costly to primates due to factors such as disease, stress, social disruption, vulnerability to poachers, and interference with rehabilitation and reintroduction.

Bringing together interdisciplinary expertise in wildlife/nature tourism and primatology, experts present and discuss their accumulated experience from individual primate sites that are open to tourists, formal studies of primate-focused tourism, and trends in nature and wildlife tourism. Chapters offer species- and site-specific assessments, weighing conservation benefits against costs, and suggest strategies for the development of informed guidelines for ongoing and future primate tourism ventures.

Primate Tourism has been written for primatologists, conservationists, and other scientists. It is also relevant to tourists and tourism professionals.

Anne E. Russon is a Professor of Psychology at Glendon College, York University, Toronto, Canada, whose research focuses on learning and intelligence in ex-captive Bornean orangutans. She is a member of the Scientific Advisory Board to the Borneo Orangutan Survival Foundation (Indonesia) and several orangutan conservation agencies. She has edited two volumes for Cambridge University Press: with D.R. Begun, *The Evolution of Thought: Evolution of Great Ape Intelligence* (2004) and with K. Bard and S. Parker, *Reaching into Thought: The Minds of the Great Apes* (1996).

Janette Wallis has published extensively on a wide range of issues in primatology, including conservation. Currently, she is the Editor of the *African Conservation Telegraph*, the newsletter of the Society for Conservation Biology – Africa Section, and the Budongo Forest Research Station's newsletter. She is on the board of directors for the Society for Conservation Biology – Africa Section, and Vice President for Conservation for the International Primatological Society. Previously, she has served as the Series Editor of the American Society of Primatologists' book series.

Primate Tourism

A Tool for Conservation?

Edited by

Anne E. Russon
Glendon College, York University, Canada

Janette Wallis
Interdisciplinary Perspectives on the Environment, The University of Oklahoma, USA

CAMBRIDGE
UNIVERSITY PRESS

CAMBRIDGE
UNIVERSITY PRESS

University Printing House, Cambridge CB2 8BS, United Kingdom

Cambridge University Press is part of the University of Cambridge.

It furthers the University's mission by disseminating knowledge in the pursuit of
education, learning and research at the highest international levels of excellence.

www.cambridge.org
Information on this title: www.cambridge.org/9781107018129

© Cambridge University Press 2014

First published 2014

Printed in the United Kingdom by TJ International Ltd. Padstow Cornwall

A catalogue record for this publication is available from the British Library

Library of Congress Cataloguing in Publication data
Primate tourism : a tool for conservation? / edited by Anne E. Russon, Glendon College, York
University; Janette Wallis, Interdisciplinary Perspectives on the Environment, The University of
Oklahoma.
   pages   cm
Includes bibliographical references and index.
ISBN 978-1-107-01812-9 (hardback)
1. Ecotourism–Environmental aspects.   2. Primates.   3. Primates–Habitat–
Conservation.   4. Wildlife conservation.   I. Russon, Anne E., author, editor
of compilation.   II. Wallis, Janette, author, editor of compilation.
G156.5.E26P74 2014
333.95′9816–dc23
2014012757

ISBN 978-1-107-01812-9 Hardback

Contents

List of contributors

Benjamin Andriamihaja
MICET
Manakambahiny
Antananarivo
Madagascar

Carol M. Berman
Department of Anthropology,
North Campus, SUNY at Buffalo,
Buffalo, NY,
USA

Stella de la Torre
Colegio de Ciencias Biológicas y Ambientales,
Universidad San Francisco de Quito,
Cumbayá, Quito,
Ecuador

David F. Dellatore
Sumatran Orangutan Society,
Orangutan Information Centre,
Medan, Sumatera Utara,
Indonesia

James S. Desmond
EcoHealth and the Jane Goodall Institute,
Golden, CO,
USA

Jennifer A. Z. Desmond
EcoHealth and the Jane Goodall Institute,
Golden, CO,
USA

Ivona Foitovà
Department of Botany and Zoology,
Masaryk University,
Brno,
Czech Republic

Michele L. Goldsmith
School of Arts and Sciences,
Southern New Hampshire University,
Manchester, NH,
USA

Jenna Guerriero
Department of Anthropology and Institute for the Conservation of Tropical Environments,
Stony Brook University,
Stony Brook, NY,
USA

Chloe Hodgkinson
Fauna and Flora International,
Cambridge,
UK

Josephine Hubbard
Department of Anthropology and Institute for the Conservation of Tropical Environments,
Stony Brook University,
Stony Brook, NY,
USA

Glen T. Hvenegaard
Department of Geography,
University of Alberta,
Augustana Campus,
Camrose, Alberta,
Canada

Consuel S. Ionica
"Francisc I Rainer" Anthropology Institute,
The Romanian Academy,
Bucharest,
Romania

Laurie Kauffman
Department of Biology,
Oklahoma City University,
Oklahoma City, OK,
USA

Stephen J. King
Department of Anthropology,
University of Massachusetts,
Amherst, MA,
USA

Christopher Kirkby
TReeS, Tambopata, Peru,
Tambopata Reserve Society ñ Research and Monitoring Studies Unit
(TReeS-RAMOS),
Puerto Maldonado, Madre de Dios,
Peru

Hiroyuki Kurita
Division of Cultural Property,
Board of Education, Oita City,
Niage-machi, Oita
Japan

Heather C. Leasor
Australian National University,
Deakin, ACT,
Australia

Jin-Hua Li
School of Life Sciences,
Anhui University,
Hefei, Anhui,
China

Elizabeth J. Macfie
IUCN SSC Primate Specialist Group,
Bankend,
Abbey St Bathans,
Duns, Scottish Borders,
Scotland
UK

Oliver J. Macgregor
Australian National University,
Deakin, ACT,
Australia

Deborah L. Manzolillo Nightingale
Sarit Centre,
Nairobi,
Kenya

Megan D. Matheson
Psychology and Primatology,
Central Washington University,
Ellensburg, WA,
USA

Eleanor J. Milner-Gulland
Imperial College of Science and Technology,
Silwood Park Campus,
Ascot,
Berkshire,
UK

Michael P. Muehlenbein
Evolutionary Physiology and Ecology Laboratory,
Anthropology Department,
Indiana University,
Bloomington, IN,
USA

Hideshi Ogawa
School of International Liberal Studies,
Chukyo University,
Kaizu-cho,
Toyota, Aichi,
Japan

Anne E. Russon
Psychology Department,
Glendon College,
York University,
Toronto, ON,
Canada

Robert M. Sapolsky
Departments of Biological Sciences, Neurology and Neurological Sciences,
Neurosurgery,
Stanford University School of Medicine,
Stanford, CA,
USA

Shirley C. Strum
Anthropology Department,
University California San Diego,
La Jolla, CA,
USA
and
Uaso Ngiro Baboon Project,
Nairobi,
Kenya

Adi Susilo
Nature Conservation and Rehabilitation Center,
Forestry Research and Development Agency,
Bogor,
Indonesia

Corri D. Waitt
Department of Zoology,
Oxford University,
Oxford,
UK

Janette Wallis
Interdisciplinary Perspectives on the Environment (IPE),
The University of Oklahoma,
Norman, OK,
USA

Elizabeth A. Williamson
IUCN SSC Primate Specialist Group,
Psychology, School of Natural Sciences,
University of Stirling,
Stirling, Scotland,
UK

Patricia C. Wright
Department of Anthropology and Institute for the Conservation of Tropical Environments,
Stony Brook University,
Stony Brook, NY,
USA

Part I

Introduction

1 Reconsidering primate tourism as a conservation tool: an introduction to the issues

Anne E. Russon and Janette Wallis

Introduction

This book aims to assess the conservation effects of nature tourism. In particular, our focus is on tourism to visit nonhuman primates and their habitats. Although humans are also primates, for convenience, we refer to nonhuman primates as "primates" and nature tourism to visit them as "primate tourism."

Using nature tourism as a conservation tool is not new. It has been advocated since the 1800s, from the view that nature tourism is an impact-free activity that will lead visitors to value nature and help fund its protection (Honey, 1999/2008). Hopes have been especially high for nature tourism that aspires to "ecotourism," broadly referring to responsible travel to natural areas that conserves the environment and improves the well-being of local people (e.g. Ceballos-Lascurain, 2000; Honey, 1999/2008). It is now well known that even the most ecologically responsible nature tourism is not always the impact-free activity envisioned and it has often failed to deliver on its conservation promises, especially to the conservation of the wildlife and natural areas visited (Higham, 2007). Evidence now shows that adverse effects of nature tourism are widespread (Butynski, 2001; Higham 2007; Knight & Cole, 1995). One result is an increase in calls for evaluative research on whether nature tourism is generating conservation benefits for the wildlife and natural areas visited, what conservation costs it incurs, and whether its benefits to their conservation outweigh its costs (e.g. Higham, 2007).

In this context, primate tourism merits attention. First, it has proven to be very popular and lucrative, largely because of primates' biological and behavioral similarities to humans (Honey, 1999/2008; Kinnaird & O'Brien, 1996; Lanyero, 2011; Wollenberg et al., 2011). Now a major form of the human–primate interface, it is considered one of the most important issues facing primatology in the twenty-first century (Fuentes et al., 2007; Paterson & Wallis, 2005). Second, the same similarities that generate high tourist interest often generate threats to the primates visited, notably in the form of habitat competition or disease transmission. Of the primate species surviving today, 55% are at risk of extinction (IUCN, 2013), so primate conservation is increasingly of great concern. Third, primate tourism takes many forms,

Primate Tourism: A Tool for Conservation?, ed. Anne E. Russon and Janette Wallis. Published by Cambridge University Press. © Cambridge University Press 2014.

ranging from safari-like tracking adventures into remote areas to drop-in day visits to see primates living in their natural habitat but "ready-to-view" at temples, monkey parks, or rehabilitation sites. An overview of primate tourism, then, offers insights into tourism effects that are shared across primates as well as those specific to particular primate taxa, sites, or forms of tourism, and into the causal factors involved.

As with nature tourism in general, empirical evaluations of primate tourism's effects on primate conservation were relatively limited until recently. Our goal in this book is to work to improve this situation by (1) presenting empirical assessments of the impacts of primate tourism on primate conservation on a sample of the world's primates, (2) weighing, to the extent possible, its conservation costs against its conservation benefits, and (3) developing recommendations for improving the net benefits of primate tourism to primate conservation. This chapter sets the stage with overviews of primate traits important in tourism, primate tourism's history, broad issues in assessing tourism's impacts, and this book's contents.

Why primates

Although primate tourism may share many patterns found in other nature tourism, it may also have distinctive features. Primates are mammals, and distinguished within the mammals by a collection of ecological, biological, and behavioral traits that include their habitat (primarily tropical to subtropical forest and savanna), diet, anatomy, life histories (relatively long lifespans, slow reproduction, slow ontogeny), sociality, large brains, high potential for independent and social learning, and high behavioral plasticity. These traits contribute to primate tourism's popularity with tourists and to some of its typical consequences.

- Primates' tropical–subtropical concentration leads to high sympatry and competition with humans for habitat (Fuentes, 2006).
- Physiological and dietary similarities result in a high risk of humans' infecting primates with human diseases. Some human diseases have devastated entire populations of great apes that are threatened with extinction (Huijbregts *et al.,* 2003; Leroy *et al.,* 2004).
- Primates' high learning potential, long lifespans, slow ontogeny, and social lifestyles combine to produce powerful learning capacities. Primates can learn how to interact with humans through tourists and refine their knowledge and skills over many years. One important result can be intensifying their involvement with humans, which has increased their crop- and garbage-raiding, aggressiveness, and vulnerability to poachers (Banks *et al.,* 2003; Fuentes, 2006; Kemnitz *et al.,* 2002; Knight, 2009a; Unwin & Smith, 2010). A second is that tourism effects can spread beyond the individual primates that learned them to conspecifics, via social learning. A third is that tourism's effects on the primates visited can change substantially over time. Problems that develop slowly have sometimes remained undetected – and unmanaged – for years. Examples are baboons developing

adult-onset diabetes due to prolonged garbage eating and some primate species becoming dangerously aggressive and violent toward humans due to long-term provisioning for tourism (Banks *et al.*, 2003; Kemnitz *et al.*, 2002; Knight, 2009a; Zhao & Deng, 1992).
• Most remaining primate habitat is in developing countries (Fuentes, 2006), which influences accessibility to tourists, infrastructure, local residents' interest in participating in primate tourism and opportunities to do so, governmental perspectives, and instabilities that affect tourism (economic, socio-political, environmental).

Primates also vary greatly in size, from tiny mouse lemurs to massive gorillas, and in lifestyle, from semi-solitary to highly social. In diet, most are primarily plant eaters but some are insectivores or generalists. They vary in mating and reproduction patterns, intelligence, sensitivities, and aggressiveness. Most inhabit the tropics or subtropics, but a few inhabit temperate zones. Primates are distributed around the globe, so they are subject to a wide variety of human cultural, political, and economic contexts. All of these factors combine to further diversify primate tourism's characteristics. Primate tourism itself varies in form, from small group primate tracking that achieves some of the criteria for ecotourism to mass tourism based on staged primate viewing at scheduled times and places. Additional primate viewing occurs in "safari" drive-through experiences in Africa, where a variety of large terrestrial mammals (including some primate species) provide a less focused but more inclusive tourism experience. Accordingly, the effects of primate tourism may vary considerably across primate species and sites.

An overview of primate tourism stands to be useful in identifying effects that are common to the order as well as those that are species- and site-specific. The risk of infecting primates with human disease is common, but its severity can vary according to the primate species, type of disease, or site location. Likewise, the effects of provisioning can vary with the primate species involved and site-specific management. Identifying shared and distinctive effects should facilitate identification of causal factors and problem mitigation.

The development of primate tourism

Primate tourism has at least two historical roots: safari-like adventures and primate provisioning linked with long-term human–primate sympatry (Butynski & Kalina, 1998; Fuentes, 2006, 2010). Safari-like treks to find and track wildlife targeted spectacular primates such as the great apes, often to shoot them as trophies or capture them for western zoos. Primate safaris shifted to viewing as visitor interest shifted to protecting wildlife. Viewing provisioned primates developed out of the practice of local peoples' regularly feeding sympatric primates, either for religious or cultural reasons (e.g. protected monkeys near Buddhist or Hindu temples: Fuentes *et al.*, 2005, 2007; Zhao, 2005) or to deter crop-raiding (e.g. Barbary and Japanese macaques: Fuentes *et al.*, 2007; Unwin & Smith, 2010; Kurita, this volume).

Provisioning for religious reasons greatly facilitated primate tourism; macaque tourist sites throughout Asia, for example, are often near Hindu and Buddhist temples (Fuentes *et al.*, 2007).

Modern primate tourism emerged in the 1950s as interest rose in viewing wild primates. It developed around the two viewing formats already used in wildlife tourism: provisioning primates to draw them to tourist areas and habituating them to human presence within their normal ranges to facilitate tracking them. Provisioning fosters habituation but also concentrates primates at specific places and times, so tourists can view them easily, up close, and on schedule (Knight, 2009a). Early examples are viewing provisioned Japanese macaques at Japan's Takasakiyama monkey park (1953: Yamagiwa, 2010), tracking mountain gorillas in Uganda's Mgahinga Game Reserve (1955: Butynski & Kalina, 1998), and viewing provisioned Barbary macaques on Gibraltar (1960: Fuentes *et al.*, 2007).

Primate tourism quickly proved popular. Primate conservation was not prominent among its early aims in either form. Income appeared to be the main motive in early tourism to track eastern lowland gorillas, based on the large tourist groups and poor organization reported (Fawcett *et al.*, 2004; Weber, 1993). Sightseeing, educating Japanese people about Japanese macaques, offering visitors opportunities to "play" with the monkeys by feeding them, and reducing crop-raiding were among the initial aims of Japan's monkey parks in offering macaque viewing at scheduled feedings (Knight, 2005; Kurita, this volume).

Provisioning- and habituation-based primate tourism tend to involve different species and audiences, partly because of their roots. Provisioning-based tourism focuses on primates that fare relatively well in human contexts, notably macaques, and makes viewing wild primates as easy as viewing captives, so it is likely to attract visitors who may be more focused on the social, recreational, and entertainment aspects of their visit than the educational and conservation ones (Parker & Ballantyne, 2012) and to resemble mass more than ecotourism. Habituation-based tourism tends to involve primates living in large expanses of natural habitat relatively remote from humans (e.g. apes, prosimians, proboscis monkeys) and trekking into primates' natural habitat, so it is more akin to ecotourism. Some primate-viewing opportunities also arise in broader habituation-based drive-through "safaris" to see large African mammals (e.g. game drives through the Serengeti National Park). Although primates are rarely the focus of such tourism, most of these locations include primate species that add to the overall experience. A large troop of baboons never fails to attract the attention of tourist vehicles.

By the 1970s, primate tourism was increasing in popularity. For instance, by 1972, 40 free-ranging Japanese monkey parks had opened in Japan after the first opened in the 1950s (Kurita *et al.*, this volume). With the growth of modern primatology and awareness that populations of some primate species were shrinking rapidly, primate conservationists and researchers began promoting and developing primate tourism as a strategy for securing support and funds for their protection. Most famous is regulated mountain gorilla tracking, which the Mountain Gorilla Project (now the International Gorilla Conservation Program, IGCP) launched in 1979 as a conservation measure to help save mountain gorillas from extinction (Vedder & Weber, 1990). Provisioning-based tourism to visit rehabilitant orangutans returning to forest life was launched around the same time to support wild orangutan conservation by educating visitors and generating conservation funds; it focused on rehabilitants to protect vulnerable wild populations from the added stresses of tourism (Aveling & Mitchell, 1982; Borner, 1976; Frey, 1978).

Some primate tourism sites experienced problems and took steps to alleviate them. Provisioning and habituation both contributed to these problems by bringing primates within nuisance range as well as within viewing range (Knight, 2009a). Problems for the primates visited and their habitat were prominent where tourism was based on provisioning and tourist volume was high (e.g. Japanese macaques, rehabilitant orangutans). At one Japanese macaque tourism site, there was a serious risk of tourists spreading human diseases to the macaques because they fed the macaques and came too close to them. These tourist behaviors also caused dangerous human-directed theft and aggression, and overprovisioning; the overprovisioning caused macaque overpopulation which increased social competition, crop-raiding, and natural habitat damage (Kurita *et al.*, 2008). Tourism with rehabilitant orangutans generated similar problems and further undermined the rehabilitation process by encouraging rehabilitants' dependency on humans (Frey, 1978; MacKinnon, 1977; Rijksen, 1978, 1982; Rijksen & Rijksen-Graatsma, 1975). In both cases, tourist management changes were recommended; some were instituted.

The 1980s and 1990s saw a dramatic growth in primate tourism (e.g. Fuentes *et al.*, 2007; Hartup, 1994; Nakamura & Nishida, 2009). Tourist numbers increased at existing primate sites, new primate sites were opened for tourism, and primate-viewing options expanded to include boats and blinds (e.g. Fuentes *et al.*, 2007; Klailova, Hodgkinson, & Lee, 2010; Knight, 2009a; Macfie & Williamson, 2010; Mugisha, 2008; Nishida & Nakamura, 2008). Its growth followed worldwide trends in tourism (Higham, 2007), but was also inspired by the conservation success of mountain gorilla tourism. Rwanda's gorilla tourism was credited with saving mountain gorillas from the brink of extinction in the 1980s by increasing their importance, protecting them and their national park, and generating substantial revenues for the country and local people (Harcourt & Stewart, 2007; Macfie & Williamson, 2010).

Conservation became more common as a rationale for primate tourism. Researchers continued to be instrumental in promoting and developing conservation-oriented tourism with wild primates (e.g. Nishida & Nakamura, 2008; Russell, 1995; Wright & Andriamihaja, 2002, 2003). Governments of several primate-habitat countries developed tourism for the income it generates and promoted their primates as tourist attractions, partly to fund their country's nature conservation. Rwanda made mountain gorillas a national symbol and featured them on its passports, visas, and bank notes (Williamson, 2001). Uganda started mountain gorilla and chimpanzee tourism (Lloyd, 2002; Moyini, 2000). Madagascar launched its national park system, with nature tourism as a substantial source of financial support and lemurs as its main attraction (Wollenberg *et al.*, 2011; Wright & Andriamihaja, 2002, 2003). Malaysia used orangutans to head its 1990 tourism campaign and was developing and promoting orangutan tourism facilities by the late 1990s (Bennett, 1998; Kaplan & Rogers, 2000). The importance of incorporating local communities into nature conservation also grew, and primate tourism offered a means of alleviating local poverty and providing financial and other incentives for protecting nearby natural areas (Archabald & Naughton-Treves, 2001; Hodgkinson, 2009; Horwich & Lyon, 1987).

Concerns about primate tourism's adverse effects also grew. The rise in concerns probably reflected increased tourism pressures that intensified the human–primate interface, pressures to prioritize economic benefits, the gradual buildup of slow-developing problems, and difficulties in instituting effective tourist controls (see Figure 1.1). Problems recognized include infecting the primates visited with human diseases, aggravating versus alleviating primate–human conflict (crop-raiding, attacks), artificially distorting the primates' reproduction and social dynamics, altering other facets of their behavior, and undermining rehabilitation. Mountain gorilla tourism came under criticism as a conservation tool, notably because of tourists' stressing the gorillas they visited and potentially infecting them with human diseases (Butynski & Kalina, 1998; Homsy, 1999; Palacios *et al.*, 2011). Malaysian resorts marketed tourism to visit captive orangutans they labeled "rehabilitants," apparently capitalizing on the caché of conservation for its income-generating potential (Brend, 2001). Lack of or ineffective tourist controls and tourist education were considered largely responsible for the persistence or intensification of the problems with rehabilitant orangutan tourism identified in the 1970s (Rijksen & Meijaard, 1999; Sajuthi *et al.*, 1991). Such concerns sparked empirical studies of the effects of primate tourism on the primates visited and their habitat (e.g. Cochrane, 1998; de la Torre *et al.*, 1999, 2000; Lloyd, 2002; Russell, 1995; Zhao & Deng, 1992).

Growing concerns over the potential for primate tourism to have adverse effects also spurred efforts to mitigate them. For example, an intensive mountain gorilla vaccination program was undertaken after the gorillas suffered an outbreak of measles to which gorilla tourism could have contributed (Hastings *et al.*, 1991). Persistent problems with rehabilitant orangutan tourism motivated the Indonesian government to order its orangutan rehabilitation projects to stop tourist operations (Rijksen & Meijaard, 1999). To reduce its Japanese macaques' soaring population size and attacks on humans, Japan's Takasakiyama monkey park reduced its official provisions (1965) and prohibited visitor feeding (1993) (Kurita *et al.*, 2008). Some of these changes were effective, others were not. Better managed provisioning eventually reduced the macaque population size and attacks on humans (Kurita, this volume; Kurita *et al.*, 2008). The move to stop rehabilitant orangutan tourism in Indonesia failed: all sites that have been offering rehabilitant orangutan tourism since the 1970s continue to do so.

Whether primate tourism benefited primate conservation eventually came into question. Some mountain gorilla experts concluded that the conservation benefits of mountain gorilla tourism outweighed its costs (Harcourt, 2001; Williamson, 2001)

Figure 1.1a   Unsafe tourist viewing practices. A baboon attempts to open a car door along a road
in Cape Town, South Africa, as tourists look on. Baboons ranging in this area are
over-habituated and readily raid any tourist vehicles that stop for closer viewing. In the
following Figure 1.1b, a tourist sits within 5 m of a chimpanzee at Gombe National
Park, Tanzania. In both situations, humans are not maintaining a safe distance from the
primates. (© J. Wallis.)

Figure 1.1b

but others were uncertain (Butynski & Kalina, 1998). Some orangutan experts likewise considered that rehabilitant orangutan tourism could operate to benefit or at least not undermine orangutan conservation (Aveling, 1982; Payne & Andau, 1989) while others considered that its benefits did not offset its costs to orangutans' health and rehabilitation (Rijksen & Meijaard, 1999).

Primate tourism has continued to grow in the twenty-first century. Several sites have reported substantial increases in tourist numbers, on the order of 6–20% per annum (e.g. Fuentes *et al.*, 2007; UBOS, 2011; Wright *et al.*, this volume; Wyman & Stein, 2010). Tourists can now visit three of the four gorilla subspecies (Macfie & Williamson, 2010), at least 11 macaque species (Fuentes *et al.*, 2007; Kinnaird & O'Brien, 1996; Serio-Silva, 2006), and orangutans at 13 "rehabilitant" and 11 wild sites (Russon & Susilo, this volume). Among the exceptions are Japanese monkey parks: many are now closed or fenced in because they have fallen out of fashion and their practices caused serious problems for the monkeys, natural habitat, visitors, and local people (Knight, 2009b). Continuing growth increases the range and intensity of taxonomic, ecological, socio-political, and economic factors at play in generating tourism's effects on the primates and sites visited.

Primate tourism sites experiencing this growth undoubtedly face greater challenges. Deterioration of good tourism practices is one result: notable are exceeding the maximum allowed number and size of tourist groups visiting primates, approaching closer than the minimum tourist–primate viewing distance, feeding or bothering the primates, and primates being visited by tourists who are ill (Berman *et al.*, 2007; de la Torre *et al.*, 2000; Fuentes, 2010; Fuentes *et al.*, 2007; Goldsmith, 2000; Muehlenbein *et al.*, 2010; Nakamura & Nishida, 2009; Nishida & Nakamura, 2008; Ruesto *et al.*, 2010; Sandbrook & Semple, 2006; Unwin & Smith, 2010; Zhao, 2005). Among the most serious concerns are tourist practices that jeopardize the primates' health. Close tourist–primate proximity is of great concern because it increases stress, opportunities for primate aggression and theft, and risks of disease transmission; it is hard to avoid, however, given its importance to tourist satisfaction (Knight, 2009a; Tapper, 2006). High tourist numbers can undermine the primates' health by increasing stress (Berman *et al.*, 2007; Butynski & Kalina, 1998; Guerriero *et al.*, 2009; Hubbard, 2011; Ruesto, *et al.*, 2010), increasing the risk of human disease transmission (Goldsmith, 2000; Muehlenbein *et al.*, 2010; Woodford, Butynski, & Karesh, 2002), and altering reproductive performance and mortality (Morelli *et al.*, 2009; Treves & Brandon, 2005; Unwin & Smith, 2010).

At the same time, the twenty-first century has also seen achievements in primate tourism favorable to primate conservation, notably recommendations for alleviating more of the primate tourism problems documented and corresponding improvements in practices. IUCN recent best practice guidelines for great ape tourism qualify as a major achievement, given their comprehensive nature and the great apes' vulnerability to extinction (Macfie & Williamson, 2010). These and other conservation-oriented recommendations are beginning to be implemented. To reduce direct disease transmission risks, some sites have raised the minimum tourist distance for chimpanzee and gorilla viewing to 10 m from 5 m and 7 m, respectively, and at least one site requires tourists to wear facial masks while visiting great apes (Macfie & Williamson, 2010; Nishida & Nakamura, 2008; Pusey *et al.*, 2008). To reduce the behavioral and health problems caused by provisioning for tourism, more sites have stopped official primate provisioning or improved its management, now discourage or prohibit tourists and guides feeding primates, and/ or have improved their enforcement of primate feeding regulations (e.g. Fuentes *et al.*, 2007; Kurita *et al.*, 2008; Pusey *et al.*, 2008; Ruesto *et al.*, 2010; Wallis, pers. obs.). Other important achievements include a growing body of empirical studies on primate tourism's effects, including more of the long-term studies needed to identify effects that are slow to develop (e.g. Knight, 2009b; Kurita *et al.*, 2008; Muehlenbein *et al.*, 2012). While these studies tend to show a mix of conservation benefits and costs similar to that reported in the 1980s and 1990s, they are building a much stronger understanding of the positive and the adverse consequences of primate tourism for primate conservation, the causal factors involved, and the effectiveness of the methods used.

It is clear that primate tourism can generate important benefits for primate conservation but equally clear that it can also generate serious costs. Like other forms of nature tourism, primate tourism may offer net benefits for primate conservation insofar as its risks and costs can be prevented, reversed, or controlled (Maekawa *et al.*, 2013). This can be complex and costly, with success depending largely on how tourists are managed (Macfie & Williamson, 2010; Maekawa *et al.*, 2013). Management complexities include risks generated by primate tourism that may be unavoidable, notably primates' susceptibility to human disease and to intensifying their involvement with humans, and variability in the effects of some tourist practices as a function of the primate species visited, location, and socio-political context. A further consideration is that many of the conservation benefits cited accrue at the level of the primate species or parent population, the national park or park system, or national and local economies. The conservation costs identified are often incurred at the local level, that is they affect the individual primates, habitats, and local communities visited. An important question is whether using primate tourism as a conservation measure effectively "robs Peter to pay Paul" and, if so, how to weigh the costs and benefits involved.

Mountain gorilla tourism illustrates the dichotomies that can be involved and the difficult situation facing conservationists. On the one hand, mountain gorilla tourism has been pivotal in saving this species from extinction: by 2010 the total mountain gorilla population had grown to 864 from its estimated lowest in 1978, around 252–285, and the income that mountain gorilla tourism generates helps fund protection for the gorillas and their habitat, other national parks (e.g. Uganda), and national and local economies (Gray *et al.*, 2011; Weber & Vedder, 1983). On the other hand, one epidemic of a lethal human disease introduced to a gorilla tourist group could potentially eradicate the entire population.

This book

With the aim of working toward balanced views of primate tourism's conservation costs and benefits, this book offers recent assessments of the effects of primate tourism on primate conservation across a sample of the world's primates. The book project developed out of a symposium on this topic at the 2004 Congress of the International Primatological Society, Torino, Italy and grew to include additional chapters on more species.

We focused on tourism with primates living free in native habitat, that is as wildlife or nature tourism, because tourists typically construe visiting primates as a nature experience (Fuentes *et al.*, 2007; Knight, 2009a) and this is the form commonly promoted as a conservation tool. We therefore included tourism to view free-ranging rehabilitant orangutans and Japanese macaques in Japan's monkey parks, but did not include tourism to view primates that do not range freely in native habitat (e.g. theme or non-native "safari parks", zoos, sanctuaries) even though it can support primate conservation. We did not focus on ecotourism because primate tourism does not necessarily operate as such or aim to do so.

Our focus is the realities of primate tourism as a conservation measure, so we invited primate specialists to present empirical assessments of the effects of primate tourism on the primates they study. We sought contributions on primate species from all the major primate families (prosimians, Old and New World monkeys, apes), most major primate regions (South and Central America, Africa, Asia), and many primate-habitat countries. We asked our contributors to consider the effects of primate tourism on the conservation of the primates visited and their habitats as well as on primate conservation more broadly.

Most of our chapters represent single-site case studies. We were very fortunate to include several that offer the long-term evidence needed to assess tourism effects that are slow to develop, and several that compare tourism's effects across several sympatric primate species. We also include three chapters that review major broad issues: economics, disease, and tourism guidelines. The final chapter aims to integrate our contributors' material, as a step toward more balanced views of the role of primate tourism in primate conservation.

References

Archabald, K. and Naughton-Treves, L. (2001). Tourism revenue-sharing around national parks in Western Uganda: early efforts to identify and reward local communities. *Environmental Conservation*, 28 (2): 135–149.

Aveling, R. J. (1982). Orang utan conservation in Sumatra, by habitat protection and conservation education. In: L. E. M. de Boer (ed.), *The Orang Utan: Its Biology and Conservation*, The Hague: Dr. W. Junk, pp. 299–315.

Aveling, R. J. and Mitchell, A. (1982). Is rehabilitating orangutans worthwhile? *Oryx*, 16: 263–271.

Banks, W., Altmann, J., Sapolsky, R., *et al.* (2003). Serum leptin levels as a marker for a Syndrome X-like condition in wild baboons. *Journal of Clinical Endocrinology and Metabolism*, 88: 1234–1240.

Bennett, E. L. (1998). *The Natural History of Orang-Utan*. Kota Kinabalu, Sabah: Natural History Publications (Borneo).

Berman, C. M., Li, J. H., Ogawa, H., Ionica, C., and Yin, H. (2007). Primate tourism, range restriction and infant risk among *Macaca thibetana* at Mt. Huangshan, China. *International Journal of Primatology*, 28: 1123–1141.

Borner, M. (1976). Sumatra's orang-utans. *Oryx*, 13 (3): 290–293.

Brend, S. (2001). It shouldn't happen to an ape. *BBC Wildlife*, June. www.responsibletravel.com (accessed July 27, 2005).

Butynski, T. M. (2001). Africa's great apes. In: B. Beck, S. Stoinski, M. Hutchins *et al.* (eds.), *Great Apes and Humans: The Ethics of Coexistence*. Washington, DC: Smithsonian Institution Press, pp. 3–56.

Butynski, T. M. and Kalina, J. (1998). Gorilla tourism: A critical look. In E. J. Milner-Gulland and R. Mace (eds.), *Conservation of Biological Resources*. Oxford: Blackwell Press, pp. 294–313.

Ceballos-Lascurain, H. (2000). Interview by Ron Mader "Ecotourism Champion: a Conversation with Hector Ceballos-Lascurain." www.planeta.com/ecotravel/weaving/hectorceballos.html (accessed Sept. 16, 2010).

Cochrane, J. (1998). *Organisation of Ecotourism in the Leuser Ecosystem*. Unpublished report to the Leuser Management Unit.

de la Torre, S., Snowdon, C. T., and Bejarano, M. (1999). Preliminary survey of the effects of ecotourism and human traffic on the howling behaviour of red howler monkeys, *Alouatta seniculus*, in Ecuadorian Amazon. *Neotropical Primates*, 7: 84–86.

de la Torre, S., Snowdon, C. T., and Bejarano, M. (2000). Effects of human activities on pygmy marmosets in Ecuadorian Amazon. *Biological Conservation*, 94: 153–163.

Fawcett, K., Hodgkinson, C., and Mehlman, P. (2004). *An Assessment of the Impact of Tourism on the Virunga Mountain Gorillas: Phase I – Analyzing the Behavioral Data from Gorilla Groups Designated for Tourism*. Rwanda: Unpublished report, Dian Fossey Gorilla Fund International.

Frey, R. (1978). Management of orangutans. In *Wildlife Management in Southeast Asia, Biotrop*. Special Publication 8, pp. 199–215.

Frothmann, D. L., Burks, K. D., and Maples, T. L. (1996). Letter to the editor: African great ape ecotourism considered. *African Primates*, 2: 52–54.

Fuentes, A. (2006). Human-nonhuman primate interconnections and their relevance to anthropology. *Ecological and Environmental Anthropology*, 2 (2): 1–10.

Fuentes, A. (2010). Natural cultural encounters in Bali: Monkeys, temples, tourists, and ethnoprimatology. *Cultural Anthropology*, 25 (4): 600–624. DOI: 10.1111/j.1548–1360.2010.01071.x.

Fuentes, A., Shaw, E., and Cortes, J. (2007). Qualitative assessment of macaque tourist sites in Padangtegal, Bali, Indonesia, and the Upper Rock Nature Reserve, Gibraltar. *International Journal of Primatology*, DOI 10.1007/s10764–007–9184-y.

Fuentes, A., Southern, M., and Suaryana, K. (2005). Monkey forests and human landscapes: Is extensive sympatry sustainable for *Homo sapiens* and *Macaca fascicularis* on Bali? In J. Paterson and J. Wallis (eds.), *Commensalism and Conflict: The Primate-human Interface*. Norman, OK: American Society of Primatology Publications. pp. 168–195.

Goldsmith, M. L. (2000). Effects of ecotourism on behavioral ecology of Bwindi gorillas, Uganda: Preliminary results. *American Journal of Physical Anthropology*, Supp. 30: 161.

Gray, M., Fawcett, K., Basabose, A., *et al.* (2011). *Virunga Massif Mountain Gorilla Census – 2010 Summary Report*. Unpublished report, International Gorilla Conservation Program.

Guerriero, J., Larney, E., King, S., and Wright, P. C. (2009). *The influence of tourism on height and activity budget of golden bamboo lemurs (Hapalemur aureus) in Ranomafana National Park, Madagascar*. Undergraduate Research and Creative Activities (URECA), Stony Brook University, Stony Brook, NY [abstract].

Harcourt, A. H. (2001). The benefits of mountain gorilla tourism. *Gorilla Journal*, 22: 36–37.

Harcourt, A. H. and Stewart, K. J. (2007). *Gorilla Society: Conflict, Cooperation and Compromise Between the Sexes*. University of Chicago Press.

Hartup, B. K. (1994). Community conservation in Belize: Demography, resource uses, and attitudes of participating landowners. *Biological Conservation*, 69: 235–241.

Hastings, B. E., Kenny, D., Lowenstine, L. J., and Foster, J. W. (1991). Mountain gorillas and measles: ontogeny of a wildlife vaccination program. *Proceedings of the American Association of Wildlife Veterinarians*, pp. 198–205.

Higham, J. E. S. (2007). Ecotourism: competing and conflicting schools of thought. In J. E. S. Higham (ed.), *Critical Issues in Ecotourism: Understanding a Complex Tourism Phenomenon*. Oxford: Elsevier. pp. 1–19.

Hodgkinson, C. (2009). *Tourists, Gorillas and Guns: Integrating Conservation and Development in the Central African Republic*. Unpublished PhD Dissertation, University College London.

Homsy, J. (1999). *Ape Tourism and Human Diseases: How Close Should We Get? A Critical Review of the Rules and Regulations Governing Park Management and Tourism for Wild Mountain Gorillas (Gorilla gorilla beringei)*. Nairobi, Kenya: Unpublished report, Consultancy for the International Gorilla Conservation Program.

Honey, M. (1999/2008). *Ecotourism and Sustainable Development: Who owns Paradise?* Washington, DC: Island Press.

Horwich, R. H. and Lyon, J. (1987). Development of the 'Community Baboon Sanctuary' in Belize: an experiment in grass roots conservation. *Primate Conservation*, 8: 32–34.

Hubbard, J. (2011). *Tourists, Friend or Foe: The Effects of Tourism on Propithecus edwardsi in Ranomafana National Park*. Madagascar. Undergraduate Research Paper for Study Abroad.

Huijbregts, B., De Wachter, P., Ndong, L. S., Akou, O., and Akou, M. E. (2003). Ebola and the decline of gorilla *Gorilla gorilla* and chimpanzee *Pan troglodytes* populations in Minkebe Forest, north-eastern Gabon. *Oryx*, 37: 437–443.

IUCN (2013). *The IUCN Red List of Threatened Species. Version 2013.1*. www.iucnredlist. org (accessed Sept. 25, 2013).

Kaplan, G. and Rogers, L. (2000). *The Orangutans: Their Evolution, Behavior, and Future*. Cambridge, MA: Perseus Publishing.

Kemnitz, J. W., Sapolsky, R. M., Altmann, J., *et al.* (2002). Effects of food availability on serum insulin and lipid concentrations in free-ranging baboons. *American Journal of Primatology*, 57: 13–19.

Kinnaird, M. F. and O'Brien, T. G. (1996). Ecotourism in the Tangkoko Dua Saudara Nature Reserve: Opening Pandora's Box? *Oryx*, 30: 65–73.

Klailova, M., Hodgkinson, C., and Lee, P. (2010). Behavioral responses of one western low-land gorilla (*Gorilla gorilla gorilla*) group at Bai Hokou, Central African Republic, to tourists, researchers and trackers. *American Journal of Primatology*, 72: 897–906.

Knight, J. (2005). Feeding Mr. Monkey: cross-species food "exchange" in Japanese monkey parks. In J. Knight (ed.), *Animals in Person: Cultural Perspectives on Human-Animal Intimacies*. Oxford: Berg. pp. 231–253.

Knight, J. (2009a). Making wildlife viewable: habituation and attraction. *Society and Animals*, 17: 167–184.

Knight, J. (2009b). *Herding Monkeys to Paradise: How Macaque Troops are Managed for Tourism in Japan*. Leiden: Brill.

Knight, R. L. and Cole, D. N. (1995). Wildlife responses to recreationists. In: R. L. Knight and K. J. Gutzwiller (eds.), *Wildlife and Recreationists*, Washington, DC: Island Press, pp. 51–70.

Kurita, H., Sugiyama, Y., Ohsawa, H., Hamada, Y., and Watanabe, T. (2008). Changes in demographic parameters of *Macaca fuscata* at Takasakiyama in relation to decrease of provisioned foods. *International Journal of Primatology*, 29: 1189–1202.

Lanyero, F. (2011). UWA lowers gorilla tracking fees. *Daily Monitor*, www.monitor.co.ug/News/National/-/688334/1164758/-/c1i12wz/-/index.html (accessed Sept. 19, 2011).

Leroy, E. M., Rouquet, P., Formenty, P., *et al.* (2004). Multiple Ebola virus transmission events and rapid decline of Central African wildlife. *Science*, 303: 387–390.

Lloyd, J. (2002). *Chimpanzee Ecotourism in Uganda: Workshop Planning Document*. Unpublished MSc Thesis, Oxford-Brookes University, UK.

Macfie E.J. and Williamson, E.A. (2010). *Best Practice Guidelines for Great Ape Tourism*. Gland, Switzerland: International Union for Conservation of Nature and Natural Resources.

MacKinnon, J. R. (1977). Rehabilitation and orangutan conservation. *New Scientist*, 74: 697–699.

Maekawa, M., Lanjouw, A., Rutagarama, E., and Sharp, D. (2013). Mountain gorilla tourism generating wealth and peace in post-conflict Rwanda. *Natural Resources Forum*, 37: 127–137.

Morelli, T. L., King, S., Pochron, S. T. and Wright, P C. (2009). The rules of disengagement: takeovers, infanticide and dispersal in a rainforest lemur, *Propithecus edwardsi*. *Behaviour*, 146 (4–5): 499–523.

Moyini, Y. (2000). *Analysis of the Economic Significance of Gorilla Tourism in Uganda*. Report, International Gorilla Conservation Programme (IGCP), Kampala, Uganda.

Muehlenbein, M. P., Ancrenaz, M., Sakong, R. *et al.* (2012). Ape conservation physiology: Fecal glucocorticoid responses in wild *Pongo pygmaeus morio* following human visitation. *PLoS ONE*, 7 (3): e33357.

Muehlenbein, M. P., Martinez, L. A., Lemke, A. A., *et al.* (2010). Unhealthy travelers present challenges to sustainable primate ecotourism. *Travel Medicine and Infectious Disease* 8 (3): 169–175.

Mugisha, A. (2008). Potential interactions of research with the development and management of ecotourism. In: R. W. Wrangham and E. Ross (eds.), *Science and Conservation in African Forests: The Benefits of Long-term Research*. Cambridge University Press. pp. 115–127.

Nakamura, M. and Nishida, T. (2009). Chimpanzee tourism in relation to the viewing regulations at the Mahale Mountains National Park, Tanzania. *Primate Conservation*, 24 (1): 85–90.

Nishida, T. and Nakamura, M. (2008). Long-term research and conservation in the Mahale Mountains, Tanzania. In: R. W. Wrangham and E. Ross (eds.), *Science and Conservation in African Forests: The Benefits of Long-term Research*. Cambridge University Press. pp. 173–183.

Palacios, G., Lowenstine, L. J., and Cranfield, M. R. *et al.* (2011). Human metapneumovirus infection in wild mountain gorillas, Rwanda. *Emerging Infectious Disease*, 17 (4): 711–713.

Parker, J. and Ballantyne, R. (2012). Comparing captive and non-captive wildlife tourism. *Annals of Tourism Research*, 39 (2): 1242–1245.

Paterson, J. and Wallis, J. (eds.) (2005). *Commensalism and Conflict: The Human-primate Interface*. Norman, OK: American Society of Primatologists.

Payne, J. and Andau, P. (1989). *Orang-Utan: Malaysia's Mascot*. Kuala Lumpur: Berita Publishing Sdn. Bhd.

Pusey, A. E., Wilson, M. L., and Collins, D. A. (2008). Human impacts, disease risk, and population dynamics in the chimpanzees of Gombe National Park, Tanzania. *American Journal of Primatology*, 70 (8): 738–744.

Rijksen, H. D. (1978). *A Field Study of Sumatran Orang Utans (Pongo pygmaeus abelii, Lesson 1872), Ecology, Behavior and Conservation*. Wageningen, the Netherlands, Mededlingen Landbouwhogeschool: H. Veenman and Zonen B.V.

Rijksen, H. D. (1982). How to save the mysterious 'man of the forest'? In: L.E.M. de Boer (ed.), *The Orang Utan: Its Biology and Conservation*. The Hague: Dr. W. Junk Publishers, pp. 317–341.

Rijksen, H. D. and Meijaard, E. (1999). *Our Vanishing Relative: The Status of Wild Orang-Utans at the Close of the Twentieth Century*. Kluwer Academic Publishers.

Rijksen, H. D. and Rijksen-Graatsma, A. G. (1975). Orangutan rescue work in North Sumatra. *Oryx*, 13: 63–73.

Ruesto, L. A., Sheeran, L. K., Matheson, M. D., Li, J. H., and Wagner, R. S. (2010). Tourist behavior and decibel level correlate with threat frequency in Tibetan macaques (*Macaca thibetana*) at Mt. Huangshan, China. *Primate Conservation*, 25: 99–104.

Russell, C.L. (1995). The social construction of orangutans: An ecotourist experience. *Society and Animals*, 3 (2): 151–170.

Sajuthi, D., Karesh, W., McManamon, R. *et al.* (1991). Medical aspects in orangutan reintroduction. *Proceedings of the International Conference on "Conservation of the Great Apes in the New World Order of the Environment"*, Indonesia.

Sandbrook, C. and Semple, S. (2006). The rules and the reality of mountain gorilla *Gorilla beringei beringei* tracking: How close do tourists get? *Oryx*, 40 (4): 428–433.

Serio-Silva, J. C. (2006). Las Islas de los Changos (the monkey islands): The economic impact of ecotourism in the region of Los Tuxtlas, Veracruz, Mexico. *American Journal of Primatology*, 68: 499–506.

Tapper, R. (2006). *Wildlife Watching and Tourism: A Study on the Benefits and Risks of a Fast Growing Tourism Activity and its Impacts on Species*. UNEP/CMS Secretariat, Bonn, Germany.

Treves, A. and Brandon, K. (2005). Tourism impacts on the behavior of black howler monkeys (*Alouatta pigra*) at Lamanai, Belize. In: J. Paterson and J. Wallis (eds.), *Commensalism and Conflict: The Primate-human Interface*. Winnipeg, Manitoba: Hignell Printing. pp. 147–167.

UBOS (2011). *Visitors to National Parks (Citizens and Foreigners), 2006 – 2010*. Uganda Bureau of Statistics.

Unwin, T., and Smith, A. (2010). Behavioral differences between provisioned and non-provisioned Barbary macaques (*Macaca sylvanus*). *Anthrozoos*, 23 (2): 109–118.

Vedder, A. and Weber, A. W. (1990). The mountain gorilla project. In A. Kiss (ed.), *Living with Wildlife: Wildlife Resource Management with Local Participation*. Washington, DC: World Bank Technical Publication, 130: 83–90.

Weber, A. W. (1993). Primate conservation and eco-tourism in Africa. In C. S. Potter, J. I. Cohen, and D. Janczewski (eds.), *Perspectives on Biodiversity: Case Studies of Genetic Resource Conservation and Development*. Washington, DC: American Association for the Advancement of Science Press. pp. 129–150.

Weber, A. W. and Vedder, A. (1983). Population dynamics of the Virunga gorillas: 1959–1978. *Biological Conservation*, 26: 341–366.

Williamson, E. A. (2001). Mountain gorilla tourism: Some costs and benefits. *Gorilla Journal*, 22: 35–37.

Wollenberg, K. C., Jenkins, R. K. B., Randrianavelona, R., *et al.* (2011). On the shoulders of lemurs: pinpointing the ecotouristic potential of Madagascar's unique herpetofauna. *Journal of Ecotourism*, 10 (2): 101–117. DOI: 10.1080/14724049.2010.511229.

Woodford, M. H., Butynski, T. M., and Karesh, W. B. (2002). Habituating the great apes: The disease risks. *Oryx*, 36 (2): 153–160.

Wright, P. C. and Andriamihaja, B. A. (2002). Making a rain forest national park work in Madagascar: Ranomafana National Park and its long-term research commitment. In: J. Terborgh, C. van Schaik, M. Rao, and L. Davenport (eds.), *Making Parks Work: Strategies for Preserving Tropical Nature*. California: Island Press. pp. 112–136.

Wright, P. C. and Andriamihaja, B. A. (2003). The conservation value of long-term research: A case study from Parc National de Ranomafana. In: S. Goodman and J. Benstead (eds.), *Natural History of Madagascar*. University of Chicago Press. pp. 1485–1488.

Wyman, M. and Stein, T. (2010). Examining the linkages between community benefits, place-based meanings, and conservation program involvement: A study within the Community Baboon Sanctuary, Belize. *Society & Natural Resources*, 23 (6): 542–556.

Yamagiwa, J. (2010). Research history of Japanese macaques in Japan. In: N. Nakagawa, M. Nakamichi, and H.Sugiura (eds.), *The Japanese Macaques*. Tokyo: Springer, pp. 3–25.

Zhao, Q. K. (2005). Tibetan macaques, visitors, and local people at Mt. Emei: Problems and countermeasures. In: J. Paterson and J. Wallis (eds.), *Commensalism and Conflict: The Human-primate Interface*. Norman, OK: American Society of Primatologists. pp. 376–399.

Zhao, Q. K. and Deng, Z. (1992). Dramatic consequences of food handouts to *Macaca thibetana* at Mount Emei, China. *Folia Primatologica*, 58: 24–31.

Part II

Asian primates

2 Tourism, infant mortality, and stress indicators among Tibetan macaques at Huangshan, China

Carol M. Berman, Megan D. Matheson, Jin-Hua Li, Hideshi Ogawa, and Consuel S. Ionica

Introduction

Primate tourism is a recent and growing trend in primate-habitat countries. Many primate tourism operations are outgrowths of community-based conservation initiatives (Hill, 2002) and have been promoted for their potential to achieve conservation goals as well as financial and educational benefits for local communities. One of the earliest and most successful initiatives is the Mountain Gorilla Project in the Virunga Mountains (Harcourt & Stewart, 2007). Gorilla tourism has been credited with bringing an important source of foreign currency to impoverished nations, educational opportunities for local inhabitants, and even increased reproduction in gorilla groups used for tourism (Harcourt & Stewart, 2007; and see Goldsmith, this volume). However, the extent to which many primate tourist operations are meeting these goals is not clear. As a result, conservationists, who were generally enthusiastic and encouraging about establishing primate tourism operations, are sounding more cautious, noting specific examples in which tourism has harmed wild primates, and pointing out that we know little about the impact of tourism on most of the populations it targets (Butynski, 2001). Most agree that we need to do much more research to better understand the ways in which primate tourism affects primate health, behavior, and reproduction. Only then will we be able to make sound recommendations that maximize conservation goals and minimize harm. This chapter reviews some approaches to assessing the effects of tourism on primate populations, presents findings on some of the negative impacts of tourism on a population of Tibetan macaques, and offers several recommendations to reduce these negative impacts both in China and elsewhere.

Management for primate tourism across sites varies considerably. In some cases, tourists follow guides to locate and observe primates where they happen to find them. In others, primates are lured or even herded to specific locations for viewing. The size of tourist groups may vary from a few to scores, and their frequency of contact may vary from less than once per week to several times per day. Provisioning by

Primate Tourism: A Tool for Conservation?, ed. Anne E. Russon and Janette Wallis. Published by Cambridge University Press. © Cambridge University Press 2014.

guides, feeding by tourists, and other interactions between humans and nonhumans may or may not be involved. In nearly all cases, however, primate tourism involves close contact with humans (Woodford et al., 2002, and references therein). As such, the strongest case for caution concerns the potential for disease transmission; several outbreaks of disease among monkeys, great apes, and humans have been attributed to close contact between nonhuman and human primates (Adams et al., 2001; Köndgen et al., 2008; Wallis & Lee, 1999; Woodford et al., 2002). Tourism may also contribute to habitat destruction and possibly disorders caused by eating garbage and tourist foods, particularly when tourist demand leads to hotels and restaurants within primate habitat (Brandon & Wells, 1992; Wrangham, 2001; in this volume see chapters by Dellatore et al., Kauffman, Sapolsky).

Most recently, primate behavioral ecologists have focused concern on the possible detrimental effects of tourism on the behavior of primates. Reports are beginning to emerge that close contact with large numbers of humans, including tourists, can cause stress in wild primates (Butynski, 2001; Kinnaird & O'Brien, 1996; Maréchal et al., 2011), avoidance of tourist areas (Boinski & Sirot, 1997; de la Torre, this volume; Kinnaird & O'Brien, 1996), and problems associated with over-habituation and hyperaggression (Boinski & Sirot, 1997; Grossberg et al., 2003; Small, 1995; Zhao 2005). More subtle effects may include changes in usage of the habitat, activity patterns (de la Torre et al., 2000; Goldsmith et al, 2006; Griffiths & van Schaik, 1993; Kinnaird & O'Brien, 1996; Klailova et al., 2010; Koganezawa & Imaki, 1999; Leary & Fa, 1993; Treves & Brandon, 2005), or communicative behavior (de la Torre et al. 1999; Johns, 1996). Numerous chapters in this volume report similar findings. For example, in one study, pygmy marmosets (Cebuella pygmaea) in groups that were exposed to relatively more tourists and tourist-related activities avoided the lower strata of the forest and engaged in less social play than those in less exposed groups, as they apparently attempted to avoid contact with humans. These changes could be serious if they detrimentally affect the abilities of group members to hunt, avoid predation, compete with or exchange members with other social groups, garner an adequate diet, or develop normal social behavior. Notably, more highly exposed pygmy marmoset groups produced fewer viable infants than less highly exposed groups (de la Torre et al., 2000), although it is not clear exactly why or even whether this was due to tourism per se.

Such pioneering studies clearly justify our caution. Some conservationists have become pessimistic about the potential for primate tourism to achieve conservation goals. Others suggest that it may be possible, with further research and education, to refine the practices used by primate tourism operations in ways that eliminate harmful practices and encourage the development of practices that provide rewarding experiences to tourists without placing their objects of interest in jeopardy. Since a single tourist operation usually involves a variety of co-occurring changes for the primates involved (e.g. increased exposure to humans who behave in a variety of ways; inadvertent introduction to new foods; changes in the distribution and density of foods; and exposure to new pathogens, building noises, vehicles, manufactured goods, and human refuse), we need to distinguish which factors are responsible for any detrimental changes in behavior and reproduction and which are more nearly neutral or even beneficial (Kauffman, 2010). Given that different primate species have diverse, flexible, and variable responses to their environments, we also need to ascertain whether particular kinds of practices have consistent effects on a variety of species (Kauffman, 2011; Kauffman & Boinski, 2008; Westin & Kauffman, 2011) or even on different groups of the same species. Some species, such as Japanese macaques (Kurita, this volume), long-tailed macaques (Fuentes, 2010), and gorillas (Harcourt *et al.*, 1983) have apparently enjoyed increased reproductive success when subjected to tourism, whereas others (e.g. pygmy marmosets: de la Torre *et al.*, 2000; black howlers: Treves *et al.*, unpublished data, cited in Treves & Brandon, 2005; and in this volume see de la Torre, Wright *et al.*) may not have been as fortunate.

Discovering how tourism practices affect behavior and reproductive success is difficult, but there are a number of approaches that can be taken, each with its own advantages and limitations. First, when baseline data are available, it may be possible to compare behavior and reproductive success before and during tourism. This has the advantage of identifying and tracking long-term behavioral changes in the same group, perhaps even in the same individuals in the same microhabitat. However, it may be difficult to distinguish the effects of tourism from other changes, anthropogenic or not, that may have coincided with the advent of tourism. It may also be difficult to disentangle the effects of specific tourism practices if they co-occurred with one another. On the other hand, the use of data covering multiple years in which specific tourism practices and other types of changes did not completely co-vary might go a long way toward solving these problems.

Second, when groups exposed to tourism can be monitored at the same time as undisturbed groups in the same population, it may be possible to compare the two groups. This approach allows one to control for yearly fluctuations in food resources, weather conditions, and other ecological conditions that may affect behavior and reproduction, but it may be limited when the two groups live in different microhabitats (affecting food distribution, habitat structure, or predation risk) or differ in group composition. Diverse microhabitats could produce differences in reproductive rates, activity patterns, foraging efficiency, and social dynamics, including competitive patterns, independently of any effects of tourism. However, if there are several groups of each type, there may be sufficient variation among them to distinguish the effects of tourism, and perhaps specific tourism practices, from other sources of intergroup variation.

A third approach involves examining short-term (days or even hours) differences in behavior in the same social groups when tourists are present versus absent, or describing variation in behavior as a function of variation in tourist numbers, tourist behavior, distance from tourists etc., or other variables of interest. This approach is particularly useful in reconstructing how particular human practices may be linked to particular immediate behavioral responses from the animals, although it may be limited in its ability to detect behavioral effects that carry over into situations in which no tourists are present.

In this chapter we describe two complementary sets of studies that examine links between tourism and monkey behavior among Tibetan macaques (*Macaca thibetana*) in Huangshan, China. The studies used a combination of approaches to pinpoint aspects of tourism that may be harmful to the monkeys, particularly infants. The first set, described in detail in Berman *et al.* (2007), examined a social group that researchers have monitored since 1986 and that has been managed for tourism since 1992. These researchers had the impression that rates of aggression increased soon after management for tourism began. In addition, for the first time, they observed severe attacks by adults on infants and infant corpses with bite wounds. Based on this, we compared infant mortality six years before the group was managed for tourism (1986–1991), during 12 years of management for tourism (1992–2002, 2004) and during one year when management was temporarily suspended (2003). We also compared aggression rates among adults before and during tourism management and tested several hypotheses about specific factors (numbers of tourists, degree of range restriction, demographic changes, changes in alpha males) that may have been harmful to infants. The second set of studies (Jones *et al.*, 2008; Mack *et al.*, 2008; Matheson *et al.*, 2006, 2007; McCarthy *et al.*, 2009; Ruesto *et al.*, 2010; Yenter *et al.*, 2008) examined the same monkey group beginning in 2004, and compared its immediate behavioral responses to tourists and tourism practices with responses to those of a group that had only recently been subjected to tourism.

Methods

Animals and study site

Tibetan macaques live in small isolated populations in sub-tropical, montane forests consisting of mixed deciduous and broadleaf evergreen trees across central China and parts of India. The IUCN Red list classifies the species as being Near Threatened (IUCN, 2012). This classification is used for species that are not immediately threatened and are the focus of a conservation program, but would likely become threatened within five years if conservation efforts were discontinued. Like other macaques, Tibetan macaques display female philopatry, male dispersal, and linear dominance hierarchies (Berman *et al.*, 2004; Deng & Zhao, 1987; Li & Wang, 1996; Li *et al.*, 1996b; Zhao, 1996). Reproduction is seasonal with overlapping birth seasons – January to August – (Yin *et al.*, 2004) and mating seasons – July to January – (Li *et al.*, 2005).

Tibetan macaques have been studied at two sites, Mt. Emei and Huangshan. At both sites, monkeys come into contact with humans as part of a tourist experience. At Mt. Emei, a Buddhist community visited mainly for its temples, there is little regulation and no instruction of tourists. As a result, tourists regularly hand feed the monkeys and suffer frequent monkey attacks. Some of these attacks result in serious injuries to both visitors and monkeys and there have also been deaths of both visitors and monkeys (Zhao, 2005). In contrast, a primate tourism program at

Huangshan was intentionally set up to avoid the problems experienced at Mt. Emei. Tourists are restricted to viewing pavilions and rules prohibit them from feeding the monkeys. They are also sometimes given brief lectures about monkey biology and behavior. Thus the signs of disturbance we observed were not expected.

The first set of studies focused on the Yulingkeng A1 (YA1) social group at Huangshan. Researchers began to observe YA1 in 1986, six years before management for tourism began (Li, 1999). To facilitate observations in the steep terrain, the researchers provisioned the group with about 6 kg of whole corn per day in an open area by a stream. An assistant stood in the middle of the provisioning area and tossed the corn, spreading it as widely as possible. In 1992, the local government herded this macaque group to an area adjacent to their natural range in preparation for tourism (Berman & Li, 2002; Li et al., 1996a). Observation and provisioning continued as before, in a new open area that was also by a stream and virtually the same size (142 m^2 vs. 144 m^2) as the former provisioning area, while government staff built a viewing pavilion on the edge of the opening. To prevent the macaque group from returning to its original range, the staff herded it back when it strayed too far. Several staff members would get behind the group, sometimes shouting and throwing stones as they chased it back to the provisioning area. In so doing, they reduced the group's range from 7.75 km^2 to less than 3 km^2 (Li et al., 1996a). It was around this time that researchers first observed severe attacks by adults on infants and infant corpses with bite wounds.

By 1994, the viewing pavilion was complete and tourism began. Tourists climbed a stairway up a hill to the open pavilion and observed the monkeys in the provisioning area for 30–45 minutes at a time, usually at three to four set times of day. A mean of about twenty tourists and two to six staff were present at each feeding. Rules prohibited tourists from feeding, shouting, and contacting the monkeys directly, but they were unevenly enforced. The monkeys often threatened tourists and staff, and adult males occasionally jumped into the pavilion to attack them. The staff sometimes responded by throwing stones at the adult males. This pattern of management continued until 2002, when a new director further restricted the group's range to about 1 km^2 immediately surrounding the provisioning area to make the group continuously available to tourists. By the end of the year, infant mortality was very high, and the director agreed to suspend regular provisioning, range restriction, and tourism over the winter and spring of 2003. Regular provisioning, tourism, and inconsistent range restriction resumed in the later part of 2003.

Thus, between 1986 and 2004, range restriction occurred at four levels – none before 1991 and during the suspension of management in 2003; inconsistent during 1992, 1993, and 2004; consistent from 1994 to 2001; and severe in 2002. Between 1994 and 2004, the numbers of tourists varied from 8915 to 43 881 per year (mean + SD = 24 760 + 10 861). The group was provisioned throughout the study in a consistent manner, although provisioning was done more regularly after 1992.

A second study group, named Yulingkeng A2 (YA2), resulted from a fissioning of YA1 in 1996. YA2 continues to be a common destination for emigrants from YA1. In 2002, active management of this group for tourism began. As with YA1, a wooden pavilion was built within YA2's range. Tourists climbed a stairway to the pavilion, where they observed park staff provision the monkeys with corn at four set times per day. One difference between the two groups, however, was that the YA2 pavilion was closer to ground level than the YA1 pavilion; one section allowed monkeys and humans to be at eye-level with no fencing between them. Between feeding times, park staff followed the group and herded them by yelling and throwing stones when they strayed too far from the provisioning site (over the mountain ridge), and when they did not respond to loud calls given by staff in the provisioning area that served to announce the start of provisioning. On at least two occasions since tourism began, YA2 park staff have been unable to locate the group for more than one week, suggesting that their natural home range exceeded that normally enforced by park staff. In addition, the staff has attempted to exclude potentially competing groups from occupying the provisioning area of YA2. For example, in 2005, a previously unknown social group of Tibetan monkeys showed up on several occasions at the provisioning site and exchanged threats with members of YA2. Park staff intervened on YA2's behalf, driving the unknown group away by yelling and throwing stones.

We compared the two groups between 2004 and 2007. YA2 was the larger group during this period, containing five to ten adult males, seven to ten adult females, zero to one subadult males (sexually mature natal males not yet integrated into the adult hierarchy), eighteen to twenty-one juveniles, and two to seven infants. YA1 contained two to five adult males, five adult females, two to three subadult males, seven to twelve juveniles, and zero to four infants.

Data collection

We used long-term census records collected by various researchers for data on group membership, births, and deaths from 1986 to 2007. Each study group was censused at least three to four times each year before, during, and after the birth season. We used all-occurrence sampling and focal animal sampling methods (Altmann, 1974) in YA1 to collect data on aggression between adults during six time periods: (a) 10 weeks immediately before the group was translocated (12/5/1991–2/8/1992), (b) 10 weeks immediately after translocation (2/16/1992–4/26/1992), (c) seven weeks several months after translocation (9/16/1992–11/4/1992), (d) five-and-a-half months when tourism was well established (8/1/2000–1/28/2001), (e) three additional months when tourism was well established (2/27/2001–5/29/2001), and (f) seven-and-a-half months when tourism and severe range restriction were enforced (12/9/2001–7/25/2002). We chose these time periods not only because our behavioral definitions and data collection methods were comparable, but also because dominance relationships within these periods were stable (Berman et al., 2004; Ogawa, 2006). Thus, aggression rates were not affected by hierarchical instability.

Data used to examine the immediate effects of tourist presence and tourist behavior on monkey behavior were collected by faculty and students of the Joint Central Washington University – Anhui University Biodiversity and Primate Field Research

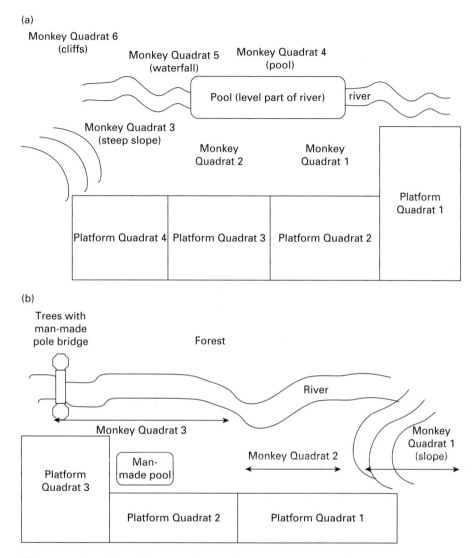

Figure 2.1   Diagrams of tourist pavilion areas for monkey groups (a) YA1 and (b) YA2.

Program during August of 2004, 2005, 2006, and 2007. During each period, data collectors watched the monkey groups from the viewing pavilions primarily during afternoon hours. We were able to collect data for YA1 both when tourists were and were not present, and both during and between provisioning events. This was not the case for YA2, because the monkeys were typically present in the provisioning area only during provisioning times. During each observation session for each group, one observer collected instantaneous point time samples of the number of people (tourists and researchers) in the pavilions at two-minute intervals (2004 to 2006), or when significant changes to densities occurred (2007), while other observers recorded data on monkey and human behavior.

We report findings from several data sets relating to the immediate effects of humans on monkey behavior, each collected with different methods. Data used to compare YA1 and YA2 on measures of monkey presence and duration in the provisioning area, rates of threatening, and rates of grooming were recorded using instantaneous scan sampling of both groups at two-minute intervals during August 2004 (Matheson et al., 2006) and August 2007 (Mack et al., 2008). For the scans, observers noted the monkeys' locations based on a quadrat system developed by the researchers (see Figures 2.1(a) and 2.1(b)). They also recorded proximity, contact, and grooming between monkeys. Between intervals, researchers recorded all occurrences of aggression and threats to humans or other monkeys. YA1 was observed across three conditions: (1) prior to provisioning with no tourists present (baseline condition); (2) prior to provisioning with tourists present (tourist condition); and (3) immediately following the start of provisioning with tourists present (food condition). YA2 could not be compared across conditions as the group was present in the provisioning area only during and immediately following the start of provisioning.

Data on human–monkey interactions come from three data sets. During August 2005, Ruesto (2010) developed an ethogram of human behaviors in the pavilion (see Table 2.1). She recorded all occurrences of human behaviors at the YA1 provisioning site when the monkeys were in the provisioning area, while a second observer recorded all occurrences of monkey threats. Ruesto also measured decibel levels at one corner of the pavilion at two-minute intervals. During August 2006, McCarthy et al. (2009) expanded and utilized Ruesto's ethogram of human behavior as well as Berman et al.'s (2004) ethogram of Tibetan monkey behavior to record monkey–human interactions using focal sampling at the YA1 provisioning site. When a monkey entered the provisioning area, s/he was chosen as a focal and one researcher spoke the monkey's behavior aloud to a second observer, who simultaneously recorded this and the behavior of humans interacting with the focal monkey. In that way, an intact sequence of monkey–human interaction components was obtained. Behaviors were considered to be part of one sequence when they occurred within five seconds of each other. Individual focal samples continued until the sequence ceased, that is the human–monkey interaction ceased for more than five seconds; then a new focal was chosen. In 2007, Jones et al. (2008) used McCarthy et al.'s methods to study human–monkey interactions at both YA1 and YA2, as well as to determine whether monkeys differentiate between different categories of humans (tourists, researchers, keepers).

Finally, data on self-directed behaviors (SDBs; a behavioral measure of stress; see Maestripieri et al., 1992; Schino et al., 1996) and tourist density were collected during August of 2005, 2006, and 2007 (Matheson et al., 2007; Yenter et al., 2008). Observers used five- to ten-minute focal samples with continuous recording of SDBs as well as general social behavior, based on Berman et al.'s (2004) ethogram.

For all data collected on monkey or human behavior, observers achieved a minimum inter-observer reliability of 90% before proceeding with formal data collection.

Table 2.1 Tourist behavioral ethogram (developed by Ruesto, 2007 and McCarthy *et al.*, 2007)

Behavior	Description
Barbed wire shake	Tourist shakes the barbed wire that borders the viewing platform railing
Dangle	Tourist dangles food, body parts, or objects over the viewing platform railing toward monkeys
Flee	Tourist turns and runs away from monkeys
Foot noise	Tourist stamps feet or kicks wall in tourist platform
Hand noise	Tourist makes noises with one or both hands (e.g. clap, snap, smack their body, or smack a book)
Mimic[a]	Tourist mimics facial expressions and/or body language of a monkey threat (e.g. eyebrow raise and stare)
Mouth noise[b]	Tourist makes noise with mouth directed toward monkey
Point	Tourist points at monkeys, arm extends out of tourist platform. Pointing that did not extend out of the platform was not recorded because we could not be sure that it was visible to the monkeys
Rock[c]	Tourist pretends to throw rock at monkeys
Slap railing[d]	Tourist slaps rail or post in observation area. Slap railing was recorded separately from other hand noise because it involved a forward motion that was more clearly directed toward a monkey
Spit	Tourist spits into monkey area
Throw food	Tourist drops or throws food item into the monkey area, or directly to a monkey
Throw object[e]	Tourist drops or throws nonfood item into monkey area
Wave	Tourist waves at monkey (with hands or objects)

Notes:
[a] If mimicry includes slap, it is coded as Mimic not Slap.
[b] Whistling, kissing noises, shouts, etc.
[c] If rock is thrown, it is coded as Throw object not Rock.
[d] Slap railing may occur using hands or objects.
[e] Thrown object may include rocks.

Results and discussion

Tourism and infant mortality

Tourism did not appear to affect birth rates in the YA1 study group. A similar percentage of reproductive age females gave birth during the years in which they were (1992–2002, 2004) and were not (1986–1991, 2003) managed for tourism (unmanaged: mean ± SE = 71.2% ± 7.3; managed: 73.1% ± 4.5). However, their infants were less likely to survive during the years of tourist management. Figure 2.2 shows percentages of infants that died before reaching one year of age each year from 1986 to 2004. Infant mortality was low in the six years before management for tourism began, except in 1988, the year that a disease (which was never diagnosed) swept through the group and killed 17 monkeys in a single week, including 4 of 6

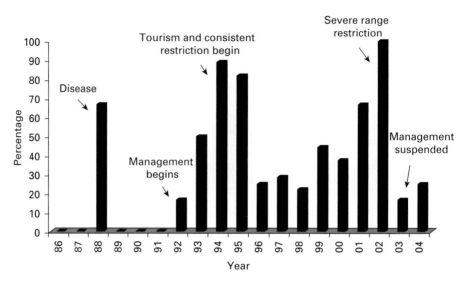

Figure 2.2   The percentage of infants that died before one year of age each year from 1986 to 2004 in YA1 (from Berman *et al.*, 2007).

infants (Li, 1999; Wada & Xiong, 1996). Infant mortality began to increase in 1992, the year the group was translocated. It fluctuated considerably during the years of management (1992–2002, 2004), peaking twice – once in 1994, the year that tourism and consistent range restriction began, and again in 2002, the year that the group's range was severely restricted. It appeared to decrease sharply in 2003, the year that management was temporarily suspended. When we compared infant mortality for individual infants born before (1986–1991), during (1992–2002, 2004), and after management (2003), the differences were statistically significant (logistic regression model $\chi^2 = 17.0$, $df = 2$, $p < 0.001$). Mortality rates were significantly higher during management than before it (during: 53/97 = 54.6%; before: 4/27 = 14.8%; $\beta = -1.9$, Exp (β) = 0.14, Wald = 11.2, $df = 1$, $p = 0.001$), but they were similar before and after (before: 4/27 = 14.8%; after: 1/6 = 16.7%).

Many of the infants that died during the tourism management years were severely injured shortly before their deaths. Figure 2.3 shows the number of infant deaths from wounding and other causes from 1999 to 2004, the years in which we had complete records. Overall, there were 47 births and 25 infant deaths between 1999 and 2004, 15 (60.0%) of which were due to wounding. Although our records were incomplete before 1999, a minimum of 7 additional infant deaths (24%) by wounding occurred between 1992 and 1998, but no infant deaths were attributed to wounding before 1992, the year that management for tourism began.

Unfortunately, we had no information for 14 of the 22 (64%) known infant deaths by wounding; infants or their corpses were found with bite wounds after the fact. In the remaining 8 cases, our information was fragmentary. Nevertheless, we had no reason to believe that infant injuries were caused by poaching, attack by

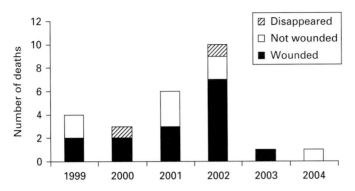

Figure 2.3   The number of infant deaths from wounds, from other causes, and from unknown causes (i.e. disappearances) each year from 1999 to 2004 (from Berman *et al.*, 2007).

tourists or staff, capture, predation, or inter-group aggression. When humans threw objects at the monkeys, they targeted adults rather than immatures. No large predators were present in Huangshan, and the few encounters we saw between groups were not aggressive in nature. Rather our observations pointed toward aggression directed by adult group members toward infants. In the most clear-cut case involving a four–month-old female infant born in 1996, observers saw the alpha male chase, grab, and bite the infant on the legs. The infant had died by the end of the day. In two other cases involving a twelve-month-old female in 1993 and a month-old female infant in 1992, we heard species-typical aggressive vocalizations shortly before a mother appeared carrying a freshly injured or dead infant. In the former case and in an additional case involving a four-month-old female in 1994, the alpha male was observed with blood on his mouth shortly after the mother appeared with her injured infant. In four remaining cases (a nine-month-old female in 2001, a four-month-old female in 2002, a month-old female in 2002, and a two-month-old female in 2002) infants appeared with wounds following an outbreak of aggression among adults in the provisioning area. Typically, the fight moved into the forest where visibility was limited. After the fight died down, the mother carried the injured or dead infant back into the provisioning area.

Our quantitative data on aggression also point toward the notion that adult aggression in the provisioning area was responsible for high levels of infant mortality during tourism management. Aggression rates among adults in the provisioning area (mean + SE = 3.22/h + 0.73) varied significantly across the six time periods (F = 7.20, df = 5, 119, p < 0.001; Figure 2.4) and correlated positively with yearly infant mortality rates for the five years (rs = 0.90, n = 5 years, p = 0.037; Figure 2.4). They were significantly lower during the first three time periods (before translocation and in the early period of management when tourists were absent and range restriction was inconsistent) than during the second three time periods (when tourism and consistent range restriction were in full force), and significantly higher during 2002 (the year that the group's range was severely restricted) than in any other

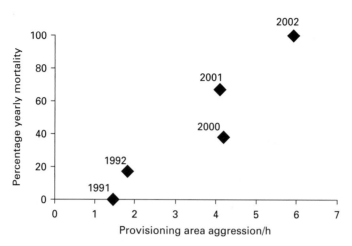

Figure 2.4   Percentage of infant mortality as a function of mean rates of aggression among adults in the provisioning area over five years (see text for details) (from Berman *et al.*, 2007).

time period. In contrast, mean rates of aggression in the forest were generally low (mean + SE = 0.33/h + 0.03), similar across all six time periods (F = 0.55, *df* = 5, 56, ns) and unrelated to infant mortality rates (*rs* = 0.70, *n* = 5 years, *p* = 0.19).

Taken together, our results confirm our suspicions that Tibetan macaque infants at Huangshan were at increased risk of death during periods of management for tourism, and they suggest that some aspect or aspects of management led to increased adult aggression in the provisioning area that led in turn to infant attacks. But which of the many changes associated with management led to an upsurge in deadly aggression, and how? To help narrow down which may have been most harmful to infants, we examined relationships between infant mortality and two specific aspects of tourist management: degree of range restriction and numbers of tourists. We also examined the possibility that management may have affected infants indirectly by affecting group demography (group size, numbers of adults, male:female sex ratios, and group fission). Fissions, in particular, can involve heightened aggression, and they were more frequent after management began (Li *et al*, 1996a). Finally, we asked whether the infant mortality we observed was consistent with the sexual selection theory of infanticide (Hausfater & Hrdy, 1984), which has been shown to increase under disturbed conditions (van Schaik, 2000), by asking whether infant mortality was greater in years in which there was a change in the alpha male position than in years in which there was no change. Although provisioning is known to increase aggression in a variety of species (e.g. Altmann, 1988; Hsu *et al.*, 2009; Lee *et al.*, 1986; Sugiyama & Ohsawa, 1982), we ruled out provisioning as the sole factor causing increases in infant mortality because the group was provisioned in a similar manner both before and during tourism management, and because previous studies of several species living under a variety of provisioning regimes suggest that provisioned groups actually experience *lower* infant mortality than unprovisioned groups (Lyles & Dobson, 1988).

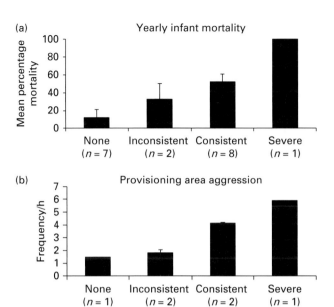

Figure 2.5 (a) Mean percentage of infant mortality and (b) mean rate of adult–adult aggression by degree of range restriction. See text for details. n = number of years (from Berman et al., 2007).

The only strong relationship we found for infant mortality was with the degree of range restriction (Figure 2.5(a)). Infant mortality rates were higher in years with higher degrees of range restriction (F = 6.0, df = 3,15, n = 19 years, p < 0.007). Indeed range restriction accounted for 54.5% of the variance in infant mortality rates. Mean rates of total adult aggression in the provisioning area were also significantly related to the degree of range restriction (F = 888.6, df = 3,4, n = 5 years, $R2$ = 100%, p < 0.025) in spite of the small sample of years for which we had aggression data; aggression was higher in years with more intense range restriction (Figure 2.5(b)). These findings suggest that range restriction may have jeopardized infants by raising levels of aggression in the provisioning area. Range restriction could have done this in at least two ways. First, range restriction most likely increased the group's dependence on provisioned food by limiting the monkeys' access to dispersed natural food resources. This change would have been expected to increase intra-group competition for the relatively more clumped provisioned food, leading to increased aggression. Indeed, the group showed more variation in measures of body fat/leanness than did an unprovisioned group (Berman & Li, 2002; Berman & Schwartz, 1988), as expected if they experienced higher levels of intra-group competition for food. While we reject provisioning as the sole cause, we suggest that range restriction may have exacerbated its effects by increasing the group's dependence on provisioned food. In this case, the effects may have been particularly severe, because the amount of provisioned food was limited and was not increased when higher degrees of range restriction were imposed.

It is also possible that the way range restriction was accomplished at Huangshan, that is through herding, which sometimes involved shouting and rock-throwing, may have been stressful to group members and led to increased aggression. The monkeys were often visibly fearful as they were driven down the slope into the provisioning area, and they frequently directed threats toward other staff and tourists waiting in the viewing pavilion. According to this hypothesis, increased aggression during periods of high degrees of range restriction may have been due to increased levels of fear-induced stress (cf Hinde, 1974, p. 261). Of course, the two hypotheses (intra-group competition and stress-inducing herding) are not mutually exclusive; both intra-group competition and herding may have contributed to higher levels of aggression under conditions of range restriction, and aggression due to intra-group competition for food may also have been mediated by stress.

Tourism, stress indicators, and behavior

Avoidance of humans

Our data suggest that both aggressive and positive social behavior varied in the presence of tourists, and possibly as a function of degree of habituation to tourist presence. There is also evidence that less well-habituated monkeys attempt to avoid proximity to humans. In 2004, YA2, the less-habituated group, was usually not present in the provisioning area outside the provisioning times and remained in the provisioning area for shorter amounts of time than was true of YA1. YA2 spent a mean of 18.7 minutes in the provisioning area, almost always following provisioning, whereas YA1 was typically visible for hours at a time. Only three out of twenty observation sessions of YA1 were ended by monkeys leaving the area during provisioning, whereas seven out of eight sessions of YA2 were ended by the monkeys leaving. We almost never observed grooming in YA2 in the provisioning area, whereas we saw it frequently in YA1 (see below). These group differences could be related to YA2's relative lack of habituation, although we cannot rule out YA2's relatively closer proximity with tourists due to the layout of the viewing area. It is also noteworthy that YA2's male dominance hierarchy was in a period of instability in 2004, likely heightening the tension within the group and making members more susceptible to human influence.

Grooming in YA1 varied across conditions in 2004, with rates ranging from 4.65 bouts per hour when tourists but not food were present to 1.3 bouts per hour during baseline observations and 0.97 bouts per hour when both tourists and food were present ($\chi^2 = 9.97$, $df = 2$, $p < 0.01$). Interestingly, most grooming bouts took place in Quadrat 3, which is adjacent to the observation platform, but above it and separated from it by a chain-link fence (see Figure 2.1(a)). Time spent in body contact was also greater in this quadrat and during the tourist condition. The tendency for YA1 to groom more while tourists were present was surprising, but could be interpreted as a stress-buffering mechanism given the protected location in which it took place. Past research has pointed to the stress-buffering potential of positive social contact (e.g. Cohen *et al.*, 1992; Gust *et al.*, 1994, 1996), and grooming in particular has a profoundly calming effect physiologically (Boccia *et al.*, 1989).

Tourist presence versus tourist behavior

In general, threat behavior was not associated with tourist density per se in either group. However, threat behavior was associated with particular human behaviors. In 2004 and 2005, tourist density was unrelated to rates of threatening by monkeys (both those directed to humans and to other monkeys), although humans (keepers, tourists, and researchers) were the most frequent target of monkey threats in both groups. On the other hand, rates of threats per observation period in 2005 were significantly correlated with both the total number of behaviors directed by humans toward monkeys ($r = 0.391$, $df = 38$, $p = 0.014$; see Table 2.1 for an ethogram of tourist behaviors) and with decibel levels on the observation platform ($r = 0.334$, $df = 38$, $p = 0.038$; Ruesto *et al.*, 2010). Thus, both the behavior of the humans and the noise generated by them may negatively affect monkeys. The most common human behavior was pointing (44.97% of 1503 human behaviors recorded), followed by "mouth noise" (16.76%), waving (16.03%), and throwing food (10.18%). Notably, pointing was performed frequently not just by tourists, but also by researchers attempting to locate monkeys.

McCarthy *et al.* (2009) followed up Ruesto's research in 2006 by looking at the sequences of behaviors within interactions between the YA1 monkeys and humans. McCarthy *et al.* found that humans were more likely to initiate sequences than were monkeys (84.6% vs. 15.4%, $n = 487$, binomial $p < 0.05$). Humans used pointing, rail slapping, fleeing, and rock showing, more often in human–monkey sequences than in sequences that consisted only of human behavior. Monkeys were likely to respond to pointing, rail slapping, and rock showing when they occurred, and humans were likely to flee in response to a monkey behavior. Pointing and rail slapping were the human behaviors that most commonly preceded monkey threats to humans. Monkeys typically responded to pointing with a facial threat but more commonly responded to rail slapping by lunging and/or ground slapping. These data suggest that particular human behaviors – including pointing by researchers – provoke aggression in monkeys.

Notably, rail slapping by humans was most often followed by lunging and/or ground slapping by monkeys. The form of these behaviors is similar, suggesting that the monkeys may perceive the humans' behavior as a threat. Anecdotally, we have often noted that humans apparently attempt to mimic monkey behavior to incite or prolong an interaction. While this is entertaining for humans, if monkeys perceive these interactions as aggression, they undoubtedly experience stress, and this may lead to aggressive interactions between monkeys and humans. Such interactions could also lead to redirection of aggression within the monkey groups, possibly contributing to the problem of infant injury and mortality.

Recently, Jones *et al.* (2008) also found that humans' monkey-directed behavior correlated positively with monkey threats to humans in both groups ($r = 0.784$, $df = 43$, $p < 0.001$ for YA1; $r = 0.743$, $df = 55$, $p < 0.001$ for YA2). Staff were responsible for some of the most antagonistic forms of behavior, for example, throwing rocks or tree stumps without provocation, which in one instance caused an injury to an adult male. Thus it is not surprising that staff also received the majority of threats (57%) from both groups, a significantly disproportionate amount given their representation among the humans present ($\chi^2 = 30.09$, $df = 3$, $p < 0.0001$ for YA1, and $\chi^2 = 1497.28$, $df = 3$, $p < 0.0001$ for YA2).

Stress indicators and human presence

While tourist density was not strongly related to rates of threatening by monkeys, human presence nevertheless appeared to contribute to the monkeys' stress levels, as indicated by self-directed behaviors (SDBs). Across both 2005 and 2006, self-scratching was overwhelmingly the most prevalent SDB (253 bouts vs. 21 self-groom bouts and 6 body-shakes in 2006; 171 vs. 23 self-grooms and 1 body-shake in 2005; yawning was not observed either year). Data for 2007 (Mack *et al.*, 2008) indicated that self-grooming rates were positively related to tourist density across all subjects in YA1 ($r = 0.353$, $df = 33$, $p = 0.04$) as well as among adult males specifically ($r = 0.345$, $df = 33$, $p = 0.04$). The rate of SDBs in general, however, appeared to vary by the location of the monkeys relative to the tourists. In 2005, we found no significant correlation between tourist densities and frequencies of SDBs in YA1 subjects across 71 focal samples.

However, exploratory analyses revealed a significant correlation specific to monkeys in Quadrat 1, the only quadrat in which monkeys are surrounded on two sides by human observers ($r = 0.378$, $df = 69$, $p < 0.01$). In YA1, Quadrats 4 to 6 are further away from the pavilion but still visible from it. Data for 2006 confirmed these results across 152 focal samples. In addition, we found a significant correlation between tourist densities and SDB frequencies for monkeys in Quadrat 2, but only among those monkeys who spent above-median amounts of time in that quadrat ($r = 0.388$, $df = 27$, $p < 0.05$). Finally there was a similar correlation for Quadrat 3 ($r = 0.667$, $df = 10$, $p < 0.05$), although individuals spent relatively little time in that quadrat and almost half of the focal individuals (6 out of 13 adults and subadults) were never observed there. We interpreted these data to mean that humans have the potential to induce stress in the monkeys, but that monkeys who are either more susceptible to stress or more stressed on a given day may simply avoid the human-proximate areas. Indeed, recent data (Yenter *et al.*, 2008) suggest that YA1 monkeys spend less time in Quadrats 1 and 2 than in the more distant Quadrats 4 to 6 (Wilcoxon test, $T = 0$, $n = 8$, $p = 0.01$).

Taken together our results suggest that human behavior is associated with elevated aggression and stress in both groups of monkeys. Monkeys appear to avoid locations within the provisioning area that are closest to humans, and in so doing modulate their stress. Nevertheless, the management practice of range restriction and herding of monkeys into provisioning areas undoubtedly limits this ability. YA2 may be particularly vulnerable, as their provisioning area is smaller and the monkeys are necessarily close to humans while in it.

General discussion

In this chapter we described two sets of studies that took a variety of approaches to understanding the effects of tourism on the behavior of Tibetan macaques at Huangshan, China. Berman and her colleagues (2007) examined infant mortality rates and aggressive behavior in a single social group over 19 years, comparing years before management for tourism began, with years of tourist management and with a year in which management was temporarily suspended. Matheson and her colleagues and students (Jones et al., 2008; Mack et al., 2008; Matheson et al., 2006, 2007; McCarthy et al., 2009; Ruesto et al., 2010; Yenter et al., 2008) examined more immediate behavioral responses to human presence and behavior in two social groups that had been exposed to tourism for different amounts of time. Our results are both complementary and congruent, providing a detailed critique of negative effects of tourism on the behavior of Tibetan macaques at Huangshan.

First, management for tourism is strongly associated with high rates of infant mortality. Although infants may not be the initial targets of aggression, infant deaths appear to be triggered by high levels of aggression among adults within the provisioning area. Second, a particular aspect of management, namely artificial range restriction, is closely associated with both infant mortality and adult aggression, and as such we hypothesize that it exacerbates the effects of provisioning on aggression. Range restriction, accomplished by herding, may lead to aggression by increasing intra-group competition for food and/or by raising levels of fear-induced stress. These effects may be particularly intense when tourists are allowed to throw highly palatable foods, such as candy, to the monkeys. Third, while aggression is not directly related to the density of humans present either immediately or over the long term, there is evidence that specific human behaviors do indeed provoke aggression in the monkeys. Monkeys frequently respond to herding, human noise, pointing, rail-slapping, and rock-showing with threats directed to the offending humans, and sometimes with lunges and physical attacks. They direct the most threats at keepers, some of whom behave antagonistically toward them. Finally, there is also evidence that human density alone is associated with increases in stress indicators such as self-scratching, self-grooming, and body shaking.

Given the negative effects of specific human behaviors on the monkeys, we suggest that management adopt stricter rules regarding human behavior and enforce them more consistently. In particular, interaction with the monkeys by tourists, staff, and observers (in the case of pointing) should be vigorously discouraged. Most seriously, we believe that the behavior of the staff themselves needs to be addressed. During 2002 and 2005 in particular, some staff invited tourists to hand feed the monkeys, sometimes selling them cans of corn for this purpose and encouraging adult male

monkeys onto the platforms to enhance photographic opportunities; these practices reportedly resumed in 2011 (E. Dunayer, personal communication). And, as
noted above, the herding of the monkeys produces fearful behavior and may lead to
aggression both within the groups and directed to humans. In 2010, the park war-
dens also began using fire-crackers thrown at the monkeys to herd them (Matheson,
personal observation). In 2011, aggressive herding caused an adult male monkey to
fall from a cliff, resulting in an apparently broken arm (E. Dunayer, personal communication). Given this treatment, and unprovoked antagonistic behavior by some
keepers, it is not surprising that monkeys show clear indications of stress and direct
disproportionate amounts of aggression toward their keepers.

The management policy at Huangshan was deliberately designed to avoid the
severe problems of unregulated interaction between tourists and Tibetan macaques
experienced at Mt. Emei, where many tourists have been seriously attacked by
monkeys seeking food handouts. Indeed some attacks have caused tourists to fall
to their deaths, and on some occasions humans have retaliated violently, seriously
injuring monkeys (Zhao, 2005). While problems of this severity have not occurred
at Huangshan, some effects of the tourist operation are nevertheless counterproductive to the aims of conservation for this species. Range restriction continues to
the present, as does uneven enforcement of rules against interaction between tourists and monkeys. Thus it should come as no surprise that infant deaths by wounding also continue to the present. Clearly, high infant mortality is unsustainable and
counterproductive to the long-term success of both tourism and the population.
We strongly recommend the following:

(1) an immediate end to artificial range restriction;
(2) the institution of a management policy that does not guarantee that tourists
will see the monkeys on any given day;
(3) a stronger emphasis on education of tourists about the value of wildlife conservation, the behavior and ecology of Tibetan macaques, and on the need for
appropriate behavior on their part before they enter the viewing pavilion;
(4) a prohibition on the sale, showing, or eating of food by humans on the staircases or in the viewing pavilions;
(5) a requirement that keepers escort groups of tourists to the platforms, to avoid
potentially aggressive interactions between monkeys and humans on the stairs
leading to the platforms;
(6) the construction of barriers (e.g. chain-link) between tourists and monkeys
in the viewing pavilion to limit interaction and the transmission of disease
(particularly given that monkeys are frequently seen with used tissues or water
bottles that tourists have dropped in the monkey areas);
(7) the use of monkey-proof trash containers along the staircases and in the
pavilions;
(8) education of keepers and observers about both appropriate methods to control monkey behavior and to prevent provoking stress as well as about behavior that humans should not use under any circumstances;

(9) education of keepers about the problems associated with hostile behavior, particularly when it is unprovoked;

(10) the institution of penalties for keepers who behave violently toward monkeys or use other inappropriate behavior; and

(11) continued research and monitoring of both long-term and immediate effects of tourist management practices on the behavior and ecology of macaques.

Ideally, we suggest the formation of an advisory committee, composed of experts in conservation, primate behavior, wildlife management, and ecology as well as park managers and representatives of the local community, to review and approve management practices periodically. We are optimistic that with these measures, a valuable natural resource, source of local income, and educational/recreational opportunity could be put back on track to achieve its original goals.

Acknowledgments

We thank the Huangshan Monkey Management Center and the Huangshan Garden Forest Bureau for permission to carry out research at Mt. Huangshan. CMB, Li, and CSI received financial support from the Leakey Foundation, the Wenner-Gren Foundation, the National Geographic Society, National Natural Science Foundation of China, Key Teacher Program of the Ministry of Education of China, the Excellent Youth Foundation of Anhui, the Mark Diamond Foundation, and the Naroll Research Fund. MDM's team received some funding from Central Washington University's Office of Graduate Studies and Research. We thank May Lee Gong, Krista Jones, Stephan Menu, Stephanie Pieddesaux, Justin Sloan, and Lei Zhang for field assistance. The research complied with protocols approved by Institutional Animal Care and Use Committees (IACUC) at the University at Buffalo and Central Washington University and adhered to the legal requirements of the People's Republic of China.

References

Adams, H. R., Sleeman, J., Rwego, I., and New J. C. (2001). Self-reported medical history survey of humans as a measure of health risk to chimpanzees (*Pan troglodytes schweinfurthii*) of Kibale National Park, Uganda. *Oryx*, 35: 308–312.

Altmann, J. (1974). Observational study of behavior: Sampling methods. *Behaviour*, 49: 227–266.

Altmann, J. (1988). Foreword. In: J. E. Fa and C. H. Southwick (eds.), *Ecology and Behavior of Food Enhanced Primate Groups*. New York: Alan R. Liss, pp. 9–13.

Berman, C. M., Ionica, C. S., and Li, J. H. (2004). Dominance style among *Macaca thibetana* on Mt. Huangshan, China. *International Journal of Primatology*, 25: 1283–1312.

Berman, C. M. and Li, J. H. (2002). Impact of translocation, provisioning and range restriction on a group of *Macaca thibetana*. *International Journal of Primatology*, 23: 283–297.

Berman, C. M., Li, J. H., Ogawa, H., Ionica, C., and Yin, H. (2007). Primate tourism, range restriction and infant risk among *Macaca thibetana* at Mt. Huangshan, China. *International Journal of Primatology*, 28: 1123–1141.

Berman, C. M. and Schwartz, S. (1988). A non-intrusive method for determining relative body fat in free-ranging rhesus monkeys. *American Journal of Primatology*, 14: 53–64.

Boccia, M. L., Reite, M., and Laudenslager, M. (1989). On the physiology of grooming in a pigtail macaque. *Physiology and Behavior*, 45: 667–670.

Boinski, S. and Sirot, L. (1997). Uncertain conservation status of squirrel monkeys in Costa Rica, *Saimiri oestdi oerstedi* and *Saimiri oerstedi citrinellus*. *Folia Primatologica*, 68: 181–193.

Brandon, K. E. and Wells, M. (1992). Planning for people and parks: design dilemmas. *World Development*, 20: 557–570.

Butynski, T. M. (2001). Africa's great apes. In: B. Beck, S. Stoinski, M. Hutchins *et al.* (eds.), *Great Apes and Humans: The Ethics of Coexistence*. Washington, DC: Smithsonian Institution Press, pp. 3–56.

Cohen, S., Kaplan, J. R., Cunnick, J. E., Manuck, S. B., and Rabin, B. S. (1992). Chronic social stress, affiliation, and cellular immune response in nonhuman primates. *Psychological Science*, 3: 301–304.

de la Torre, S., Snowdon, C. T., and Bejarano, M. (1999). Preliminary study of the effects of ecotourism and human traffic on the howling behavior of red howler monkeys, *Alouatta seniculus*, in Ecuadorian Amazonia. *Neotropical Primates*, 7: 84–86.

de la Torre, S., Snowdon, C. T., and Bejarano, M. (2000). Effects of human activities on wild pygmy marmosets in Ecuadorian Amazonia. *Biological Conservation*, 94: 153–163.

Deng, Z. Y. and Zhao, Q. K. (1987). Social structure in a wild group of *Macaca thibetana* at Mount Emei, China. *Folia Primatologica*, 49: 1–10.

Fuentes, A. (2010). Natural cultural encounters in Bali: monkeys, temples, tourists and ethno-primatology. *Cultural Anthropology*, 25: 600–624.

Goldsmith, M. L., Glick, J., and Ngabirano, E. (2006). Gorillas living on the edge: literally and figuratively. In: N. E. Newton-Fisher, H. Notman, J. D. Paterson, and V. Reynolds (eds.), *Primates of Western Uganda*. New York, NY: Springer, pp. 405–422.

Griffiths, M. and van Schaik, C. P. (1993). The impact of human traffic on the abundance and activity periods of Sumatran rain forest wildlife. *Conservation Biology*, 7: 623–626.

Grossberg, R., Treves, T., and Naughton-Treves, L. (2003). The incidental ecotourist: measuring visitor impacts on endangered howler monkeys at a Belizean archaeological site. *Environmental Conservation*, 30: 40–51.

Gust, D. A., Gordon, T. P. Brodie, A. R., and McClure, J. H. (1994). Effect of a preferred companion in modulating stress in adult female rhesus monkeys. *Physiology & Behavior*, 55: 681–684.

Gust, D. A., Gordon, T. P., Wilson, M. E., Brodie, A. R. *et al.* (1996). Group formation of female pigtail macaques (*Macaca nemestrina*). *American Journal of Primatology*, 39: 263–274.

Harcourt, A. H., Kineman, J., Campbell, G., Yamagiwa, J., and Redmond, I. (1983). Conservation and the Virunga gorilla population. *African Journal of Ecology*, 21: 139–142.

Harcourt, A. H. and Stewart, K. J. (2007). *Gorilla Society: Conflict, Cooperation and Compromise Between the Sexes*. University of Chicago Press.

Hausfater, G. and Hrdy, S. B. (1984). *Infanticide: Comparative and Evolutionary Perspectives*. Hawthorne, NY: Aldine.

Hill, C. M. (2002). Primate conservation and local communities – ethical issues and debates. *American Anthropology*, 104: 1184–1194.

Hinde, R. A. (1974). *Biological Bases of Human Social Behaviour*. New York: McGraw-Hill.

Hsu, M. J., Kao, C. C., and Agorammoorthy, G. (2009). Interactions between visitors and Formosan macaques (*Macaca cyclopis*) at Shou-Shan Nature Park, Taiwan. *American Journal of Primatology*, 71: 214–222.

IUCN (2012). *The IUCN Red List of Threatened Species, Version 2012.2*. www.iucnredlist.org (accessed April 4, 2014).

Johns, B. G. (1996). Responses of chimpanzees to habituation and tourism in the Kibale forest, Uganda. *Biological Conservation*, 78: 257–262.

Jones, A. M., Matheson, M. D., Sheeran, L. K., Li, J. H. and Wagner, R. S. (2008). Aggression and habituation toward humans in two troops of Tibetan macaques (*Macaca thibetana*) at Mt. Huangshan, China. *American Journal of Primatology*, 70 (Suppl. 1): 61.

Kauffman, L. (2010). Local predictors of primate response to tourists in the Central Suriname Nature Reserve. *American Journal of Physical Anthropology*, 141 (Suppl. 50): 139–140.

Kauffman, L. (2011). Creating sustainable primate-based tourism: a view from the Central Suriname Nature Reserve. *American Journal of Physical Anthropology*, 144 (Suppl. 52): 183.

Kauffman, L. and Boinski, S. (2008). Do primates see ecotourists as potential predators? *American Journal of Physical Anthropology*, 135 (Suppl. 46): 128.

Kinnaird, M. F. and O'Brien, T. G. (1996). Ecotourism in the Tangkoko DuaSudara Nature Reserve: opening Pandora's box? *Oryx*, 30: 65–73.

Klailova, M., Hodgkinson, C., and Lee, P. C. (2010). Behavioral responses of one western lowland gorilla (*Gorilla gorilla gorilla*) group at Bai Hokou, Central African Republic, to tourists, researchers and trackers. *American Journal of Primatology*, 72: 897–906.

Koganezawa, M. and Imaki, H. (1999). The effects of food sources on Japanese monkey home range size and location, and population dynamics. *Primates*, 40: 177–185.

Köndgen, S., Kuhl, H., N'Goran, P. K., Walsh, P. D., Schenk, S. *et al.* (2008). Pandemic human viruses cause decline of endangered great apes. *Current Biology*, 18: 260–264.

Leary, H. O. and Fa, J. E. (1993). Effects of tourists on Barbary macaques at Gibraltar. *Folia Primatologica*, 61: 77–91.

Lee, P. C., Brennen, E. J., Else, J. G., and Altmann, J. (1986). Ecology and behavior of vervet monkeys in a tourist lodge habitat. In: J. G. Else, and P. C. Lee (eds.), *Primate Ecology and Conservation*. Cambridge University Press, pp. 229–235.

Li, J. (1999). *The Tibetan Macaque Society: a Field Study*. (Partially translated by Lei Zhang.) Hefei, China: Anhui University Press.

Li, J. H. and Wang, Q. S. (1996). Dominance hierarchy and its chronic changes in adult male Tibetan macaques (*Macaca thibetana*). *Acta Zoologica Sinica*, 42: 330–334.

Li, J. H. Wang, Q. S. and Han, D. S. (1996a). Fission in a free-ranging Tibetan macaque group at Huangshan Mountain, China. *Chinese Science Bulletin*, 41: 1377–1381.

Li, J. H. Wang, Q. S. and Li, M. (1996b). Migration of male Tibetan macaques (*Macaca thibetana*) at Mt. Huangshan, Anhui Province, China. *Acta Theriologica Sinica*, 16: 1–6.

Li, J. H. Yin, H., and Wang, Q. S. (2005). Seasonality of reproduction and sexual activity in female Tibetan macaques *Macaca thibetana* at Huangshan, China. *Acta Zoologica Sinica*, 51: 365–375.

Lyles, A. and Dobson, A. (1988). Dynamics of provisioned and unprovisioned primate populations. In: J. E. Fa and C. H. Southwick (eds.), *Ecology and Behavior of Food-enhanced Primate Groups*. New York: Alan R. Liss, Inc., pp. 167–198.

Mack, H., Matheson, M. D., Sheeran, L. K., Li, J. H. and Wagner, R. S. (2008). Grooming behavior of Tibetan macaques (*Macaca thibetana*) in the presence of tourists at Mt. Huangshan, China. *American Journal of Primatology*, 70 (Suppl. 1): 59.

Maestripieri, D., Schino, G., Aureli, F., and Troisi, A. (1992). A modest proposal: Displacement activities as an indicator of emotions in primates. *Animal Behaviour*, 44: 967–979.

Maréchal, L., Semple, S., Majolo, B., Quarro, M., Heistermann, M., and MacLarnon, A. (2011). Impacts of tourism on anxiety and physiological stress levels in wild male Barbary macaques. *Biological Conservation*, 144: 2188–2193.

Matheson, M. D., Hartel, J., Whitaker, C., *et al.* (2007). Self-directed behavior correlates with tourist density in free-living Tibetan macaques (*Macaca thibetana*) at the Valley of the Wild Monkeys, Mt. Huangshan, China. *American Journal of Primatology*, 69 (Suppl. 1): 41–42.

Matheson, M. D., Sheeran, L. K., Li, J. H., and Wagner, R. S. (2006). Tourist impact on Tibetan macaques. *Anthrozoos*, 19: 158–68.

McCarthy, M. S., Matheson, M. D., Lester, J. D., Sheeran, L. K., Li, J. H., and Wagner, R. S. (2009). Sequences of Tibetan macaque (*Macaca thibetana*) and tourist behaviors at Mt. Huangshan, China. *Primate Conservation*, 24: 145–151.

Ogawa, H. (2006). *Wily Monkeys: Social Intelligence of Tibetan Macaques*. Kyoto University Press.

Ruesto, L. A., Sheeran, L. K., Matheson, M. D., Li, J. H., and Wagner, R. S. (2010). Tourist behavior and decibel level correlate with threat frequency in Tibetan macaques (*Macaca thibetana*) at Mt. Huangshan, China. *Primate Conservation*, 25: 99–104.

Schino, G., Perretta, G., Taglioni, A. M., Monaco, V., and Troisi, A. (1996). Primate displacement activities as an ethopharmacological model of anxiety. *Anxiety*, 2: 186–191.

Small, M. (1995). Hanuman's troops, servants of the gods. *Garuda Air Magazine*, Jan, 26–28.

Sugiyama, Y. and Ohsawa, H. (1982). Population dynamics of Japanese monkeys with special reference to the effect of artificial feeding. *Folia Primatologica*, 39, 238–263.

Treves, A. and Brandon, K. (2005). Tourist impacts on the behavior of black howling monkeys (*Alouatta pigra*) at Lamanai, Belize. In: J. Paterson and J. Wallis (eds.), *Commensalism and Conflict: The Human-primate Interface*. Norman, OK: American Society of Primatologists, pp. 146–167.

van Schaik, C. P. (2000). Infanticide by males: the sexual selection hypothesis revisited. In: C. P. van Schaik and C. H. Janson (eds.), *Infanticide by Males and its Implications*. Cambridge University Press, pp. 361–387.

Wada, K. and Xiong, C.-P. (1996). Population changes of Tibetan monkeys with special regard to birth interval. In: T. Shotake and K. Wada (eds.), *Variations in the Asian Macaques*. Tokyo: Tokai University Press, pp. 133–145.

Wallis, J. and Lee, D. R. (1999). Primate conservation: the prevention of disease transmission. *International Journal of Primatology*, 20: 803–826.

Westin, J. and Kauffman, L. (2011). Tourism in Suriname: do monkeys view tourists as predators or conspecifics? *American Journal of Physical Anthropology*, 144 (Suppl. 52): 309–310.

Woodford, M. H., Butynski, T., and Karesh, W. B. (2002). Habituating the great apes: the disease risks. *Oryx*, 36: 153–160.

Wrangham, R.W. (2001). Moral decisions about wild chimpanzees. In: B. Beck, S. Stoinski, M. Hutchins, *et al.* (eds.), *Great Apes and Humans. The Ethics of Coexistence*. Washington, DC: Smithsonian Institution Press, pp. 230–244.

Yenter, T. A., Matheson, M. D., Sheeran, L. K., Li, J. H. and Wagner, R. S. (2008). Self-directed behaviors, tourist density and proximity in a free-living population of Tibetan macaques (*Macaca thibetana*) at an ecotourism destination in Anhui Province, China. *American Journal of Primatology*, 70 (Suppl. 1): 52.

Yin, H., Li, J. H., Zhou, L., *et al.* (2004). Birth of Tibetan monkey (*Macaca thibetana*) in winter. *Acta Theriologica Sinica*, 24: 19–22.

Zhao, Q. K. (1996). Etho-ecology of Tibetan macaques at Mount Emei, China. In: J. Fa and D. G. Lindburg (eds.), *Evolution and Ecology of Macaque Societies*. Cambridge University Press, pp. 263–289.

Zhao, Q. K. (2005). Tibetan macaques, visitors, and local people at Mt. Emei: Problems and countermeasures. In: J. Paterson and J. Wallis. (eds.), *Commensalism and Conflict: The Human-primate Interface*. Norman, OK: American Society of Primatologists, pp. 376–399.

3 Provisioning and tourism in free-ranging Japanese macaques

Hiroyuki Kurita

Introduction

Primate-focused tourism in Japan began with provisioning after World War II. Researchers from Kyoto University formed the Primate Research Group in 1951 and searched extensively for good sites to study Japanese macaques (*Macaca fuscata*) in their natural habitats (Yamagiwa, 2010). The macaques were frightened of humans, however, because of past hunting and chasing pressures from the farmland inhabitants. Therefore, researchers tried provisioning to habituate macaques to humans and this finally succeeded at Koshima and Takasakiyama in 1952 (Yamagiwa, 2010). Provisioning allowed researchers to conduct long-term observational studies with individual identification that produced many important findings on the elements of macaque social organization, such as dominance rank among individuals and kin groups, and cultural behaviors (reviewed by Yamagiwa, 2010).

These scientific findings, along with newspaper articles and TV programs on Japanese macaques and their behavior, made these macaques interesting to people other than researchers and led to the establishment of commercial free-ranging monkey parks. Such monkey parks, where visitors could observe provisioned free-ranging Japanese macaques up close, were established in Japan from the 1950s (Mito & Watanabe, 1999; Nakagawa *et al.*, 2010; Yamagiwa, 2010). They became an important recreational activity for the Japanese people at a time when the country as a whole was recovering from World War II. Mito and Watanabe (1999) identified five factors that led to the popularity of such monkey parks in Japan at that time: (1) there was little opportunity for recreation after World War II; (2) such monkey parks needed little initial investment; (3) visitors could observe wild macaques without being separated from them by fences or cages; (4) park managers felt their jobs were worthwhile because people enjoyed visiting free-ranging monkey parks; and (5) the development of primatology in Japan increased Japanese people's interest in the social structure and behavior of Japanese macaques. People who visited monkey parks were delighted to learn about new findings by primatologists. The provisioning that allowed visitors to observe these macaques easily, however,

Primate Tourism: A Tool for Conservation?, ed. Anne E. Russon and Janette Wallis. Published by Cambridge University Press. © Cambridge University Press 2014.

Figure 3.1   Map of Japan with boundaries of prefectures and locations referred to in this chapter.

eventually caused drastic increases in macaque population sizes and this in turn led to agricultural and forest damage.

The aim of this chapter is to assess the past, present, and future roles of tourism in free-ranging monkey parks in the conservation of Japanese macaques, especially relative to the provisioning that has been involved.

Establishment of provisioning and monkey parks

The provisioning of Japanese macaques began at Koshima and Takasakiyama in 1952 (Figure 3.1). It stimulated the interest of researchers as well as the Japanese people and the tourism industry in Japan. Oita City Municipal Office began provisioning free-ranging Japanese macaques in 1952 and opened the Takasakiyama Natural Zoo in 1953 to attract tourists. The zoo's aims included allowing tourists to observe macaques at close range (Yamagiwa, 2010). Japanese macaques and their social life were introduced to the public through the media, in newspapers, journals, films, and TV programs. As a result many tourists paid to visit Takasakiyama for monkey watching (Yamagiwa, 2010). Forty-one free-ranging monkey parks were established between 1952 and 1972; 19 of them, including Koshima and Takasakiyama, opened in the 1950s and 17 more in the 1960s (Mito & Watanabe, 1999). Japanese macaques were native to 30 of these parks and translocated from their original habitats at 11 others (Mito & Watanabe, 1999). The origin of the

Table 3.1 Reproductive parameters of mature females aged 5 years or more in non-provisioned and provisioned Japanese macaque populations

	Yakushima	Kinkazan	Ryozen	Shiga B2	Arashiyama	Katsuyama
Condition[a]	NP	NP	NP	NP	P	P
(1)	0.270	0.353	0.336	0.349	0.538	0.495
(2)	0.25	0.227	0.277	0.533	0.103	0.102
(3)	1.5	1.59	—	—	1.15	1.29
(4)	2.24	2.37	—	—	1.46	1.58

[a] NP: non-provisioned population; P: provisioned population.
(1) Birth rate (births/female/year); (2) infant mortality within one year; (3) average interbirth interval following the death of infants within one year after birth (yr); (4) average interbirth interval following surviving infants (yr).
Source: Cited from Takahata *et al.* (1998).

Miyajima population, for example, was 47 animals from a troop captured as agricultural pests in Shodoshima (Kanaiduka, 2002). The establishment of many free-ranging monkey parks and the related newspaper and TV coverage were the basis for informing Japanese people about the behaviors and society organization of Japanese macaques.

Although one important purpose of establishing free-ranging monkey parks was the promotion of tourism, another was preventing crop-raiding. Because macaques were damaging agricultural crops around Takasakiyama, Mr. Ueda, then the Mayor of Oita, began provisioning macaques at Takasakiyama in 1952 not only to promote tourism but also to prevent crop damage. Provisioning aimed to draw monkeys away from crop fields by attracting them to the zoo's feeding ground.

Population dynamics of provisioned macaque populations

The provisioned foods provided to Japanese macaques contain more soluble carbohydrate (e.g. sweet potato and wheat), crude protein (soybean), or lipid (peanut) but less crude fiber than natural resources, so provisions are more digestible (Iwamoto, 1988). For example, Japanese macaques can extract 1.72 times more energy per gram of food from wheat than they can from natural leaves (Iwamoto, 1988). Provisioning with these higher-energy/lower-fiber foods dramatically improved the provisioned monkeys' nutritional levels and, as a result, increased their population size greatly. In several provisioned populations, reproductive parameters superior to those of non-provisioned populations were found (Table 3.1).

At Takasakiyama, in 1950, before provisioning started, 166 macaques were counted (Itani *et al.*, 1964). Standardized macaque censuses were established at Takasakiyama in 1970 and have been conducted annually since (Kurita *et al.*, 2008).

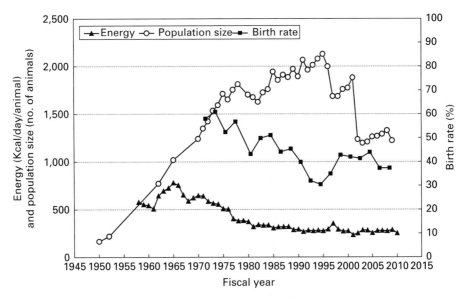

Figure 3.2 Changes in energy (kcal) of food given to macaques (1958–2010), population size of macaques at Takasakiyama (1950–2009), and birth rate (%) of macaques for every 2 years at Takasakiyama (1971–2009). *Sources.* Energy – 1958–1987: Sugiyama and Ohsawa (1988); 1988–1999: Takasakiyama Management Committee (2001); 2000–2010: Takasakiyama Natural Zoo, unpublished data. Population size – 1950–2001: Kurita *et al.*, 2008; 2002–2009: Takasakiyama Natural Zoo (2009). Birth rate – TakasakiyamaNatural Zoo (1971–2009).

They showed that the macaque population grew rapidly after the establishment of the park because of heavy provisioning (Figure 3.2).

Impacts of provisioning-induced population increase on Japanese macaque conservation

Provisioning-induced increases in the population size of Japanese macaques at monkey parks have had strong impacts on Japanese macaque ecology and human life. Provisioning resulted in serious ecological damage because increasing the macaque population led to increased macaque feeding on natural vegetation. For example, the high density of macaques in and around Takasakiyama seriously damaged the forest (Kurita *et al.*, 2008). To reverse the macaque population growth rate, caretakers at Takasakiyama have decreased provisioned foods since 1965 (Kurita *et al.*, 2008; see Figure 3.2). Over the period when the amount of provisioned foods was decreased (1973 to1989), edible trees for macaques also decreased in Takasakiyama forest. Yokota (1993) compared composition ratios of vegetation at Takasakiyama Forest between 1973 and 1989 for the four types of forest where macaques often foraged for natural food. For this period, he found no change or

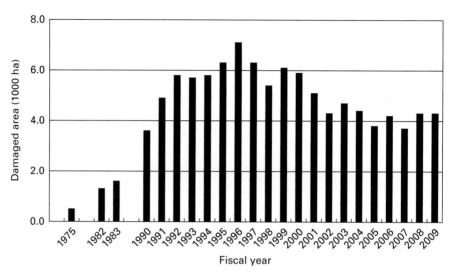

Figure 3.3   Area of agricultural damage caused by Japanese macaques in Japan. Modified from Muroyama and Yamada (2010) with permission, adding data from annual statistical reports on agricultural damage caused by wildlife in Japan (Ministry of Agriculture, Forestry, and Fisheries, Japan, 2007–2009).

only slight decreases in the ratios of the number of patches of three of these four forest types to all forest patches (*Quercus glauca – Cinnamomum camphora, Q. serrata, Q. acutissima,* and *Castanopsis – Quercus*) but substantial decreases (8.5%) in artificial forests or fields such as *Cryptomeria japonica* and *Chamaecyparis obtuse,* bamboo forest, orchard, rice fields, and weed fields. In addition, Yokota and Ono (1993) reported an instance in which soil that had been trodden on by too many macaques prevented the growth of seedlings and roots in Takasakiyama Forest. For macaques that ranged there, such serious damage to vegetation led to a deterioration of the habitat more broadly, decreases in the amount of food available to them from natural resources, and degradation of their living environment. In addition, the deterioration of vegetation also affected agricultural activities because it resulted in monkeys migrating into crop fields. At Takasakiyama, individuals from the large macaque troops emigrated from their original habitat inside the natural park and invaded agricultural fields and started crop-raiding.

It was not only individuals from provisioned macaque troops that raided crops. Serious crop-raiding by Japanese macaques was exacerbated by several other social factors throughout Japan. In 1947, because the macaque population was estimated to have suffered serious declines in various parts of Japan, hunting regulations were introduced (Muroyama & Yamada, 2010). In addition, as a consequence of a general change in Japan's economic structure after World War II, the human population declined in the countryside and the ratio of elderly to total residents increased in rural communities. One result, in recent decades, is that rural communities no longer have the workforce or the funds to protect their crops from raiding wildlife (Muroyama & Yamada, 2010). As a consequence of poorly controlled crop-

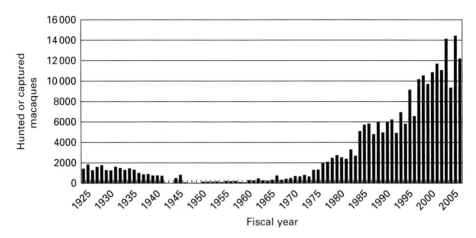

Figure 3.4   Number of Japanese macaques removed by hunting (up to 1948) and captured as nuisance animals (after 1948) or as subjects of population control under the Specified Wildlife Conservation and Management Plan (after 2001). Modified from Muroyama and Yamada (2010) with permission, adding data from annual statistical reports on wildlife (Ministry of the Environment, Japan, 2005–2007).

raiding and feeding on readily digestible food, the distribution and population size of macaques living around human settlements increased. This in turn increased their crop-raiding in many areas, and led to culling of macaques as agricultural pests (Figures 3.3 and 3.4).

Another bad consequence of provisioning has been increases in physical accidents between macaques and humans, because provisioning led to a reduction of the distance between them. Visitors were initially allowed to feed monkeys at Takasakiyama. This was effective in attracting people to the monkey park, but also further enriched the monkeys' nutritional condition and increased the number of accidents involving visitors and macaques (Sugiyama, 1993). For this reason, the park's food-selling shop was closed in 1993 and, since then, the feeding of macaques by visitors has been strictly prohibited. As a result, few incidents have occurred recently.

Accidents involving visitors and macaques have also occurred at places other than free-ranging monkey parks. At Nikko, north of Tokyo, there were incidents of tourists who were feeding macaques being attacked by macaques on roads (Koganezawa, 2002). At Hakone, macaques were sometimes run over by cars because tourists were feeding them from cars (Okano, 2002). Thus, reducing the distance between humans and macaques because of feeding has negative effects on both macaques and humans. These problems are similar to those that have resulted from habituating other wild primates for tourism (e.g. in this volume, see chapters by Berman *et al.*, Dellatore *et al.*, Goldsmith, Russon, & Susilo, Williamson & Macfie). What is necessary is humans keeping a suitable distance from wildlife and refraining from interfering with them (Muroyama, 2000). Keeping a suitable distance can also help both humans and monkeys avoid disease transmission.

Effects of ceasing or reducing provisioning on macaque populations

Because provisioning caused a marked increase in macaque population sizes, in 1965, the Oita City Municipal Office began to decrease the amount of food provided to macaques at Takasakiyama Natural Zoo. As a result, the physical condition of adult females decreased, the primiparous age increased, the birth rate decreased, and the infant mortality rate increased, such that the annual population growth rate dropped substantially from the 1970s (Kurita et al., 2008) (see Table 3.2, Figure 3.2). Decreasing the amount of provisioned foods or ceasing provisioning altogether also decreased reproductive parameters in some other populations (Table 3.2).

Decreasing provisioning could temporarily increase macaques' exploitation of forests and crop-raiding because hunger could lead them to forage in natural forests or farmers' fields more intensively. However, decreasing the macaque population growth rates by decreasing their reproductive rates ultimately relieved both problems by decreasing the impact of the macaques' foraging. Yamada, who began surveying Takasakiyama forest in 2000, for instance, found that the forest's condition has remained stable since 2000 (Yamada, 2008). The decrease in macaque population density may also lead to decreased crop-raiding in areas that surround provisioned populations. Ceasing provisioning altogether worsened some forms of conflict between macaques and humans. According to Okano (2002), one macaque troop changed its home range area because provisioning ceased, which caused it to increase crop-raiding at Hakone.

Provisioning by free-ranging monkey parks has led to increased macaque population sizes, and these larger macaque populations cannot satisfy their nutritional needs from natural resources alone. Soumah and Yokota (1991), who studied the feeding behavior of Takasakiyama macaques in 1987 and 1988, estimated that high-ranking females obtained about three-quarters of their daily digested energy from provisioned foods and one-quarter from natural foods, and low-ranking females obtained approximately equal proportions from provisioned and natural food sources. Without provisioning, many of these macaques would move from the parks into nearby agricultural areas, resulting in large-scale damage to agricultural crops. The size of provisioned populations clearly needs to be managed to protect both monkeys and humans.

Therefore, to manage provisioned macaque populations effectively, that is to lower their reproductive rates and prevent them ruining the forest and farmers' crops, existing provisioning needs to be continued to some extent and for some period. It should not be stopped or drastically reduced suddenly: both these options threaten the survival of the macaques by causing severe food shortages which in turn lead to increased rather than decreased conflict with humans. Critical to managing the size of provisioned Japanese macaque populations is controlling birth rates. Gradually decreasing provisioned foods combined with monitoring the macaques' nutritional condition could help reduce birth rates; temporary birth control is another possibility (Sugiyama, 1997; Sugiyama et al., 1995). Clearly, new provisioning should not be started.

Table 3.2 Comparison of reproductive parameters before and after the decrease in the amount of provisioned foods among provisioned, free-ranging Japanese macaque populations

Population	Reproductive parameter			
	Mortality (%) within one year after birth	Birth percentage (%) of females aged 5 years and over	Population density (no./km²)	Annual population growth rate (%)
Takasakiyama				
	4.5 (1956–62)	54.2 (1971–80)	55.3 (1950)	13 (1952–62)
	24.2 (1987–93)	45.6 (1981–90)	413.7 (1970)	4.0 (1965–70)
		39.1 (1991–2000)	568.0 (1980)	3.2 (1970–80)
			631.3 (1990)	1.1 (1980–90)
			591.7 (2000)	−0.65 (1990–2000)
Ryozen[a]				
1969–73	14.58[c]	59.26 (5–19 years-old)		13.4[e]
1974–81	27.66[c]	33.58 (6–19 years-old)		4.9[e]
Koshima[b]				
1952–63	19	47.0	70.0 (1952)[d]	9.6 (1952–71)[d]
1964–71	19	62.5	386.7 (1972)[d]	
1972–77	43	20.9	293.3 (1977)[d]	−5.4 (1972–77)[d]

[a] Macaques had been provisioned until 1973; [b] 1952–63: infrequent provisioning period, 1964–71: intensive provisioning period, 1972–77: severely restricted provisioning period; [c] mortality within 2 years after birth; [d] calculated using the population size or the island area, about 30 ha; [e] males older than 6 years were excluded.

Source: Cited from Kurita *et al.* (2008).

Discussion

Many of the free-ranging monkey parks established in the 1950s and 1960s in Japan went out of business around the 1970s. Mito and Watanabe (1999) gave three reasons for such closures: (1) a marked increase in feeding costs linked with increasing macaque populations resulting from provisioning; (2) increasing management costs, such as compensation for agricultural damage caused by macaques; and (3) decreasing numbers of visitors because the novelty of monkey parks was wearing off. Although the Takasakiyama Natural Zoo has survived as a business, it has experienced all of the above problems. The number of annual visitors was approximately 0.6 million in 1954. It increased gradually to 1.0 million in 1960 and peaked at 1.9 million in 1965. Annual visitor numbers began to decrease after 1965 and have recently been under 0.3 million (Oita City 2008; unpublished data).

Some have emphasized the negative effects of provisioning macaques at free-ranging monkey parks but presented no constructive proposal for improving them.

Simply closing these monkey parks, however, can lead to decreases in survival rates of the monkeys and/or increases in crop-raiding because of food shortages. These parks have made and can continue to make positive contributions to understanding and managing Japanese macaques. Ecotourism is considered an effective measure for promoting the conservation of Japanese macaques (Yamagiwa, 2010), and monkey parks could make contributions similar to those proposed for ecotourism. While the most important mission of many free-ranging monkey parks has been promoting sightseeing, these parks have also played an important role in enabling the research that built understanding of Japanese macaques and they have increased the importance of their educational mission (Kurita, 2007). Free-ranging monkey park staff explain Japanese macaques' behavior, life history, mother–offspring relationship, diet and feeding strategy, and other aspects of their lives to visitors. They also explain to visitors how to behave around free-ranging animals at the monkey park and/or in the forest and how to handle conflicts with macaques. This makes free-ranging monkey parks among the best locations for people to learn about the relationships among humans, wildlife, and the environment. After a history of several decades, most of the free-ranging monkey parks in Japan have now generated large amounts of data on demographic parameters (e.g. Itoigawa et al., 1992; Koyama et al., 1992; Kurita et al., 2008; Watanabe et al., 1992). They have also acquired the knowledge needed to educate visitors about the ecology and behavior of macaques and the expertise and facilities to handle macaque–human problems when they arise. Free-ranging monkey parks could devise plans to become places of nature and science education, which would support macaque conservation through a better understanding of Japanese macaque ecology and behavior.

Some monkey parks have already begun nature and science education (e.g. Kanaiduka, 2002; Oita City, 2003) and they could provide educational programs on a much larger scale (Kurita, 2007). At Takasakiyama, in 1983, zoo staff began to publish bulletins about Takasakiyama macaques and give public lectures (Kurita, 2008). The Takasakiyama Natural Zoo established the "Takasakiyama Members Club" in 1994 and began a series of bulletins for members and meetings on the ecology and behavior of Japanese macaques at Takasakiyama (Kurita, 2008; Oita City, 2003). It provides students with experiential education about the ecology and behavior of Japanese macaques and about the jobs of the zoo staff. In 2010, zoo staff began visiting elementary schools to give talks on Takasakiyama macaques. These educational activities should serve the conservation of Japanese macaques not only through improving understanding of their ecology and human–animal relationships but also through having a sense of closeness toward wildlife living close to humans.

The accumulated ecological data from free-ranging monkey parks are also very helpful for macaque conservation. Information on Japanese macaques' diet is useful in selecting tree species to plant in deteriorated forests, and information on relationships between the amount of provisioned foods, how provisions are offered, and macaque migration into crop fields should be helpful in managing provisioned troops to prevent crop-raiding.

The Takasakiyama area is part of the Setonaikai National Park, and the Japanese macaques at Takasakiyama and their habitat are designated and maintained as a national "Natural Monument" because of their value as wildlife living in urban suburbs. National Parks and national Natural Monuments are based on Japan's Natural Park and Cultural Property Protection Laws, which regulate various kinds of land development and human activities. Protecting Japanese macaques at Takasakiyama requires preserving the Takasakiyama forest as their habitat. The above laws mean that development within the Takasakiyama area, such as constructing houses and cutting down trees, is strictly regulated. Although this tourism has contributed to understanding and protecting the area's wildlife and their habitat, it often needs development such as construction and maintenance of roads. Therefore, legal regulations are needed to balance conserving wildlife habitats with the minimum development needed to sustain human support for their conservation.

Interference with wildlife to enable tourism, notably provisioning and habituation to humans that accompanies it, has had negative impacts on wildlife conservation and wildlife–human relationships here and elsewhere (in this volume, see chapters by Dellatore, Goldsmith, Kauffman, Russon & Susilo, Sapolsky, Williamson & Macfie). For visitors and area residents alike, to ease such negative impacts, it is important to maintain an appropriate distance from Japanese macaques and other wildlife. Achieving this requires educating people about appropriate behavior around wildlife, enforcing appropriate behavior, and regulating human development legally. One of the most important missions for free-ranging monkey parks now may be taking a leading role in this education.

Acknowledgments

I especially thank Professors Anne E. Russon and Janette Wallis for giving me the opportunity to write this chapter. I express my special thanks to Emeritus Professor Yukimaru Sugiyama and anonymous reviewers for their valuable comments on an earlier draft. Sincere thanks are also due to the Division of Tourism, Oita City and Takasakiyama Natural Zoo for permission to use their unpublished data, and Professor Yasuyuki Muroyama for his helpful information. I adhered to the guidelines for studies on free-ranging animals made by the Primate Research Institute, Kyoto University.

References

Itani, J., Tokuda, K., Furuya, Y., Kano, K., and Shin, Y. (1964). Social composition in wild Japanese monkeys at Takasakiyama. In: J. Itani, J. Ikeda, and T. Tanaka (eds.), *Wild Japanese Monkeys in Takasakiyama*. Tokyo: Keiso-Shobo, pp. 3–41 (in Japanese).
Itoigawa, N., Tanaka, T., Ukai, N., *et al.* (1992). Demography and reproductive parameters of a free-ranging group of Japanese macaques (*Macaca fuscata*) at Katsuyama. *Primates*, 33: 49–68.

Iwamoto, T. (1988). Food and energetics of provisioned wild Japanese macaques (*Macaca fuscata*). In: J. E. Fa and C. H. Southwick (eds.), *Ecology and Behavior of Food-Enhanced Primate Groups*. New York: Alan R. Liss, pp. 79–94.

Kanaiduka, T. (2002). Attempt of eco-museum – Miyajima in Hiroshima Prefecture. In: T. Oi and K. Masui (eds.), *Natural History of Japanese Macaques*.Tokyo: Tokai University Press, pp. 193–212 (in Japanese).

Koganezawa, M. (2002). People feeding, monkeys begging – Nikko, Tochigi Prefecture. In T. Oi and K. Masui (eds.), *Natural History of Japanese Macaques*. Tokyo: Tokai University Press, pp. 78–92 (in Japanese).

Koyama, N., Takahata, Y., Huffman, M. A., Norikoshi, K. and Suzuki, H. (1992). Reproductive parameters of female Japanese macaques: Thirty years data from the Arashiyama troops, Japan. *Primates*, 33: 33–47.

Kurita H. (2007). Preliminary study on the knowledge of junior high school students about Japanese macaques: Recommendations for science education at monkey parks. *Primate Research*, 23: 17–23 (in Japanese with an English summary).

Kurita, H. (2008). The history of Takasakiyama Management Committee. In: H. Kurita (ed.), *Takasakiyama Management Committee Report*. Oita City: Oita, pp. 1–6 (in Japanese).

Kurita, H., Sugiyama, Y., Ohsawa, H., Hamada, Y., and Watanabe, T. (2008). Changes in demographic parameters of *Macaca fuscata* at Takasakiyama in relation to decrease of provisioned foods. *International Journal of Primatology*, 29: 1189–1202.

Ministry of Agriculture, Forestry, and Fisheries, Japan. (2007–2009). *Statistics on Agricultural Damage caused by Wildlife in Japan* (yearly report) [Zenkoku no Yaseichoju ni yoru Nosakumotsu Higaijokyo] (in Japanese).

Ministry of the Environment, Japan. (2005–2007). *Wildlife Statistics* (yearly report) [Choju Kankei Tohkei] (in Japanese).

Mito, Y. and Watanabe, K. (1999). *History of Japanese Macaques with Men*. Tokyo: Tokai University Press (in Japanese).

Muroyama, Y. (2000). Japanese monkeys in villages. In: Y. Sugiyama (ed.), *Primate Ecology: Dynamics of Environments and Behaviors*. Kyoto University Press, pp. 225–247 (in Japanese).

Muroyama, Y. and Yamada, A. (2010). Conservation: present status of the Japanese macaque population and its habitat. In: N. Nakagawa, M. Nakamichi, and H. Sugiura (eds.), *The Japanese Macaques*. Tokyo: Springer, pp. 143–164.

Nakagawa, N., Nakamichi, M., and Sugiura, H. (2010). Preface. In: N. Nakagawa, M. Nakamichi and H. Sugiura (eds.), *The Japanese Macaques*. Tokyo: Springer, pp. v–xv.

Oita City. (2003). *Four Seasons of Takasakiyama V*. Oita: Oita City (in Japanese).

Oita City. (2008). *The History of Takasakiyama*. Oita: Oita City (in Japanese).

Okano M. (2002). Monkeys in hot spring resorts – Seisho Kanagawa Prefecture and Atami Shizuoka Prefecture. In: T. Oi and K. Masui (eds.), *Natural History of Japanese Macaques*. Tokyo: Tokai University Press, pp. 155–176 (in Japanese).

Soumah, A. G. and Yokota, N. (1991). Female rank and feeding strategies in a free-ranging provisioned troop of Japanese macaques. *Folia Primatologica*, 57: 191–200.

Sugiyama, Y. (1993). Problems caused by population increase and decrease of amount of provisioned foods. In: Takasakiyama Management Committee (ed.), *The Management of Japanese Macaques and Environment at Takasakiyama*. Oita: Board of Education, Oita City, pp. 41–50 (in Japanese).

Sugiyama, Y. (1997). Population control of artificially provisioned Japanese monkeys and birth control. *Primate Research*, 13: 91–94 (in Japanese).

Sugiyama, Y., Iwamoto, T., and Ono, Y. (1995). Population control of artificially provisioned Japanese monkeys. *Primate Research*, 11: 197–207 (in Japanese with English summary).

Sugiyama, Y. and Ohsawa, H. (1988). Population dynamics and management of baited Japanese monkeys at Takasakiyama. *Primate Research,* 4: 33–43 (in Japanese).

Takahata, Y., Suzuki, S., Agetsuma, N., Okayasu, N., Sugiura, H., Takahashi, H., *et al.* (1998). Reproduction of wild Japanese macaque females of Yakushima and Kinkazan Islands: A preliminary report. *Primates*, 39: 339–49.

Takasakiyama Management Committee. (2001). *On the Management of Japanese Macaques and the Nature*. Oita: Board of Education, Oita City (in Japanese).

Takasakiyama Natural Zoo. (1971–2009). *Reports of Japanese Monkeys at Takasakiyama*. Oita: Takasakiyama Natural Zoo (Mimeo in Japanese).

Watanabe, K., Mori, A., and Kawai, M. (1992). Characteristic features of the reproduction of Koshima monkeys, *Macaca fuscata fuscata*: A summary of thirty-four years of observation. *Primates*, 33: 1–32.

Yamada, T. (2008). Effects of Japanese macaques on the forest at Takasakiyama. In: H. Kurita (ed.), *Takasakiyama Management Committee Report*. Oita: Oita City, pp. 13–18 (in Japanese).

Yamagiwa, J. (2010). Research history of Japanese macaques in Japan. In N. Nakagawa, M. Nakamichi, and H. Sugiura (eds.), *The Japanese Macaques*. Tokyo: Springer, pp. 3–25.

Yokota, N. (1993). Forest as food resources for macaques. In: Takasakiyama Management Committee (ed.), *The Management of Japanese Macaques and Environment at Takasakiyama*. Oita: Board of Education, Oita City, pp. 19–37 (in Japanese).

Yokota, N. and Ono, Y. (1993). Changes of vegetation by the increase of macaque population. In: Takasakiyama Management Committee (ed.), *The Management of Japanese Macaques and Environment at Takasakiyama*. Oita: Board of Education, Oita City, pp. 38–40 (in Japanese).

4 Proboscis monkey tourism: can we make it "ecotourism"?

Heather C. Leasor and Oliver J. Macgregor

Introduction

Since the 1970s, organizations working for nonhuman primate conservation have emphasized the importance of working with the people in closest contact with nonhuman primates and their forests by attempting to give them a stakeholder share in the benefits of primate preservation and protection (Wetlands, 2001; WWF, 2001). One method for encouraging local protection of nonhuman primates, while maintaining economic balance in the community, is making primates in natural settings a focus for "ecotourism."

This chapter discusses a case study of primate tourism that focuses on visiting proboscis monkeys (*Nasalis larvatus*) along the Lower Kinabatangan River in Sabah, Malaysia. The proboscis monkey is an endangered species and a focal species for conservation and tourism in Borneo. The Lower Kinabatangan area is home to proboscis monkeys that have been studied through four PhD projects: Boonratana in 1992, Tadahiro in 2000, Leasor in 2003, and Matsuda in 2005. Tourism has been developing in the area since the mid-1990s. If tourism in this area is to contribute to conserving the area's natural habitat, proboscis monkeys, and other resident wildlife, then it should operate as ecotourism. It is often promoted as such by the Malaysian Tourism Board. However, numerous signs suggest that much of this tourism does not qualify as legitimate ecotourism.

If tourism in the Lower Kinabatangan is to become ecotourism, we must first examine the extent to which it currently satisfies the defining criteria for ecotourism. In this chapter, we outline what these criteria are, and then discuss studies on the area's proboscis monkeys and the impact that tourism has had on them over the last ten years. We use data from research on proboscis monkeys and local tourism operations in the region, including the first author's PhD research from May 2003 until July 2004 on how the tourism industry interacts with and affects the area's primates, especially its proboscis monkeys. On this basis, we suggest possible avenues for improvement by involving local people, tourism agencies, and the burgeoning sanctuary system.

Primate Tourism: A Tool for Conservation?, ed. Anne E. Russon and Janette Wallis. Published by Cambridge University Press. © Cambridge University Press 2014.

Ecotourism

Ecotourism is important because it has been seen as a panacea for conservation, potentially generating significant revenue for conservation while minimally altering or impacting the species or habitats visited (Blamey, 2001). There is no copyright on the word ecotourism nor is there a patent on any approach to it (Orams, 2001). Fennell's (2001) analysis of 85 ecotourism definitions found much variability. A universally functioning operational definition has yet to be achieved, but many authors have found three to six tenets that make up most ecotourism definitions (Donohoe & Needham, 2006). The definition used here conforms to these tenets and was developed from reviewing work by prominent authors in the field (Blamey, 2001; Bornemeier *et al.,* 1997; Burton, 1998; Ceballos-Lascurain, 1996; Gilbert, 1997; Goodwin, 1996; Hvenegaard & Dearden, 1998; Tickell, 1994; Treves & Brandon, 2005). Here, ecotourism is defined as having the following traits:

I. It involves low-impact travel to relatively undisturbed or uncontaminated natural areas.
II. It does not alter the integrity of the ecosystem.
III. It has the specific objective of studying, admiring, and enjoying the scenery and its wild plants and animals, as well as any existing cultural manifestations (both past and present) found in these areas.
IV. It includes a component of educative and interpretive programs that foster environmental and cultural understanding, appreciation, and conservation.
V. It is "ecologically sustainable," involving appropriate returns to the local community and long-term conservation of the resource.
VI. It contributes to the maintenance of species and habitats.
VII. It produces economic opportunities that make conservation of natural resources beneficial to local people, local community groups, and local projects.
VIII. It aids conservation either directly through a contribution to conservation and/or indirectly by providing sufficient revenue to the local community to enable local people to value, and therefore to protect, their wildlife heritage area as a source of income.
IX. It is conducted with small groups of tourists, ideally 6–12 and definitely less than 20 tourists per guide at one site at one time, to allow guides to control the groups, to minimize disturbances to wildlife, and to increase tourists' appreciation.

Fulfilling all these criteria is a difficult task. A tourism venture might generate social and economic benefits to the community but not benefit the environment; as such, it would not qualify as ecotourism but might be managed as sustainable tourism (Dearden, 1997). Further, because there is no consensus on the definition for ecotourism, the practices that are defined as ecotourism take many differing forms, and tourism operators are frequently in the difficult position of not knowing which rules to follow. This definitional uncertainty also means that the ecotourism label is vulnerable to being co-opted by a wide variety of tourism operations without the need to provide authentication.

To ensure that ecotourism practices do not diverge from commonly accepted eco-tourism principles, we need ongoing reviews of any tourism operation to assess the environmental and social impact and carrying capacity. Such reviews should involve baseline surveys of flora, fauna, the local communities, and tourist participants followed by periodic surveys to monitor change (Bornemeier *et al.*, 1997). These reviews need to be conducted on a two- to five-year basis so that problems can be identified as they arise and solutions implemented before it is too late (Bornemeier *et al.*, 1997). Ecotourism requires a management strategy that uses the best available knowledge in a rapid manner to correct activities that adversely impact sustainability (Weaver, 2005). Weaver (2005) says the key to sustainable management of ecotourism is not finding one solution that offers the total absence in perpetuity of any negative results, but a management system that will rapidly and adequately deal with negative impacts.

Tourism in the Lower Kinabatangan area

Tourism was officially recognized as an industry in Sabah, the more easterly of Malaysia's two Bornean states, in 1986 (Fletcher, 1997). The Kinabatangan has been an attraction since the late 1980s, starting with a small number of infrequent adventuresome tourists followed by more general tourists by 1997 (Fletcher, 1997). Sabah is home to 10 primate species, including the proboscis monkey (*Nasalis larvatus*); at least 50 mammal species, including elephant and Sumatran rhinoceros; over 200 species of birds; and a diverse array of flora and aquatic life, making it an ideal location to view a diverse array of wildlife (Wetlands, 2002). Throughout Sabah, many newly developed tour operations are identifying themselves as eco-tourism. For example, the Sabah Wildlife Department website and Sabah tourist board describe the Lower Kinabatangan as a world-renowned ecotourism destination and experience (Bagul, 2009; Sabah, 2008, 2009a, 2009b).The area around Sukau, a village on the Lower Kinabatangan River, is one of the key areas targeted by the influx of nature adventure tourists or ecotourists and small-scale tourism development (Wetlands, 2002) (Figure 4.1).

Around Sukau, commercial-scale tourism started in 1991 with the opening of Wildlife Expeditions' lodge (Vaz, 1993). Since the late 1990s, with no official policy to manage or control development, five ecotourism lodges have sprung up in the area surrounding the Menanggul River, each offering its own tourism boats and wildlife tours (Fletcher, 1997). Fletcher (1997) noted that these lodges were developed by a few urban developers and that tourism was increasing at a number of sites in the Lower Kinabatangan area. When Fletcher made this statement there were only five lodges, but she cautioned that the area was already under pressure from overuse. As of 2006 there were eight lodging options in Sukau, ranging from lodges to homestays, with a ninth a possibility.

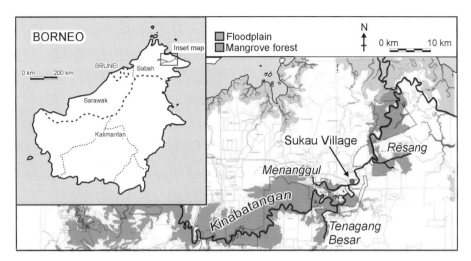

Figure 4.1 Map of the study area: Sukau Village and 3 main tourist rivers (Menanggul, Resang, and Tenagang Besar) along the Kinabatangan River. Base map sourced from WWF, Malaysia.

In 1989, a tourism feasibility study was conducted from which recommendations and a proposal for a sanctuary were drafted. Following this, Sabah's Ministry of Tourism and Environmental Development proposed the Kinabatangan Wildlife Sanctuary, originally a 50 000 ha patchwork of forest reserves and wildlife sanctuaries in key habitats along the Kinabatangan floodplain (Fletcher, 1997; Ministry of Tourism and Environmental Development & DANCED, 1996; Vaz, 1993). During the initial stages of planning, a WWF program called Partners for Wetlands was working in the area to develop an uninterrupted corridor from the coast to upland forests for wildlife, nature-based tourism, and local forest and farming industries (Hai, 2000). Other NGOs such as HUTAN, the parent organization of the Kinabatangan Orangutan Conservation Project (KOCP), and Red Ape Encounters (RAE) have also been active in the area. August 11, 2005 finally saw 26 103 ha in 17 parcels of land gazetted as the Kinabatangan Wildlife Sanctuary. It was put under Section 9 of the state Wildlife Conservation Ordinance, increasing the Wildlife Department's jurisdiction over the area (Sario, 2005).

Proboscis monkeys

Proboscis monkeys are sexually dimorphic, large-bodied colobines (leaf-eating monkeys) endemic to Borneo. They typically live in one-male groups, comprising one adult male with multiple adult and subadult females, juveniles, and dependent infants (Figure 4.2). Two other group types can exist: all-male groups and non-breeding groups. Non-breeding groups are all-male groups that are joined by non-breeding adult or subadult females; generally, these groups are short lived. Group

Figure 4.2   One-male group of proboscis monkeys on the Menanggul River, Sabah, with inset of the adult male of the group.

sizes vary from site to site but one-male groups are typically between 3 and 32 (Boonratana, 2000; Burton *et al.*, 1995; Kern, 1964; Leasor, 2010; Salter *et al.*,1985; Yeager, 1991b, 1995; Yeager & Blondal, 1991).

The range size of proboscis monkeys may be limited by their specialized diet of young leaves, seeds, and unripe fruit. They need to range in habitats like riverine forest, where tree species have high enough protein levels and relative abundance to support their nutritional needs (Bennett & Sebastian, 1988; Bernard, 1997a). They generally range on both sides of a river but high human boat traffic that prevents the monkeys from swimming to the opposite shore can alter this (Bernard, 1996; Meijaard & Nijman, 2000). Proboscis monkey groups living in riverine forests usually return to a river's edge each evening to sleep. This behavior has made the proboscis monkey a popular species for tourism, because of the ease with which they can be viewed from the comfort of a boat.

Assessing the impact of tourism in Sukau

Our research team collected data on tourism impacts for 14 months (May 2003–July 2004).

Methods

We conducted monthly censuses to monitor proboscis monkeys and the tourists visiting them along the Menanggul, Tenagang Besar, and Resang Rivers, as well as other tributaries, lakes, and the Lower Kinabatangan River. Proboscis monkey groups are known to return to the water's edge at dusk to sleep and potentially to congregate with other groups. Due to their crepuscular nature, we followed the common practice of censusing them from boats along rivers, starting in the evening from approximately 15:00 or 16:00 to18:30 and returning to recensus the same route the following dawn from 05:30 or 06:00 to 08:00, before any groups had begun to move inland. This afforded two opportunities to count the members of each group and assess their behaviors. When a proboscis monkey group was encountered, researchers recorded the number of individuals by age/sex class, their activities (using a 31-behavior ethogram), and other data not presented here. Simultaneously, researchers recorded information on tourist boats present (number, distance from proboscis monkey groups) and boat occupants (number, activities, loudness, inter-action with the monkeys).

On weeks between censuses, full-day follows of focal groups were conducted on each of the three key rivers. During follows, scan data were collected every 10 min-utes on the age/sex class of all group members, their activity, height, distance from water, and group spread; *ad libitum* observations were recorded between scans. Also, at every tourist encounter with the focal group, researchers recorded data on the boats present and their occupants and behavioral and positional data for all proboscis monkey individuals visible, using the same method followed during censuses.

Local surveys were conducted with all households in the Sukau area (total 107) covering the locals' engagement with the tourism industry; their feelings and knowledge regarding wildlife and forest; their attitudes toward tourists, oil palm plantations (OPP), and government agencies; where they would like to see tourism progressing; and their role in the future of the area.

Tourist surveys were conducted at all seven lodges and the homestay program in Sukau. Surveys asked tourists about their expectations of the cultural and wildlife experiences and their feelings regarding issues and changes relevant to ecotourism practices. Twenty surveys were obtained from each lodge except the homestay pro-gram, from which 20 willing participants were not found within the survey time-frame. In total, 149 surveys were completed.

Results and discussion

General ecological impact of tourism in Sukau

Tourism operations have already altered the integrity of the ecosystem by build-ing tourist facilities on former proboscis monkey habitats. This habitat destruction means that tourism in the area has violated ecotourism criteria I and II.

All tourists travel by boat so tourism increases local river traffic, especially on frequently visited rivers such as the Menanggul, thereby potentially decreasing water quality and increasing noise pollution. Increased river traffic may disrupt proboscis monkeys and other species during their daily activities (Bernard, 1996; Bismark & Iskandar, 1996; Leasor, 2010; Meijaard & Nijman, 2000). Our local surveys, *ad libitum* observations, and systematic boat counts showed an increase in river traffic compared to Boonratana's (pers. comm.) findings for the Menanggul in the 1990s that is linked to the introduction of tourism and the construction of multiple large OPP. This increase and disturbance could violate ecotourism criterion II.

Tourism impacts on the local community

Fletcher (1997) found that locals' involvement in tourism, especially in Sukau, was in low-paying, menial jobs at the lodges and that most lodge staff came from urban centers. Surveys in our study indicated that Fletcher's findings on the local job situation had not changed, except for the introduction of one locally operated homestay program and one bed and breakfast. The percentage of local Sukau residents that wanted to be involved with tourism (e.g. working for agencies, as guides, or in land rental, product or produce trade, or other related jobs) was very high but the percentage who were actually involved with or gaining financial benefit from tourism was low (Table 4.1). More job opportunities should be available, given the number of tourism operations in the area, but few locals were given employment opportunities in tourism, so little financial benefit was returning to the local community. These conditions violate terms V, VII, and VIII of the ecotourism definition.

The local community's responses to our survey questions on the value of the forest, forest products, rivers, and wildlife were all positive. The respondents valued and wanted to protect or preserve the forest, which could be useful for achieving ecotourism criterion VIII. Their responses also indicated a loss of culture in the area over time and disapproval of inappropriate aspects of the tourists' behavior or activities. Examples included the Westernization of younger generations after viewing tourism; tourists' inappropriate attire when entering or passing the village; and tourists photographing local people carrying out their daily tasks, bathing included, without asking their permission.

While photography of the local community may be seen as a way of admiring existing cultural manifestations, in line with ecotourism criterion III, it can also be experienced or seen as demeaning and exploitative of local community members. The fact that tourists were failing to understand that their actions were culturally insensitive indicates that Sukau area tourism was falling short of achieving ecotourism criterion IV, which concerns cultural awareness and education. Part of the reason was that few educational opportunities were available in the area to provide tourists with a deeper understanding of the ecosystem, the local culture, and conservation. Based on our tourist surveys, some of the lodges provided limited information and one presented an educational program in the evening, but most tourist programs were just guided tours of the area during which any animals seen were identified.

Table 4.1 Responses of Sukau village residents to some of the survey questions. Original responses were on a seven-point scale; they are summarized here. Cells show the percentage of respondents choosing highest and lowest scale values

Village resident responses	Highest positive response (%)	Lowest negative response (%)
Work with tourism	15	85
Desire to work with tourism	80	20
Rent land to tourism agencies	10	90
Want more trade or money from tourism	75	5
Value wildlife	90	1
Value proboscis monkeys	85	2
Value forest	69	2
Feel that cultural traditions were more intact prior to tourism	80	2
Desire tourists to dress more respectfully	90	1
Want tourists to know about the culture of the area	65	2
Feel that proboscis monkeys bother the village	1	90

Tourism impacts on proboscis monkeys

Due to tourism and OPP expansion, the area available to proboscis monkeys had reduced noticeably since Boonratana's study in 1992 (see Table 4.2, census distance). The portion of each river that we censused was the portion that was, at that time, declared or promised as protected; this included portions of the river that were forested, but not necessarily all forested areas. The reduction in census distances between these two studies also reflects the reduction in the length of the forested portion of some rivers. At the time of this study, the Menanggul River had been cleared at its mouth for two tourism lodges and a Jabatan Hidupan Liar (JHL, the Sabah Wildlife Department) building, and clearing was beginning farther up the river. All these areas were forested in Boonratana's study and used by his focal proboscis monkey groups. We surveyed only the first 900 m of the Resang River because the rest had been converted to OPP. We surveyed the first 3.45 km of the Tenagang Besar River because this was the only area with protection status; of this, only one bank was forested past 1.3 km and the other was converted to oil palms.

Although census distances were much shorter in our study than the earlier one, our overall proboscis monkey counts were similar and proboscis monkey densities were higher (Table 4.2). The changes in forest structure have undoubtedly compromised the integrity of the ecosystem and have not maintained species habitats, thus violating ecotourism criteria II and VI.

People did not use all the rivers in the current study or all areas on those rivers equally. Tourist agencies favored evening wildlife cruises along the first 3 km of

Table 4.2 Comparisons of census lengths and proboscis monkey population densities along main rivers between this study in 2003–2004 and Boonratana's study in 1992. Columns show the length of the river censused, best count of proboscis monkeys seen during a single census, average group size for that river, and estimated monkey density

River	Study	Census distance (km)	Best count of proboscis monkeys	Average proboscis monkey group size	Proboscis monkeys/km
Menanggul	Boonratana	10	146	17.0	14.6
	Leasor	5	108	12.6	21.6
Resang	Boonratana	6	44	ND	7.3
	Leasor	0.9	56	15.5	62.2
Tenagang Besar	Boonratana	12	95	ND	7.91
	Leasor	3.45	97	12.8	28.1

ND denotes no data available for comparison.

the Menanggul. These tourists were not always well behaved, toward each other or the primates, but no protocols existed for multiple boats or encounters with wild primates. We observed tourist boats on the Menanggul blocking each other and pushing the research boat to the center of the river while the research team was observing a focal group. We observed cases when tourist boats approached macaques and elephants close enough to touch them.

Noise levels from people and boats were also an issue, but a hard one to quantify. Based on this study, when loud noises or loud engines were approaching, most primates, proboscis monkeys included, became vigilant toward the source of the noise. Larger groups of tourists, on average, were louder than smaller ones, even if tourists were speaking at a normal conversational volume. Larger groups also require larger boats, which are usually noisier. Similar noise levels resulted from multiple smaller boats, each carrying four to ten individuals, congregating at one location. The noise levels associated with large numbers of tourists concentrated around monkey groups loosely violate ecotourism criteria I, IV, and IX. The surveys indicated that most of these tourist boats were not locally owned or operated, violating ecotourism criterion VII.

The intensity of tourist activity was negatively correlated with the observed frequency of proboscis monkeys for all rivers. The number of people on the rivers was negatively correlated with both the number of proboscis monkeys and the number of proboscis monkey groups visible. The number of boats on the rivers similarly correlated negatively with the number of proboscis monkeys visible (Figure 4.3). Thus the presence of large numbers of boats or people altered proboscis monkeys' natural behavior or ranging, violating ecotourism criterion VI. Findings indicate that the greatest visibility of proboscis monkey groups or individual monkeys occurred when less than 20 people or 4 boats were in the vicinity (Figure 4.3). A Multivariate General Linear Model was developed to analyze the relationships between multiple independent variables, including boat numbers and people numbers, and behavioral data from different age categories of proboscis monkeys. This model showed

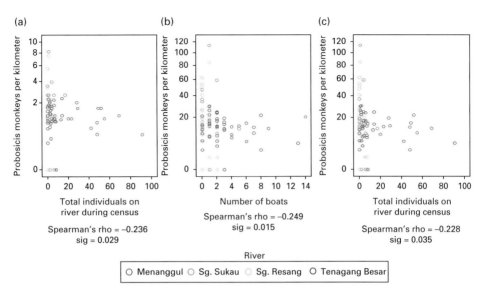

Figure 4.3 Number of proboscis monkey groups/km vs. number of people on river (a), proboscis monkeys/km vs. number of boats (b), proboscis monkeys/km vs. number of people on river (c).

a significant negative relationship between the number of people present and the frequency of proboscis monkey infants exhibiting social behavior ($t = -7084$, $p < 0.001$). This indicates that when large numbers of boats or people were present, proboscis monkeys typically retreated or hid from view, another alteration of natural behavior that violates ecotourism criterion VI. The high boat numbers and tourist numbers witnessed violate ecotourism criterion IX.

For each river, for one-male groups, behavior varied with boat presence at any time of day. From data collected during either morning or evening hours of full-day follows, group behavioral profiles varied with boat numbers. We used chi-squared analyses and modeling tests for times when any boat other than the research boat was near a group, irrespective of the total number of boats, to see whether boat presence was enough to alter the group's behavior. Data from both morning and evening hours showed a significant difference in group behavior when boats were present compared to when they were absent (see Table 4.3).

The frequencies of behaviors exhibited by one-male groups when boats were present or absent were compared to investigate what caused these significant differences. When boats were present there appeared to be more agonistic behavior, except on the Menanggul and the Tenagang Besar during the morning hours: at this time of day on those rivers only, the monkeys traveled more when boats were present. Instead of responding agonistically to boats, these monkey groups may have been moving away from them.

Comparisons of proboscis monkey behavior during censuses when four boats and no boats were present produced several interesting findings (Figure 4.4). The presence of four boats was chosen for this comparison because this number of boats was relatively common for tourism operations on the rivers studied. Agonistic behaviors

Table 4.3 Chi-squared analyses of proboscis monkey behavioral profile as influenced by boat presence and time of day. All results are significant ($p < 0.001$)

	Menanggul River		Resang River		Tenagang Besar River	
	χ^2	df	χ^2	df	χ^2	df
Boat presence vs. absence, all times of day	806.80	64	234.28	12	313.49	12
Boats presence vs. absence, morning hours, one-male groups only	149.89	24	46.34	8	79.97	12
Boats presence vs. absence, evening hours, one-male groups only	476.99	56	55.90	4	52.21	8

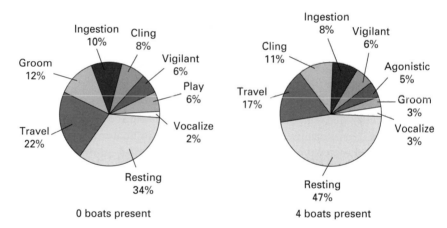

0 boats present   4 boats present

Figure 4.4   Comparison of proboscis monkey behavior when four boats vs. no boats were present during census on the Menanggul River.

occurred when four boats were present but not when boats were absent. Play and grooming rates were higher when boats were absent than when four boats were present. These data provide good evidence that proboscis monkey behavior changes when the number of boats present approaches the upper end of the range of boat numbers found at proboscis monkey groups, thus violating ecotourism criterion VI.

Comparisons of proboscis monkey behavior between our study and other studies are difficult because behavioral categories differ. Matsuda *et al.* (2009) recorded three categories (rest, eat, other) but did not include vigilance and other behaviors important to our study. Our study and Boonratana's used 11–14 behavioral categories that differ mainly in lumping or splitting (e.g. urination and defecation vs. bodily function, clinging including vs. excluding suckling). One important difference is how vigilance was recorded: in our study vigilance was mutually exclusive of other behaviors, that is, it was recorded only when a monkey was exhibiting no other behavior. Boonratana (1993) recorded vigilance when it occurred simultaneously

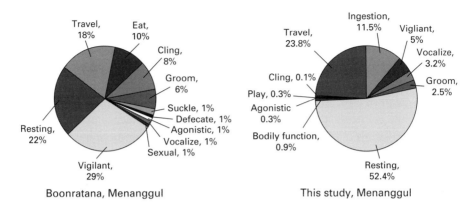

Figure 4.5 Comparison of proboscis monkey behaviors on the Menanggul River: Boonratana's 1992 (left) and this study's 2003–2004 (right) findings.

with a limited number of other behaviors as well as on its own. Partly for this reason, he found much higher vigilance rates than we did (Figure 4.5). The frequency of swimming by proboscis monkey groups, however, can be compared between studies.

Our study found lower rates of proboscis monkeys swimming and river crossing than other proboscis monkey studies, which could be a result of tourism activities. Travel in our study was broken down into travel within a tree, between trees, on the ground, and in the water (swimming, intentional or not). Most proboscis monkey travel is arboreal, with only a limited amount of travel on the ground. In this study, of all the travel on all three focal rivers during full-day follows ($n = 7704$), 94% was between or within trees, 0.25% on the ground, 0.25% in the water, and 5.5% undetermined. During Boonratana's initial habituation, his focal one-male group swam the Menanggul River five times in a three-hour period; after habituation it swam on at least 31.2% of the 93 days and swam the 150 m-wide Kinabatangan at least twice in a year (Boonratana, 1993). Matsuda *et al.* (2008) witnessed their focal female cross the Menanggul River 90 times, 93% arboreal ($n = 84$) and 7% ($n = 6$) swimming. It is unclear whether swimming in that study resulted from falling into the water from a failed arboreal crossing or deliberate choice. We witnessed 14 strictly arboreal instances of group river crossings on the Menanggul River and 8 instances when individuals swam after missing a jump and landing in the water. This does not include instances when groups crossed small tributaries where the canopy was completely closed above the water course. Proboscis monkeys attempted to cross where the canopy was closed or nearly closed, and would travel to such crossing points preferentially. We never witnessed proboscis monkeys deliberately entering the water, as reported in Tanjung Puting National Park (Yeager, 1991a, 1991b). In our study, all swimming was initiated while attempting to jump from tree branch to tree branch across the river.

Tourist boats were unregulated in regard to viewing distance or number of boats at a site, thus large blockades of boats frequently formed a formidable flotilla either across the river, along the opposite bank, or under the proboscis monkeys. No proboscis monkey studies that reported multiple episodes of swimming noted high amounts of boat traffic. Bernard (1997b) witnessed no swimming in the Klias Peninsula, a site known to have hunting and large amounts of boat traffic for fishing. Yeager (1992) found over the years of her study that increased boat traffic made it difficult for groups to swim across the river to reach new feeding areas. Over the course of this study, we observed that the presence of boats interrupted river crossings or caused groups to abort crossing and move inland. The infrequency of proboscis monkeys swimming the Menanggul River during our study could therefore be due to higher boat traffic in the area since the time of Boonratana's study.

Recommendations for tourism

The current study aimed to address alterations to the socioecology of proboscis monkeys when viewed by tourists. We assessed the extent to which tourism meets ecotourism criteria, focusing on proboscis monkeys. Based on information collected, we suggest how tourism can be changed to bring it closer to the ecotourism ideal.

Based on the ecotourism definition used in the chapter and findings from the Lower Kinabatangan area tourism, including its impacts on local ecology, local communities, and most importantly proboscis monkeys, the tourism being implemented and developed in the area is not truly ecotourism. This does not mean we should give up on the tourism, the site, or the species involved. Rather we should alter current practices, legislation, and guidelines. This section discusses recommendations for realigning Lower Kinabatangan tourism to something more akin to ecotourism. As noted earlier, not all the changes we detected result from tourism and thus not all alterations can be achieved through tourism without contributions from external forces.

Tourism education in Sukau

Ecotourists expect more information during their nature experience than mass tourists, although this information and its delivery should not be overly heavy or scientific, and guides should be knowledgeable in general and specialized in their particular type of tour group (Gilbert, 1997). Learning can at a minimum foster satisfaction and basic understanding of host cultural or natural attractions, and at best be a transformative experience leading to a deep understanding of the host community and ecosystem which, in turn, leads to a more ethical and environmentally responsible lifestyle in the future (Weaver, 2005).

The new JHL building erected in 2003/04 at the mouth of the Menanggul River, the main tourist river, contributed to destroying proboscis monkey habitat. Given that reality, it is an ideal location for a visitor education/information center that, if used properly, could help create an effective interface between tourists, government, and local communities. For instance, it is a great location for pre-cruise educational purposes. There could be presentations prior to or after the time when proboscis monkeys return to the river, scheduled to space the tourists entering the river more evenly. The building could house informative displays about local peoples and cultures or other areas of Sabah. Implementation could follow models used in some North American national parks and mountain gorilla parks in Uganda and Rwanda, where a ranger or local representative gives a brief talk on what to expect while visiting, the rules and regulations for wildlife observation, and appropriate behavior. Codes of behavior can be developed on the basis of readily accessible great ape tourism guidelines (Macfie & Williamson, 2010; Williamson & Macfie, this volume). This would bring things in better alignment with ecotourism criteria III and IV.

Based on personal experience and survey responses, very few people in the Sukau community spoke languages other than their own dialect and Malaysian. Non-local guides spoke adequate but not fluent English and very few spoke other foreign languages. This limits the tourist experience for non-English speakers. Based on this study, additional languages important in Sukau tourism are Dutch, Japanese, French, Spanish, and German. There is a great opportunity to involve Malaysian universities, possibly even from mainland Malaysia, as an avenue to enlightening the "mainland" view of outlying states like Sabah. These are all ways of working toward achieving ecotourism criterion IV.

Training is crucial at the field level; it can determine whether a tourism project is a success or failure. Well-trained guides and personnel have better skills, knowledge, and thus dedication to a project (Bornemeier et al., 1997). A joint project of the UMS, the Darwin Initiative, and NGO HUTAN/KOCP runs guide-training courses periodically. Taman Negara National Park and Universiti Pertanian Malaysia had a good guide-training program in the mid-1990s that could be used as a Malaysian template for appropriate training techniques; this has a two fold advantage of keeping training based in Malaysia and fostering relations between peninsular Malaysia and Sabah (Shuib, 1997). Other potential sources of training are the Malaysian Tour Guides' Council and a Malaysian nature society that runs nature tour guide courses (Nor & Wayakone, 1997). Training credentials like these and a one-week on-site training program should be prerequisites for working as guides in the Lower Kinabatangan; the local program is important to familiarize guides with the physical boundaries and lay of the land, local culture, and procedures. Shuib (1997) found that local residents can be superior guides because of their local knowledge of the forest and interesting anecdotes. Programs similar to these could be implemented if properly supported at local and national levels. This could assist with achieving ecotourism criteria III, V, VII, and VIII.

Contributing to the local community

Our study found some tension in the local community with tourist incursion. Cultural clashes could potentially result from differences in behavior and mores between the local community and the tourists. A cultural program to educate the children of the village and tourists about the area's rich culture could help to maintain the strong cultural heritage of the Lower Kinabatangan and the river people (see Strum & Manzolillo Nightingale, this volume). UMS has conducted a few unpublished studies to document the culture and traditional stories of the area. If these were brought to the fore to help local people feel culturally proud, the people might be in a better position to handle the influx of new experiences that the tourist culture brings without losing their own. This could also assist in the creation of education programs about the local culture for the tourists. Efforts like this would better fulfill ecotourism criteria I, V, VII, and VIII.

Primate–human encounters

One way to mitigate human–primate encounters could be limiting the river routes, distances, and frequency of tourist travel so that the surrounding areas remain undisturbed primate refuges. Based on a chimpanzee model, an observer distance of 40 m is optimal to maintain natural behavior and protect against disease transmission (Cowlishaw & Dunbar, 2000). It is not always possible, because of river width and curvature, to be in visual contact with primates and at least 40 m away. For the Kinabatangan area, a longer section of the Menanggul River should be protected to give the monkeys places of escape if they do not wish to encounter humans. Tourists should not be allowed to travel farther than 4–5 km upriver by boat, for instance, so that proboscis monkey groups have sufficient opportunities to escape deeper into the sanctuary, and the sanctuary itself should be extended another 1–2 km beyond the 5 km mark. To safeguard the proboscis monkeys, conservation authorities must protect sufficiently large areas of habitat with effective law enforcement (Meijaard & Nijman, 2000). This could assist with satisfying ecotourism criterion VI.

Maintaining safe distances between humans and nonhuman primates, not allowing provisioning, and proper waste disposal would lessen the risk of tourist–monkey disease transmission. Mountain gorilla tourism has very strict guidelines, which include only one-hour visits per day by at most 8–10 people to a few groups of habituated mountain gorillas (Homsy, 1999; Macfie & Williamson, 2010). Regulations this extreme may not be needed with boat-based proboscis monkey tourism of the sort practiced at Sukau. After sunset, however, proboscis monkey viewing should be minimal because extended visits and/or spotlighting could provoke a group to alter sleeping locations. The minimum distance rule in gorilla tourism is currently 7–10 m, to protect against the transmission of airborne diseases (Macfie & Williamson, 2010). Similar distances should be respected for boat tourism, paying attention to both horizontal and vertical distance (Homsy, 1999). This could assist with satisfying ecotourism criterion VI.

Noise levels caused by tourists can be controlled with tourist education. Boat noise can be controlled by using canoes or boats with electric trolling motors (Yeager & Blondal, 1991). Canoe travel would have to originate at the mouth of a tributary because the current on the Kinabatangan is too strong to negotiate and too dangerous for tourists.

Boat pollution can be reduced by regular basic boat maintenance (and other forms of transport maintenance), which also keeps the boats running efficiently and is more cost effective. A practical measure is training a group of locals or JHL staff in basic boat maintenance. Use of solar recharged batteries to operate motors, already an occasional practice at one lodge, and 4-stroke fuel-efficient engines on the main river could be promoted by JHL and the tourism board. This could help clean up the noise, air, and water pollution that can disrupt wildlife behavior.

Based on both the number of boats and tourists seen throughout our study and on the conversations with managers and locals, there seemed to be no check on the number or size of tourist groups visiting the area. Low numbers would enhance educational and viewing opportunities. Ideally, fewer than two tour boats should be near any given proboscis monkey group, and tourist groups limited to ten members (including boatman and guide) are most appropriate for ecotourism (see Figure 4.3). Both limitations could mitigate ecotourism criterion II.

In theory, proper boat placement combined with maintaining appropriate distances could mitigate some of the wariness that proboscis monkeys exhibit when boats gather near them. Further monitoring of proboscis monkey swimming, vigilance, ranging, and other behaviors relative to the number and placement of boats should be undertaken in developing measures to minimize tourism impacts on behavior. This could help to meet ecotourism criterion IX.

Conclusion

It is possible to bring the proboscis monkey-focused tourism currently in practice closer to the guidelines for ecotourism. Surveys carried out in this study showed that tourists desire a change and that the local community is open to change and could benefit from it. Further monitoring studies need to be conducted periodically to measure tourism-related change in proboscis monkeys, including monitoring boats and primate behavior. River traffic from both the estates and tourists must be assessed further to develop appropriate management strategies, especially relative to the tourism sector. Behavioral monitoring of proboscis monkeys should be continued to enable comparisons over time, paying attention to swimming, vigilance, and other behaviors when boats are nearby. The key issues raised in this chapter are education, protection, and appropriate returns to the community. Sukau area nature tourism will not be able to fulfill the criteria for ecotourism, but it could certainly come closer to them and be sustainable into the future. The key factor is that proboscis monkeys must continue to thrive in the area, because they are a strong factor in tourists choosing the site for travel.

Acknowledgments

The authors would like to thank the editors for their assistance in formulating this chapter and to the compilers of the book for the inclusion of this chapter. This project was made possible by grants from: Wildlife Conservation Society, Primate Conservation Inc., Columbus Zoo and Aquarium, The Faculty of Arts at The Australian National University, and Paul and Lola Leasor. We are grateful for the permission to carry out this research from Jabatan Hidupan Liar (Sabah Wildlife Dept.), Ketua Kampung, JKKK, Ketua Police, Pemaju Masyarakat, Hutan/REA/WCU/KOCP, Puan dan Tuan Penduduk Kampung, and research ethics clearance and other permissions from The Australian National University. We would like to thank the field assistants: Jonny Walsh, Mansor Choy Bin Ismail (Choy), Mat Sarudin Bin Abdul Karim (Udin), Sabli Bin Guyau (Sabli), and Adian Syah Bin Asmar (Adi). Statistical assistance was provided by the ANU Statistical consulting unit.

Researchers obtained prior ethical approval for both surveys (from the Australian National University (ANU), the Malaysian Economic Planning Unit (EPU), the Sabah Wildlife Department, and village and lodge authorities) and permission from participants at the time of the survey. All survey responses were confidential. All primate interactions were in accordance with ethical requirements of the Malaysian EPU, the Sabah Wildlife Department, and the ANU.

References

Bagul, A. H. B. P. (2009). *Success of Ecotourism Sites and Local Community Participation in Sabah* (PhD Thesis). Victoria: University of Wellington.

Bennett, E. and Sebastian, A. C. (1988). Social organization and ecology of proboscis monkeys (*Nasalis larvatus*) in mixed coastal forest in Sarawak. *International Journal of Primatology*, 9 (3): 233–255.

Bernard, H. (1996). *The Distribution and Abundance and some Aspects on Behaviour and Ecology of the Proboscis Monkey (Nasalis larvatus) in the Klias Peninsula, Sabah.* Unpublished MSc thesis. Reading, Whiteknights: The University of Reading.

Bernard, H. (1997a). Some aspects of behaviour and feeding ecology of the proboscis monkey, *Nasalis larvatus*, based on a brief study in the Klias Peninsula, Sabah. *Borneo Science*, 3: 65–84.

Bernard, H. (1997b). A study on the distribution and abundance of proboscis monkey (*Nasalis larvatus*) in the Klias Peninsula. *The Journal of Wildlife Management and Research Sabah*, 1: 1–12.

Bismark, M. and Iskandar, S. (1996). *The Population and Distribution of Proboscis Monkey (Nasalis larvatus) at the Kutai National Park East Kalimantan.* Jakarta: United Nations Educational, Scientific and Cultural Organization UNESCO/UNDP/GOI Project INS/93/004 Kutai National Park Management Support.

Blamey, R. K. (2001). Principles of ecotourism. In: Weaver D, (ed.), *Encyclopedia of Ecotourism*. Wallingford: CABI Pub. pp. 5–22.

Boonratana, R. (1993). *The Ecology and Behaviour of the Proboscis Monkey (Nasalis larvatus) in the Lower Kinabatangan, Sabah.* Unpublished PhD Thesis. Mahidol University. Nakhom Pathom.

Boonratana, R. (2000). Ranging behavior of proboscis monkey (*Nasalis larvatus*) in the Lower Kinabatangan, Northern Borneo. *International Journal of Primatology*, 21 (3): 497–518.

Bornemeier, J., Victor, M. and Durst, P. (1997). *Ecotourism for Forest Conservation and Community Development: Proceedings of an International Seminar, Held in Chiang Mai, Thailand, 28–31 January, 1997*. Bangkok, Thailand: FOA/RAP Publications: 1997/26. Report nr RECOFTC Report No. 15.

Burton, F. (1998). Can ecotourism objectives be achieved? *Annals of Tourism Research*, 25 (3): 755–758.

Burton, F., Snarr, K. A. and Harrison, S. E. (1995). Preliminary report on *Presbytis francoisi leucocephalus*. *International Journal of Primatology*, 16 (2): 311–327.

Ceballos-Lascurain, H. (1996). *Tourism, Ecotourism and Protected Areas*. Gland, Switzerland: IUCN.

Cowlishaw, G. and Dunbar, R. (2000). *Primate Conservation Biology*. London: University of Chicago Press.

Dearden, P. (1997). Carrying capacity and environmental aspects of ecotourism. In: J. Bornemeier, M. Victor, and P. Durst (eds.), *Ecotourism for Forest Conservation and Community Development: Proceedings of an International Seminar, Held in Chiang Mai, Thailand, 28–31 January, 1997*. Bangkok, Thailand: FAO/RAP Publication: 1997/26; RECOFTC Report No 15. pp. 44–60.

Donohoe, H. M. and Needham, R. D. (2006). Ecotourism: The evolving contemporary definition. *Journal of Ecotourism*, 5 (3): 192–210.

Fennell, D. A. (2001). A content analysis of ecotourism definitions. *Current Issues in Tourism*, 4 (5): 403–421.

Fletcher, P. (1997). The Lower Kinabatangan: The importance of community consultations in ecotourism development. In: J. Bornemeier, M. Victor, and P. Durst (eds.), *Ecotourism for Forest Conservation and Community Development: Proceedings of an International Seminar, Held in Chiang Mai, Thailand, 28–31 January, 1997*. Bangkok, Thailand: FAO/RAP Publication: 1997/26; RECOFTC Report No 15. pp. 213–219.

Gilbert, J. (1997). *Ecotourism Means Business*. Wellington, NZ: GP Publications.

Goodwin, H. (1996). In pursuit of ecotourism. *Biodiversity and Conservation*, 5: 277–291.

Hai, T. C. (2000). *Land Use and the Oil Palm Industry in Malaysia. Abridged Report Produced for the WWF Forest Information System Database*. Kuala Lumpur: WWF. Report nr MY0057 MYS406/98.

Homsy, J. (1999). *Ape Tourism and Human Diseases: How Close Should We Get? A Critical Review of the Rules and Regulations Governing Park Management and Tourism for the Wild Mountain Gorilla, Gorilla gorilla beringei*. Nairobi: International Gorilla Conservation Programme (IGCP).

Hvenegaard, G.T. and Dearden, P. (1998). Ecotourism versus tourism in a Thai National Park. *Annals of Tourism Research*, 25 (3): 700–720.

Kern, J. (1964). Observations on the habits of the Proboscis Monkey, *Nasalis larvatus* (Wurmb), made in the Brunei Bay area, Borneo. *Zoologica*, 49: 183–192.

Leasor, H. C. (2010). *Effects of Ecotourism on the Proboscis Monkey (Nasalis larvatus) in the Lower Kinabatangan, Sabah, Malaysia*. Unpublished PhD thesis. Canberra: Australian National University.

Macfie, E. J. and Williamson, E. A. (2010). *Best Practice Guidelines for Great Ape Tourism*. Gland, Switzerland: IUCN/SSC Primate Specialist Group (PSG). www.primate-sg.org/best_practice_tourism.

Matsuda, I., Tuuga, A., Akiyama, Y. and Higashi, S. (2008). Selection of river crossing location and sleeping sites by proboscis monkeys (*Nasalis larvatus*) in Sabah, Malaysia. *American Journal of Primatology*, 70: 1097–1101.

Matsuda, I., Tuuga, A. and Higashi, S. (2009). The feeding ecology and activity budget of proboscis monkeys. *American Journal of Primatology*, 71: 478–492.

Meijaard, E. and Nijman, V. (2000). Distribution and conservation of the proboscis monkey (*Nasalis larvatus*) in Kalimantan, Indonesia. *Biological Conservation*, 92: 15–24.

Ministry of Tourism and Environmental Development & DANCED. (1996). *Sabah Biodiversity Conservation Project, Malaysia: Kinabatangan Multi Disciplinary Study*. Kota Kinabalu: Ministry of Tourism and Environmental Development, Sabah.

Nor, D. S. M. and Wayakone, S. (1997). Ecotourism in Malaysia. In: J. Bornemeier, M. Victor, and P. Durst (eds.), *Ecotourism for Forest Conservation and Community Development: Proceedings of an International Seminar, Held in Chiang Mai, Thailand, 28–31 January, 1997*. Bangkok, Thailand: FAO/RAP Publication: 1997/26; RECOFTC Report No 15.

Orams, M. B. (2001). Types of ecotourism. In: D. Weaver (ed.), *Encyclopedia of Ecotourism*. Wallingford: CABI, pp. 23–36.

Sabah (Tourism Board) (2008). *A Boost for Eco-Tourism*. Kota Kinabalu: Sabah Tourism Board.

Sabah (Sabah Wildlife Department JHL). (2009a). *The Kinabatangan Floodplain*. Kota Kinabalu: Jabatan Hidupan Liar.

Sabah (Sabah Wildlife Department JHL). (2009b). *Sepilok Orangutan Rehabilitation Centre*. Kota Kinabalu: Jabatan Hidupan Liar.

Salter, R. E., MacKenzie, N. A., Nightingale, N., *et al.* (1985). Habitat use, ranging behaviour, and food habits of the proboscis monkey, *Nasalis larvatus* (van Wurmb), in Sarawak. *Primates*, 26 (4):436–451.

Sario, R. (2005). Stronger laws in place to protect Lower Kinabatangan area. 26 September 2005. *The Star Online*. http://thestar.com.my.

Shuib, A. (1997). The training of local interpretive guides in ecotourism at Taman Negara National Park, Malaysia. In: J. Bornemeier, M. Victor, and P. Durst (eds.), *Ecotourism for Forest Conservation and Community Development: Proceedings of an International Seminar, Held in Chiang Mai, Thailand, 28–31 January, 1997*. Bangkok, Thailand: FAO/RAP Publication: 1997/26; RECOFTC Report No 15.

Tickell, C. (1994). Foreword. In: E. Cater and G. Lowman (eds.), *Ecotourism a Sustainable Option?* Brisbane: John Wiley & Sons, pp. ix–x.

Treves, A. and Brandon, K. (2005). Tourism impacts on the behavior of black howling monkeys (*Alouatta pigra*) at Lamanai, Belize. In: J. D. Paterson and J. Wallis (eds.), *Commensalism and Conflict: The Human-primate Interface*. Norman, OK: The American Society of Primatologists, pp. 147–167.

Vaz, J. (ed.). (1993). *The Kinabatangan Floodplain: An Introduction*. Kota Kinabalu: WWF.

Weaver, D. (2005). Comprehensive and minimalist dimensions of ecotourism. *Annals of Tourism Research*, 32 (2):439–455.

Wetlands (Partners for Wetlands). (2001). *Tourism in the Kinabatangan-a Potential Tool for Conservation*. Special Report Quarterly March 2001. Kota Kinabalu: WWF Partners for Wetlands.

Wetlands (Partners for Wetlands). (2002). *Malaysia: Vision for the Lower Kinabatangan Floodplain Unveiled*. Kota Kinabalu: WWF Partners for Wetlands.

WWF (2001). *Malaysia: Making Land Use Sustainable in the Lower Kinabatangan Floodplain*. Kota Kinabalu: WWF Partners for Wetlands.

Yeager, C. P. (1991a). Possible anti-predator behavior associated with river crossings by proboscis monkeys (*Nasalis larvatus*). *American Journal of Primatology*, 24: 61–66.

Yeager, C. P. (1991b). Proboscis monkey (*Nasalis larvatus*) social organization: intergroup patterns of association. *American Journal of Primatology*, 23: 73–86.

Yeager, C. P. (1992). Changes in proboscis monkey (*Nasalis larvatus*) group size and density at Tanjung Puting National Park, Kalimantan Tengah, Indonesia. *Tropical Biodiversity*, 1 (1): 49–55.

Yeager, C. P. (1995). Does intraspecific variation in social systems explain reported differences in the social structure of the proboscis monkey (*Nasalis larvatus*)? *Primates*, 36 (4): 575–582.

Yeager, C.P. and Blondal, T. K. (1991). The proboscis monkey of Brunei Bay – conservation status and ecotourism potential. *The Brunei Museum Journal*, 7 (3): 112–116.

5   Orangutan tourism and conservation: 35 years' experience

Anne E. Russon and Adi Susilo

Introduction

Orangutan tourism has operated continuously since the 1970s, on both islands that orangutans inhabit (Borneo and Sumatra) and in both countries that govern them (Indonesia and Malaysia). As such it offers a rich example of primate tourism and an important one for conservation, especially from the long-term perspective (Butler, 1980; Catlin *et al.*, 2011; Duffus & Dearden, 1990). Both Bornean and Sumatran orangutans are endangered, like all the great apes, and protected internationally from trade for commercial purposes (CITES, 2012, Appendix I; IUCN, 2011). Since the 1970s, conservationists have promoted great ape tourism as a lucrative and effective strategy for securing their survival (GRASP, 2005; Litchfield, 2001; Woodford *et al.*, 2002) – close-contact tourism included, typically meaning proximity up to and including physical contact. For orangutans, however, the greatest survival threat is humans and much of that threat comes from disease, hunting, and capture, all of which are enabled by proximity (Woodford *et al.*, 2002; Yeager, 1999). With the risk of extinction and the costs of human proximity being so high for orangutans, examining how conservation and tourism interests mesh is critical.

This chapter aims to assess the impact of orangutan tourism on orangutan conservation, both its expected benefits and its known costs. We could not weigh the benefits against the costs in a balanced fashion because the data needed are not available. We therefore focused on weighing the experiences reported against the conservation expectations promoted.

History of orangutan tourism

The history of tourism with orangutans is important to contextualize assessments. Great apes have been a focus for tourism since colonial times, originally in the form of hunting safaris for killing "trophy" wildlife. As interest grew in seeing wild animals alive in their natural habitat instead of dead on a wall and concerns grew for their conservation, great ape tourism was refashioned as low-impact visits to experience wildlife in its natural habitat. Modern orangutan tourism was launched in this spirit in the

Primate Tourism: A Tool for Conservation?, ed. Anne E. Russon and Janette Wallis. Published by Cambridge University Press. © Cambridge University Press 2014.

1970s, but with rehabilitant rather than wild orangutans, that is orangutans rescued from illegal captivity and in the care of projects charged with helping them recover and resume forest life. At that time, experts considered that wild orangutans should be left undisturbed for conservation reasons and had little tourist appeal anyway, as they live in remote areas, are hard to find and observe, and are too solitary, slow-moving, and phlegmatic to be interesting (Aveling, 1982; Frey, 1978; Rijksen, 1982).

The four rehabilitation projects that have operated since the 1970s were the first to accept tourists: Sepilok (da Silva, 1965), Camp Leakey in Tanjung Puting National Park (Galdikas-Brindamour & Brindamour, 1975), Bohorok (also called Bukit Lawang) in Gunung Leuser National Park (Borner, 1976; Frey, 1978), and Semenggoh (Aveling & Mitchell, 1982). They did so in the expectation that tourism would advance conservation education about orangutans and generate funding support for their main conservation activities, returning ex-captive orangutans to native forests. All four still operate today, along with several newer orangutan rehabilitation projects (Russon, 2009), and they remain the most heavily involved in orangutan tourism.

Typical tourist experiences involved viewing rehabilitants during daily feedings at feeding sites and forest walks into areas with free-ranging rehabilitants, presumably to show a glimpse of their forest life. This quickly proved popular, probably because it offers safe and easy viewing: rehabilitants typically tolerate observers, are easily lured to feedings at times and sites that suit tourists, and are mostly lively, sociable youngsters (Russon, pers. obs.; Sepilok Orangutan Appeal, 2003). The tourist influx was already high at Bohorok and Sepilok by the late 1970s (5000 and 17 000 annually: Aveling, 1982; Aveling & Mitchell, 1982). These experiences remain the basis of rehabilitant tourism today.

Wild orangutan tourism has been growing since the mid-1980s, partly in conjunction with rehabilitant tourism (Bohorok and Tanjung Puting), partly in conjunction with orangutan research, and partly independently. Starting in 1984, for example, Earthwatch marketed tours to assist orangutan research in Tanjung Puting, where rehabilitant and wild orangutans range together (Russell, 1995). In 1989, Malaysia chose the orangutan as its mascot and had its Minister of Culture and Tourism announce the choice. In 1990 Malaysia used orangutans to head its tourism campaign; it had few if any visitor facilities in wild orangutan habitat at that time, but was developing and promoting them by the late 1990s (Bennett, 1998; Kaplan & Rogers, 2000; Payne & Andau, 1989). As of 2012, tourists can visit wild orangutans at ten sites in Borneo (four in Malaysia, six in Indonesia) and at one site in Sumatra. Tours typically offer guided forest trips (boating, trekking), list orangutans as one of many possible sights (e.g. other wildlife, indigenous people), and emphasize that orangutan sightings cannot be guaranteed.

Assessing orangutan tourism

Our assessments are based on previous evaluations of orangutan tourism and surveys of tours featuring orangutans. Most evaluations date from the late 1970s or late 1990s and represent informed opinions of experienced conservationists, researchers, and rehabilitation project managers; few are based on formal studies. Our surveys sampled internet-advertised tours featuring orangutans in 2005 and 2012, identified by searching the term *orangutan* (all spellings) combined with terms such as *adventure, tour*, or *ecotour*. We excluded captive orangutan shows at theme parks or zoos and lengthy volunteer or work visits (over three weeks), as not clearly representing tourism. These searches identified 124 organized tours in 2005 and 122 similar tours in 2012. For each tour we recorded the agents, orangutan targets (wild, rehabilitant, both; sites visited), orangutan selling points (e.g. close encounters, feeding), duration (days), visitor controls (health, behavior), and conservation donations. For the 2005 tours, we also recorded prices (USD estimate, per-person double occupancy, inclusions, and exclusions).

Our internet samples are clearly biased. They exclude tours not advertised on the internet (e.g. brief local tours) or geared to non-English speakers. The 2005 sample may over-represent wild relative to rehabilitant tours because we searched more intensively for wild tours (we found fewer, 55 vs. 72) and classified tours visiting both wild and rehabilitant orangutans as wild tours (they were typically marketed as such). Both samples nonetheless span the expected range of agents (Malaysian, Indonesian, international), orangutan targets (rehabilitant, wild, both; range of sites), durations (1–22 days), and prices (2005: $22–$7980 US).

Rehabilitant orangutan tourism

The earliest evaluations identified serious problems with rehabilitant orangutan tourism by the late 1970s. One of the first problems they identified was excessive tourist–rehabilitant close contact, which increased the risks of spreading tuberculosis, hepatitis B, and other infectious human diseases to rehabilitants (Frey, 1978; MacKinnon, 1977; Rijksen, 1978, 1982; Rijksen & Rijksen-Graatsma, 1975). Indirectly, this put wild orangutans at risk of human diseases. Early centers rehabilitated and released ex-captives in forests with resident wild orangutans, in the belief that this would bolster shrinking wild orangutan populations, but could provide their rehabilitants only primitive quarantine and medical care (Harrisson, 1961; Rijksen & Rijksen-Graatsma, 1975). Tourism also undermined the process of rehabilitation by encouraging ex-captives to increase rather than decrease their orientation to humans and to remain human-dependent rather than resume independent forest life.

Some experts, citing Bohorok as an example, argued that rehabilitant tourism could and had been managed effectively for economic and educational benefits (Aveling, 1982; Frey, 1978). Others argued that the tourist controls and benefits promised were rarely realized and the benefits gained did not offset the costs to the orangutans' health and rehabilitation (MacKinnon, 1977; Rijksen, 1982). Many recommended change, especially prohibiting tourist–orangutan contact (Aveling & Mitchell, 1982; MacKinnon, 1977; Rijksen, 1982; Rijksen & Rijksen-Graatsma, 1975).

Instead, in the 1980s, rehabilitant tourism soared in an uncontrolled fashion. By promoting orangutans as major tourist attractions, Malaysia's 1990 tourist campaign probably contributed to this (Kaplan & Rogers, 2000). Efforts to stop tourism at some problem rehabilitation centers failed because of resistance to losing tourist revenues (Aveling, 1982). By the turn of the century, annual visitors to Tanjung Puting, Semenggoh (with its newer partner site, Matang), Sepilok, and Bohorok totaled approximately 3000, 65 000, 89 600, and 50 000, respectively; daily, an average of 100 tourists (max. 300) viewed rehabilitant feedings at Bohorok (Corpuz, 2004; ICZM Malaysia, 1998; Rijksen & Meijaard, 1999; Sarawak Forestry Department, 2002). Many of the tourism problems identified in the 1970s persisted or intensified, often because lessons learned went unheeded and experts' recommendations were not implemented (Bennett & Gombek, 1992; Sajuthi *et al.*, 1991; Sugardjito & van Schaik, 1991; Tilson *et al.*, 1993).

Also near the turn of the century, formal evaluative studies substantiated early concerns (e.g. Cochrane, 1998; Rijksen, 1997; Russell, 1995). Some cited evidence that rehabilitation centers rarely, if ever, controlled tourism or informed visitors of rehabilitation issues and appropriate behavior with ex-captives and evidence that visitors transmitted infectious human diseases to rehabilitants despite preventive measures (Drewry, 1996; Rijksen & Meijaard, 1999; Russell, 1995; Singleton & Aprianto, 2001). Some condemned rehabilitation as a blight that commodifies ex-captives to attract tourist revenues and leads tourists to think they help orangutan conservation by visiting ex-captives when instead they undermine it (Lardoux-Gilloux, 1994; Leiman & Ghaffer, 1996; Yeager, 1997).

Some positive changes resulted. Advocates of rehabilitant orangutan tourism came to agree that it is not strictly compatible with rehabilitation but maintained that it could be controlled so as not to interfere (Galdikas, 1991, 1995; Payne & Andau, 1989). Three new rehabilitation projects that opened in the 1990s were closed to tourism from the outset, by design (Rijksen & Meijaard, 1999; Smits *et al.*, 1995; Sumatran Orangutan Conservation Programme, 2005). Partly because of tourism problems, the Indonesian government ordered Bohorok and Tanjung Puting to cease rehabilitation in 1995 (Pan Eco Foundation, 2000; Rijksen & Meijaard, 1999). An orangutan workshop and IUCN guidelines for tourism with great apes both recommended that no tourism be allowed with rehabilitants eligible for or already returned to forest life (Macfie & Williamson 2010; Rosen & Byers, 2002).

As of 2012, disputes over rehabilitant tourism remain unresolved. No rehabilitation projects that were open to tourism have stopped tourist visits. Sabah and Sarawak still promote tourism at their orangutan rehabilitation centers (Sabah Tourism Board, 2012; Sarawak Forestry Corporation, 2012). Bohorok and Tanjung Puting continued rehabilitation with overt tourism agendas until the turn of the century (Pan Eco Foundation, 2000; Rijksen & Meijaard, 1999) and both still promote tourism focused on viewing rehabilitants.

Rehabilitant tourism in fact appears to have increased. Our surveys found tours to visit rehabilitants at eight sites in 2005 and thirteen in 2012. Malaysian resorts hold ex-captive orangutans and market paid viewings as serving rehabilitation and conservation (Brend, 2001; Bukit Merah Laketown Resort, 2012; Rasa Ria Resort, 2012). Independent tour operators probably contribute to these increases, especially those offering rehabilitant tours in violation of regulations (Dellatore, 2007; Russon, pers. obs). Efforts are underway to implement better tourist education and management practices at Bohorok (Dellatore et al., this volume), but one agent recently promoted its Bohorok tour with a photo of a guide touching or offering food to an orangutan with a tourist close by (Indonesia Travel Plan, 2011). Tourism has also crept into three newer rehabilitation centers, Samboja Lestari, Nyaru Menteng, and Lamandau. The first two initially allowed visitors to view ex-captives under tightly controlled conditions (e.g. only those ineligible for release, via one-way windows) but were unable to sustain these controls against mounting tourism pressures (de'Gigant Tours, 2012; OrangHutan Tours, 2012; Russon, pers. obs). The third advertises tours to the area where it frees rehabilitants to forest life, recently using a photo of a visitor within touching distance of a free-ranging orangutan (Responsible Travel, 2012).

Wild orangutan tourism

Tours to see wild orangutans represented 42% of our tour sample in 2005 and 62% in 2012. As of 2012, they operate in nine national parks or other protected areas and three areas with no formal protection status. In our 2005 sample, 5% promised wild orangutan sightings, 80% offered no-guarantee opportunities to look for wild orangutans, and 81% included rehabilitant viewing, probably in case no wild orangutans were seen. These tours also involve more remote areas and rougher travel than rehabilitant tours, so they were generally longer, more costly (overall, per day), and probably more exclusive (smaller groups, specialized tourists).

Wild orangutan tours appear to be better operated and controlled than rehabilitant tours, probably in part because they are more specialized. Their conservation impacts over the long term remain to be seen. Many target rehabilitant as well as wild orangutans, including tours to Tanjung Puting and Bohorok, where orangutan tourism management has historically been poor. Officials are concerned about wild orangutan tourism in Kutai National Park: the orangutans visited are highly habituated and tourists, unless closely monitored, can and do approach within 5 m (Russon, pers. obs.). Other wild great apes have suffered from the habituation and exposure to human disease that tourism brings (in this volume see Desmond & Desmond, Goldmith, Muehlenbein & Wallis, Williamson & Macfie). Wild orangutan tours to unprotected areas limit the authority to safeguard conservation priorities (Macfie & Williamson, 2010). Tourism problems are emerging for other wild primates in areas where wild orangutans are visited (e.g. Leasor & Macgregor, this volume) and some of these could affect orangutans.

Benefits and costs of orangutan tourism for orangutan conservation

The critical question is clearly whether the benefits of orangutan tourism outweigh the recognized costs. It is important to note that most orangutan tourism has not been ecotourism. Ecotourism generally refers to minimal-impact travel to relatively undisturbed natural areas for the purpose of experiencing them and their wild-life; it should contribute to the conservation of these areas and their wildlife and improving the well-being of local people (Boo, 1990; Macfie & Williamson, 2010; TIES, 2005). While it is often difficult to draw a hard line between eco- and trad-itional tourism, especially since "eco-" labeling is often adopted for its promotional caché, orangutan tourism has typically fallen closer to traditional tourism (Fennell & Weaver, 2005; Honey & Stewart, 2002; Orams, 1995; Weaver, 2002). By focusing on rehabilitants, it has targeted humanized contexts (e.g. feeding sites), typical visits include day tours to view staged feedings, and annual visitor numbers reach tens of thousands (Rijksen & Meijaard, 1999).

Economics

Economic benefit has been a primary conservation rationale for orangutan tourism, as it is for ecotourism: tourism revenues will help fund conservation (Hvenegaard, this volume; Weaver, 2002). Rehabilitant orangutan tourism can be lucrative. In the late 1990s, estimated annual revenues were $100 000 US (Sepilok), $43 000–$80 000 US (Bohorok), and $240 000 US (Tanjung Puting) (Rijksen & Meijaard, 1999). Who profits is another question. Tourism is highly competitive and most businesses barely generate enough income to cover their costs (Cochrane, 1998). If anyone profits, it tends mainly to be large corporations that control air travel, accommoda-tion, and tour packages (Bandy, 1996; Goodwin, 1996; Lindberg, 1998).

We found few studies on economic benefits of orangutan tourism to conserva-tion. At Bohorok, only about half the foreigner entry fee, about $0.50 US (5% of annual local revenues), was allocated to rehabilitation and visitor center operations (Cochrane, 1998). Tourism income has been a major source of funds for Sepilok's rehabilitation program but entry fees are low so adequate funding requires high vis-itor numbers, which can undermine orangutans' welfare and rehabilitation (Corpuz, 2004). Tourist day entry fees are also low at other rehabilitation sites: as of 2012, they were about $1 US (Semenggoh), $3 US (Matang), $5 US (Tanjung Puting), and $10 US (Sepilok) for one foreign adult; fees for nationals are typically lower.

Indirect evidence suggests typical patterns apply. In Indonesia, many tourists visit national parks to view orangutans but the parks gain little direct economic benefit because entry fees are low and government regulated, and income is govern-ment controlled (Cochrane, 1997). Other tourist revenues typically contribute little to parks or local communities: at Komodo National Park, leakage estimates for tourist income exceeded 50% for local spending and 80% for tour prices (Walpole & Goodwin, 2000). Indonesian parks known for nature tourism (Komodo, Bunaken, Bogani, Tangkoko) have shown typical economic inequalities favoring urban and external over local interests (Ross & Wall, 1999; Walpole & Goodwin, 2000).

We assessed leakage in orangutan tourism from 84 tours in our 2005 sample that were over two days long and provided detailed itineraries. We estimated leak-age as major tour costs for domestic air travel and major hotels, that is external to destination sites and communities. We standardized cost estimates to all domestic costs as of September 2005 (for flights, ex. Kuala Lumpur, Malaysia, or Jakarta, Indonesia) using low-end internet quotes for the class of service and discounts advertised. Where similar tours offered by multiple vendors differed by less than 5% in price, we used the least expensive. Our leakage estimates for these 84 tours averaged 40% of standardized costs and exceeded 50% for 20% of tours. For 15 tours priced to include all flights, the total and international airfare averaged 37% and 27% of standardized costs, respectively. We also compared domestic costs for organized tours and independent trips to Tanjung Puting, the most popular destination for longer orangutan tours. Prices for 14-day tours started at about $4470 US (London departure, nine days in orangutan habitat), including $1000 US donation to the tour operator. We estimated the cost of a similar independent trip for two foreign tourists at $1840 US each, including all travel and daily expenses (English-speaking guide, houseboat, hotel/double occupancy, meals, and entry fees). Donation excluded, domestic costs were $1480 US higher for the tour than the independent trip. While the tour provided experiences beyond those open to independent tourists, some of which were probably provided locally, the differential paid to the tour operator is still substantial. Overall, our estimates suggest about 40% leakage in orangutan tourism and replicate the pattern of large commercial enterprises taking the bulk of revenues.

Ten vendors in our 2005 sample advertised donating a portion of tour income to orangutan conservation but only four stated amounts (10% of tour income, $1000 US). The recipients were three NGOs (Orangutan Foundation, Environmental Investigation Agency, and Sepilok Orangutan Appeal) that sponsored or operated the relevant tours. We could not assess how donations contributed to orangutan conservation.

Overall, we found little evidence that orangutan tourism generates substantial economic benefits to orangutan conservation. Our assessments did not include spin-off benefits, such as tourists' donations to orangutan conservation after their visit. Neither, however, did they include recognized costs of tourism to researchers, governments, conservation agencies, and local people, for example the lengthy process of habituating wild orangutans to viewing, developing and maintaining the tourist facilities and services needed, and the habitat damage and other local changes that result (in this volume see chapters by Desmond & Desmond, Goldsmith, Leasor & Macgregor, Strum & Manzolillo Nightingale, Wright *et al.*). Notably, as early as 1980, orangutan researchers considered that the economic and educational benefits of rehabilitant tourism, even occasional visits, did not outweigh its costs to rehabilitants' readaptation and health (MacKinnon, 1977; Rijksen, 1982).

Protection

The "use it or lose it" belief, that wildlife and wildlands are more likely to be protected if they are economically profitable as tourist attractions, has often been invoked as a benefit of tourism to conservation (Butler & Boyd, 2000; Freese, 1998;

Sherman & Dixon, 1991; Steele, 1995). This belief may hold some truth for rehabilitant tourism. All forest areas used by orangutan rehabilitation projects are legally protected but, in Indonesia, those closed to tourists have suffered much more from illegal logging than those open to tourists (BOS Scientific Advisory Board, 2003; Russon & Susilo, 1999; Singleton, pers. comm.). For wild orangutan tourism, results appear to be mixed: it may support some reforestation work but has also caused habitat damage and loss for some sympatric primates (Leasor & Macgregor, this volume).

Education

Education is the other common conservation rationale for orangutan tourism and for ecotourism in general (Fennell & Weaver, 2005; Orams, 1995, 1997; Singleton & Aprianto, 2001). If tourism can educate in ways that support conservation, it is clearly worth doing. To be effective, education should go beyond making tourists aware of conservation problems to changing their attitudes and behaviors so as to solve these problems (Orams, 1995, 1997). The experiences that orangutan tourism provides and the messages they generate are then critical to assessing the educational impact of orangutan tourism on orangutan conservation.

Close encounters with orangutans were promoted by 25% of the tours in our 2005 sample and 20% in our 2012 sample, up to and including direct physical contact. Beyond the health risks they create, close encounters encourage rehabilitants to manipulate and rob tourists. Only six vendors stated that orangutans would be viewed from a distance (at Sepilok, Semenggoh, and Bukit Merah) and only three explained why. Some sites have distance, no-contact, or no-feeding rules and/or barriers or signs to stop tourists from approaching rehabilitants, but enforcement has been weak. Reports are common of staff and guides failing to stop tourists from coming too close to orangutans and even staging close encounters (Abdul Rahman, 2001; Corpuz, 2004; Dellatore, 2007; Frey, 1996; Kemp, 2003; Rijksen & Meijaard, 1999; Russon, pers. obs; Singleton & Aprianto, 2001). Neither signs nor barriers stop rehabilitants from approaching tourists. Contact is almost inevitable once they do, and easily becomes habitual. Rehabilitants, unchecked, will cling to, kiss or bite tourists, steal or beg from them, and eat their half-eaten foods and waste, including toxic items like bleach, mosquito repellant, and detergents (Irwin, 2001; Kaplan & Rogers, 1994; Russon, pers. obs.; Singleton & Aprianto, 2001; Wong, 2005). Many tourists report having physically contacted rehabilitants (10 out of 20 tourist reports in our 2005 sample), having sought it or done little to avoid it when they knew it was prohibited and why; many considered it the highlight of their visit (Drewry, 1996). Efforts are underway at some sites to tighten regulations and enforcement (see Dellatore et al., this volume).

Staged feedings, a cornerstone of orangutan tourism, have questionable educational value. Feeding is standard in rehabilitation as a bridge to help ex-captives try forest life while unable to fend for themselves. Feedings have shifted over time to suit tourists, however, and orangutans are lured in at show times. Viewing feeding

Figure 5.1   (Upper photograph). An adult female rehabilitant orangutan and her infant sit on a tourist bench, at the daily afternoon orangutan feeding staged for tourists, within touching distance of a tour guide (2:51pm). (Lower photograph). Foreign and Indonesian tourists, their guides and rehabilitation staff sit on and near the same bench to watch the same feeding, a few minutes before the female orangutan arrived (2:33pm). "X" marks the place where the female orangutan later sat. (© A. Russon, July 29, 2004.)

promotes close encounters by drawing rehabilitants and tourists together (see Figure 5.1). Staff and guides have behaved inappropriately to show off to tourists, including teasing or threatening orangutans, making them perform tricks, and contacting them (Rijksen, 1997; Rijksen & Meijaard, 1999). Viewing feeding then sends distorted images of orangutan behavior, sociality, and forest life; shows orangutans as curiosities in a circus-like tradition; and promotes dangerously inappropriate behavior (Cochrane, 2003; Peterson, 1993; Rijksen & Meijaard, 1999; Singleton & Aprianto, 2001; Wagner, 2004).

Forest walks may offset the poor messages generated by staged feedings but open the door to others. Forest areas with free-ranging rehabilitants are freely accessible to tourists at all rehabilitation centers open to visitors. One result has been uncontrolled encounters with free-ranging rehabilitants that have ended in humans teasing orangutans; tourist–orangutan contact; orangutans stealing from tourists; and injury to humans, orangutans, or both (e.g. Anonymous, 2002a, Donaghy, 2002; Kaplan & Rogers, 1994; Kemp, 2003; Low, 2004; Rijksen & Meijaard, 1999; Wong, 2005). Guides have not solved the problem. Local guides often lack the requisite values, knowledge, or skills (Cochrane, 1998; Rijksen, 1997). Bohorok guides have taken and solicited bribes to lure rehabilitants to private encounters with visitors, resulting in visitors and orangutans being injured and orangutans killed (Dellatore *et al.*, this volume; Rijksen & Meijaard, 1999).

Tour promotional materials in our 2005 and 2012 survey samples had little educational content. Only three tours in each of our samples marketed themselves as educational and only two in our 2005 sample specified an educational component (one lecture, one visit with staff briefing). All rehabilitation sites open to tourists have information centers and Sepilok's is extensive (Corpuz, 2004). However, studies of educational practices have found common failings: information has been insufficient or so outdated as to be misleading; neglected rehabilitation and conservation issues concerning the orangutans and projects visited; failed to interest local visitors; or was difficult to access (location, opening hours; Abdul Rahman, 2001; Cochrane, 2003; Rijksen, 1997; Rijksen & Meijaard, 1999; Russon, pers. obs.; Singleton & Aprianto, 2001). Staff and guides can inform visitors but their input has been inconsistent, tourists have complained of their poor knowledge, and instruction on how to behave around orangutans has often been absent or ineffective (Cochrane, 1998; Corpuz, 2004; Low, 2004; Rijksen, 1997; Rijksen & Meijaard, 1999; Russon, pers. obs.; Sorenson, 2004).

Some tours offer educational programs. The best we found were offered by longer (14+ days) high-end tours to work on orangutan research or conservation. Russell (1995) conducted one of the few formal studies on their educational value, based on Earthwatch visitors' attitudes to and interactions with orangutans during their Tanjung Puting tour. Tour members arrived with three main perceptions of orangutans – photographic collectibles, embodiments of pristine nature, and children to be nurtured – none of which changed over the visit (Russell, 1995; Russell & Ankenman, 1996). Photography-oriented visitors overlooked much beyond their camera lens and so experienced a fragmented view of orangutans. Pristine nature seekers pursued wild orangutans in the forest, some spurning rehabilitants as less "real." Their emphasis on pristine nature risks aggravating pressures on rare and endangered species and bringing more natural areas under exploitation as they seek out wilder zones. Some who viewed young ex-captives as much like human infants sought every chance to cuddle them. While all had been informed that contact can transmit diseases or encourage inappropriate relationships and habituated orangutans could hurt humans, they seemed unable or unwilling to accept it. Drewry (1996), studying a similar tour, found that seven out of ten tour members knowingly disregarded educational material they had received about the dangers of contact in favor of holding an orangutan.

Finally, tourism may send unintended messages to unintended audiences. Two Malaysian hotels keep ex-captives and market paid visits with them in the name of conservation and rehabilitation (Banker, 2004; Brend, 2001; Bukit Merah Laketown Resort, 2012; Rasa Ria Resort, 2012). Bukit Merah offers visits ($9 US per adult) to an orangutan "Rehabilitation Island" and meeting ex-captives; previously, it sold opportunities to pose with orangutans for photographs (Agoramoorthy, 2002; Anonymous, 2002b; Brend, 2001; Bukit Merah Laketown Resort, 2012). Rasa Ria keeps ex-captives in its private nature reserve, describes its work with them as rehabilitation, charges $23 US per adult to view feedings, and recently promoted viewings with a photograph showing a guard cuddling two young orangutans while tourists watched within easy reach (Rasa Ria Resort, 2012). Both hotels have government authorization to keep ex-captives (Anonymous, 2002b; Rasa Ria Resort, 2012; Sepilok Orangutan Appeal, 2003) but endanger their health and rehabilitation with their tourism.

Health

Concerns that tourism jeopardizes the health of the primates visited have increased as evidence accumulates that humans can infect primates with human diseases, cause illness through unsuitable feeding, lower their disease resistance by stressing them, and increase their vulnerability to injury by habituating them (in this volume, see Goldsmith, Muehlenbein & Wallis, Sapolsky, Wright et al.). Human diseases known to have infected great apes were most likely introduced by local humans or researchers but tourists still pose substantial risks (Muehlenbein et al., 2010; Muehlenbein & Wallis, this volume).

Rehabilitants exposed to tourists have been diagnosed with human diseases that include typhoid, hepatitis (A, B, C), tuberculosis, scabies, measles, conjunctivitis, meningitis, various parasites, and respiratory ailments; tuberculosis, hepatitis B, and scabies have caused rehabilitant deaths (Frey, 1978; Kosasih et al., 1977; Rijksen, 1982, 1997; Rijksen & Rijksen-Graatsma, 1975; Warren, 2001; Woodford et al., 2002; Yeager, 1997, 1999). Of the orangutans that Kilbourn et al. (1998) tested, those that had been exposed to humans (Sepilok rehabilitants) were infected with common human respiratory viruses but those that had not (wild orangutans) did not have antibodies against these viruses.

Tourism can stress orangutans. Orangutans show behavioral signs of stress on encountering humans who are unfamiliar and/or behave intrusively (e.g. approach too close, talk loudly, take flash photographs) – traits that many tourists show. Rehabilitants may also show signs of stress at tourist feedings, where food competition and human crowding can be high. One recent study found physiological signs of acute but not chronic stress in wild orangutans after tourist viewings (Muehlenbein et al., 2012); it assessed only two exceptionally well-habituated individuals, however, so the generality of these findings is uncertain.

Rehabilitants have been stabbed or poisoned to death when they endangered tourists, and seriously injured or killed by electrical cables at tourist facilities (Rijksen & Meijaard, 1999; Singleton & Aprianto, 2001; Sumatran Orangutan Society, 2000). Rehabilitants from tourism sites have been caught crop-raiding (Orangutan Information Centre, 2011), a problem potentially exacerbated by the human dependency and over-habituation that tourism fosters. Rehabilitant females reproduce younger and faster than wild females, possibly because of the extra food they receive, but suffer much higher infant mortality (Kuze et al., 2011). Cannibalism is known only in female rehabilitants exposed to tourism (Dellatore et al., 2009). Daily feedings for tourists may contribute to these alterations in reproduction patterns.

Recently, efforts have been made to improve controls against the health risks caused by tourism. IUCN guidelines for great ape tourism limit visits to small groups, a distance of at least 7 m or 10 m (with or without a surgical mask, respectively) and at most one hour in duration, and recommend against tourism with rehabilitants eligible for release (Macfie & Williamson, 2010; Williamson & Macfie, this volume). Orangutan tourism guidelines further require waiting at least seven days after foreign travel before visiting orangutans (Rosen & Byers, 2002). Signs of changes in these directions are limited. Tours still promise close encounters and guides still stage them illicitly (Dellatore et al., this volume; Indonesia Travel Plan, 2011), although fewer promised close encounters in our 2012 than our 2005 sample (17% vs. 27%). Of the 34 tours in our 2005 sample promising close encounters, none explained disease risks or stated health requirements for visiting orangutans. Three of the four main rehabilitant tourist sites are in areas with resident wild orangutans and all still focus tourism on daily feedings that draw uncontrolled numbers of tourists, rehabilitants, and wild orangutans together, so health risks still apply to wild orangutans. Some observers have reported improved health controls at sites notorious for allowing visitor–rehabilitant contact (Dellatore et al., this volume; Low, 2004) but others have not (Kemp, 2003). In summary, orangutan tourism may aim to mitigate health risks but its achievements appear limited.

Behavioral rehabilitation

The second major tourism cost to ex-captive orangutans is disrupting their rehabilitation. Tourism encourages ex-captives to maintain human orientation and skills when they should be acquiring expertise for independent forest life in an orangutan community and breaking ties with humans (Frey, 1978; MacKinnon, 1977;

Rijksen, 1978, 1982, 1997; Rijksen & Rijksen-Graatsma, 1975). Some rehabilitants who are repeatedly exposed to visitors concentrate on learning to manipulate them, typically to extort food and social support, and often target tourists as easy, naïve prey (Russon, 2009; Singleton & Aprianto, 2001). Rehabilitants drawn regularly to feeding sites for tourist viewing are ranging abnormally in ways that may interfere with establishing suitable feral home ranges, distort their activity budgets, and artificially elevate feeding competition (Dellatore, 2007; Dellatore *et al.*, this volume). Bohorok rehabilitants have been returned to captivity for raiding or harassing tourist facilities (Orangutan Information Centre, 2011; Sumatran Orangutan Society, 2000).

Habitat

Costs to habitat identified around touristed orangutan sites include increased toxic pollutants and garbage, habitat degradation and destruction, wildlife disturbance, and uncontrolled development (Cochrane, 1998; Frey, 1996; Rijksen & Meijaard, 1999; Singleton & Aprianto, 2001). At Bohorok, a tourist village mushroomed until it supplanted the forest leading to the rehabilitation center (Rijksen & Meijaard, 1999). Its tourists and tourist industry caused damage along the area's roads and waterways, river pollution, bank erosion, inadequate sewage and waste disposal, and wildlife loss (overuse of nature trails scared wildlife away, poaching eradicated species sold to tourists) (McCarthy, 1999). Bohorok is a prime example of why income-generating projects in and around natural areas may not protect them, even if they curtail illegal logging or clearing of forests (Cochrane, 2003).

Other costs

Rehabilitant tourism may create new costs to orangutans because of the need to guarantee high probabilities of seeing them (Cochrane, 1998). If more orangutans are needed to sustain tourist attractions, then purchase and poaching risks grow and tourism may cause more injury, orphaning, and capturing. Providing ex-captives to Malaysian hotels comes perilously close to this situation. One Malaysian park was caught smuggling Sumatran orangutans and was required to repatriate them (Anonymous, 2005; Atan, 2005; Reuters, 2005).

Discussion and conclusion

Based on our review, orangutan tourism has not clearly generated many of the economic, protection, or educational benefits to orangutan conservation that were expected. It can be economically profitable and may have protected some rehabilitant orangutan habitat, but its income may have contributed little to orangutan conservation and its educational efforts have not proven effective. Tourism has also imposed serious costs on the orangutans visited, especially their health, safety, reproduction, and rehabilitation, that do not clearly outweigh the benefits. While our assessment is not sufficiently systematic to establish this conclusively, the patterns we detected resemble patterns seen in tourism with other primates that have likewise led to questioning its conservation value (e.g. Butynski & Kalina, 1998; Goldsmith, this volume; Goldsmith *et al.*, 2006; Litchfield, 2001; Wallis & Lee, 1999; Woodford *et al.*, 2002). For orangutan tourism, the question is then whether it can be changed to benefit orangutan conservation. To that end, we consider the problems involved and possible improvements.

Orangutan tourism problems

Problems with orangutan tourism have persisted despite repeated assessments, recommendations for change, and attempts to improve. Proper controls have been recognized as critical to making rehabilitant tourism work to conservation advantage for over 30 years (Aveling, 1982; Frey, 1978; Galdikas, 1991, 1995; Payne & Andau, 1989). They have rarely been achieved or even attempted, however, and any effectiveness was short-lived (Rijksen, 1997). As an important first step toward improvement, we consider two important contributors to chronic failure: the long-term processes affecting orangutan tourism and the priorities of controlling parties.

Tourism areas are seen as having life-cycles wherein visitors and vested interests change over time (Catlin *et al.*, 2011; Duffus & Dearden, 1990). Early on, tourists are typically knowledgeable about what they will see and careful to have low impact, and tourist-based businesses arrive to develop the area. As tourism develops, businesses generate major habitat damage; more general-interest tourists visit, who are less knowledgeable and concerned, and who may eventually destroy what they came to see. This progression is extremely difficult to prevent or alter; it can become almost irreversible as it advances and vested interests work to maintain developed resources (Duffus & Dearden, 1990). Orangutans are highly intelligent and long-lived, and they create problems that begin as soon as tourism starts and grow slowly, for example interacting with tourists and encountering human diseases. Both problems are initially minor or even invisible but can grow to dangerous levels. Such problems may be especially hard to manage because they grow unchecked until they are serious, by which time they may be beyond control.

Conservation may not maintain top priority in managing tourism. Rehabilitation projects invested in tourism are compromised: given that the consequences of tourist dissatisfaction, restricting tourist behavior, and encouraging rehabilitants to avoid humans are not in their interests. Staff charged with enforcing regulations, often from local communities, often lack the status, education, and language skills to control tourists. Cultural values and low salaries may bias them to prioritize pleasing tourists and earning tips over protecting orangutans. Businesses are economically driven, so tourist satisfaction takes top priority. Tourists may feel entitled to satisfy personal agendas, given the high prices they paid and the expectations of close encounters that are promoted (Corpuz, 2004). Rehabilitants may be little motivated to avoid tourists because of the food, other goods, and socio-emotional support they gain. Finally, all these actors tend to be guided by short-sighted interests, probably because most do not live with the consequences.

Recommendations

We are skeptical that orangutan tourism can operate to serve orangutan conservation given that these problems have persisted despite ample opportunity for change: over 35 years' continuous operation, many facilities and sites, many management and government regimes, expert tourism and orangutan advice, and (presumably) ample tourism profits. We also hesitate to suggest alternatives, given the difficulties of reconciling conservation and tourist interests in fragile natural contexts (Higham, 2007; Ives & Messerli, 1989) and the likelihood that available alternatives will deteriorate in a similar fashion. Predictable problems have already developed with wild African ape tourism that have raised concerns about its conservation value (e.g. Goldsmith, this volume). Many consider that visiting orangutans has some legitimacy and potential value, however, and that compromise is inevitable. With these caveats, we sketch major informed recommendations for operating orangutan tourism so that it contributes to conservation.

1. *Stop tourism with rehabilitant orangutans eligible for or already released to free forest life, or in forests where they range.* Specialists recommend this and it is the only option consistent with conservation interests (Macfie & Williamson, 2010; Rosen & Byers, 2002). Non-intrusive viewing of ex-captives ineligible for release is one alternative (e.g. in visitor enclosures via one-way viewing windows, or from viewing towers or enclosed walkways). Combined with good explanations, it appears to satisfy many day visitors.
2. *Design wild orangutan tourism to offer an orangutan highlight and avoid recognized tourist pressures.* The handicaps to wild orangutan tourism recognized in the 1970s remain: remote sites, primitive tourist facilities, and uncertain sightings. In our 2005 survey, tours featuring wild orangutans averaged only 53% of their time in wild orangutan habitat, 81% included visiting rehabilitants, and one advised against its own wild orangutan tour because some groups saw no orangutans and left dissatisfied. Tours with broader themes (e.g. an area's wildlife or natural habitat), no-guarantee opportunities to spot wild orangutans, and no rehabilitant visits are recommended. This option offers better conditions for vendors and avoids known tourist pressures. In our 2005 survey, 57% of tours featuring wild orangutans used this option. As of 2012, such tours appear to be increasing, so this may be an effective alternative.
3. *Design the management of orangutan tourism to prevent problems and limit changes to acceptable levels.* Plan for, specify, and enforce limits to tourism growth (e.g. number and traits of orangutans to be habituated, tourist group size, visit duration and frequency, access only to non-priority areas for orangutan conservation) (Rosen & Byers, 2002). Establish suitable checks and balances, for example regular monitoring designed for early problem detection, rapid response systems, guide/operator licensing, training, and enforcement (Cochrane, 1998).

4. *Ensure that conservation interests hold and exercise deciding authority in orang-utan tourism operations, from the outset.* This includes the decision to develop tourism, defining and enforcing tourism regulations, exercising authority consistently as soon as tourist operations start, and maintaining authority over the long term. Conservationists should collaborate with all stakeholders, local communities, and experts on orangutan biology included (McCarthy, 1999). They must be convinced by sound cost–benefit assessments that tourism's benefits to orangutan conservation significantly outweigh its risks over the short and long term. They must further be convinced that effective management practices will be instituted to counter all known orangutan tourism problems (e.g. prohibit problematic practices such as tourist feedings and feeding sites) and to maintain conservation priorities over the long term (Singleton & Aprianto, 2001). Available conservation-oriented tourism guidelines should be instituted, for instance, with attention to the best ways to ensure adherence (Macfie & Williamson, 2010; Russell, 2004; Sirakaya, 1997; Sirakaya & Uysal, 1997; Williamson & Macfie, this volume).

Acknowledgments

The research involved in developing this paper was supported by Glendon College of York University, Toronto, Canada. Thanks especially to Connie Russell for her collaboration and substantial contribution to earlier versions of this material. Thanks also to numerous colleagues for contributing information and discussion on these issues over the years, to Laura Adams for developing the 2005 sample tours featuring orangutans, to Sarah Iannicello for assistance in updating it in 2012, and to Carol Berman, Dave Dellatore, and Ian Singleton for comments on earlier versions of this manuscript.

References

Abdul Rahman, H. M. (2001). Visitor profile and satisfaction survey at Semenggoh Wildlife Rehabilitation Centre, Sarawak, Malaysia, *Hornbill*, 5. www.mered.org.uk/Hornbill/Abdul_Rahman.htm (accessed Sept. 19, 2005).

Agoramoorthy, G. (2002). Exhibiting orangutans on a natural island in Malaysia. *International Zoo News*, 49 (5): 2–3.

Anonymous. (2002a). Ape leaves tourist naked. *Regina Leader Post*, Oct. 20, 2002.

Anonymous. (2002b). Sarawak to lend resort five more orangutans. *Malaysia Star*, Oct. 20, 2002.

Anonymous. (2005). Dept to investigate species of orangutans kept at resort. *Malaysiakini. com*, May 10, 2005.

Atan, H. (2005). Orang-utans head home. *New Straits Times*, Dec. 18, 2005. www.nst.com. my/Current_News/NST/Sunday/National/20051218083034/Article/ (accessed Dec. 19, 2005).

Aveling, R. J. (1982). Orang utan conservation in Sumatra, by habitat protection and conservation education. In: L. E. M. de Boer (ed.), *The Orang Utan: Its Biology and Conservation*, The Hague: Dr. W. Junk, pp. 299–315.

Aveling, R. J. and Mitchell, A. (1982). Is rehabilitating orangutans worthwhile? *Oryx*, 16; 263–271.

Bandy, J. (1996). Managing the other of nature: Sustainability, spectacle, and global regimes of capital in ecotourism. *Public Culture*, 8: 539–566.

Banker, J. (2004). The land on natural treasures–Kota Kinabalu, Sabah. *Kansai Scene*, 46. www.kansaiscene.com/2004_03/html/travel.shtml (accessed July 27, 2005).

Bennett, E. L. (1998). *The Natural History of Orang-Utan*. Kota Kinabalu, Sabah: Natural History Publications (Borneo).

Bennett, J. and Gombek, F. (1992). *Orangutan ecotourism in Borneo*. Paper presented at the Second Biennial International Conference of the Borneo Research Council.

Boo, E. (1990). *Ecotourism: The Potentials and Pitfalls*. Washington, DC: World Wildlife Fund.

Borner, M. (1976). Sumatra's orang-utans. *Oryx*, 13 (3): 290–293.

BOS Scientific Advisory Board. (2003). *Meratus Release Forest: Recommendations*. Unpublished report to the Borneo Orangutan Survival Foundation (Indonesia), July.

Brend, S. (2001). It shouldn't happen to an ape. *BBC Wildlife*, June 2001. www.responsible-travel.com (accessed July 27, 2005).

Bukit Merah Laketown Resort. (2012). *Orangutan Island*. www.bukitmerahresort.com.my/orangutan.htm (accessed April 15, 2012).

Butler, R. (1980). The concept of a tourist area cycle of evolution: implications for management of resources. *Canadian Geographer*, 24: 5.

Butler, R. and Boyd, S. (2000). *Tourism and National Parks: Issues and Implications*. Chichester, UK: Wiley.

Butynski, T. M. and Kalina, J. (1998). Gorilla tourism: A critical look. In E. J. Milner-Gulland and R. Mace (eds.), *Conservation of Biological Resources*. London: Blackwell, pp. 280–300.

Catlin, J., Jones, R., and Jones, T. (2011). Revisiting Duffus and Dearden's wildlife tourism framework. *Biological Conservation*, 144(5): 1537–1544.

CITES. (2012). *Appendices I, II and III, valid from 3 April 2012*. www.cites.org/eng/app/appendices.php. (accessed April 15, 2012).

Cochrane, J. (1997). *Factors Influencing Ecotourism Benefits to Small, Forest–Reliant Communities: A Case Study of Bromo Tengger Semeru National Park, East Java*. University of Hull, UK. www.recoftc.org/documents/Inter_Reps/Ecotourism/Cochrane.rtf (accessed May 22, 2005).

Cochrane, J. (1998). *Organisation of Ecotourism in the Leuser Ecosystem*. Unpublished report to the Leuser Management Unit.

Cochrane, J. (2003). *Ecotourism, Conservation and Sustainability: A Case Study of Bromo Tengger Semeru National Park, Indonesia*. University of Hull, UK: Unpublished doctoral dissertation.

Corpuz, R. (2004). *"Wild Borneo" – a perception: A Study of Visitor Perception and Experience of Nature Tourism in Sandakan, Sabah, Malaysian Borneo*. International Centre for Responsible Tourism, University of Greenwich: Unpublished MSc thesis.

da Silva, G. S. (1965). The East-coast experiment. *IUCN Public N.S. 10: Conservation in tropical South East Asia*. Bangkok, pp. 229–302.

de'Gigant Tours. (2012). *Rahai'l Jungle*. www.borneotourgigant.com/3_Days_ Kahayan_ River_Weekend_Tours.html (accessed April 30, 2012).

Dellatore, D. F. (2007). *Behavioural Health of Reintroduced Orangutans (Pongo abelii) in Bukit Lawang, Sumatra Indonesia*. Oxford Brookes, UK: MSc thesis.

Dellatore, D. F., Waitt, C. D., and Foitovà, I. (2009). Two cases of mother–infant cannibalism in orangutans. *Primates*, 50: 277–281.

Donaghy, K. (2002). *Orang Utans at Sepilok*. www.wildasia.net/main/article.cfm?articleID=18 (accessed Sept. 19, 2005).

Drewry, R. (1996). *Sustainable Development, Ecotourism and Flagship Species: The Case of Orangutan Ecotourism in Indonesia*. School of Humanities, Murdoch University, Murdoch, Western Australia: Unpublished BA (Honours) thesis.

Duffus, D. A. and Dearden, P. (1990). Non-consumptive wildlife-oriented recreation: A conceptual framework. *Biological Conservation*, 53: 213–231.

Fennell, D. and Weaver, D. (2005). The ecotourism concept and tourism-conservation symbiosis. *Journal of Sustainable Tourism*, **13**(4): 373–390.

Freese, C. H. (1998). *Wild Species as Commodities: Managing Markets and Ecosystems for Sustainability*. Covelo, CA: Island Press.

Frey, R. (1978). Management of orangutans. In *Wildlife Management in Southeast Asia, Biotrop*. Special Publication 8, pp. 199–215.

Frey, R. (1996). *Report on Present Situation and Proposal for Establishment of Sustainable Management of Bohorok Orangutan Centre*. Unpublished report.

Galdikas, B. M. F. (1991). Protection of wild orangutans and habitat in Kalimantan vis a vis rehabilitation. In *Proceedings on the Conservation of the Great Apes in the New World Order of the Environment*. Jakarta: Republic of Indonesia Ministries of Forestry and Tourism, Post, and Telecommunication, pp. 87–94.

Galdikas, B. M. F. (1995). Behavior of wild adolescent female orangutans. In R.D. Nadler, B. M. F. Galdikas, L.K. Sheeran, and N. Rosen (eds.), *The Neglected Ape*, New York: Plenum, pp. 163–182.

Galdikas-Brindamour, B. and Brindamour, R. (1975). Orangutans, Indonesia's "people of the forest". *National Geographic*, 148: 444–473.

Goldsmith, M. L., Glick, J. and Ngabirano, E. (2006). Gorillas living on the edge: Literally and figuratively. In: N. E. Newton-Fisher, H. Notman, J. D. Paterson, and V. Reynolds (eds.), *Primates of Western Uganda*. New York: Springer Publishers. pp. 405–422.

Goodwin, H. (1996). In pursuit of ecotourism. *Biodiversity and Conservation*, 5, 277–291.

GRASP. (2005). *Kinshasa Declaration on Great Apes*. Kinshasa, Sept. 9, 2005. (GRASP: Great Ape Survival Project).

Harrisson, B. (1961). Orangutan: What chances of survival? *Sarawak Museum Journal*, 10: 238–261.

Higham, J. E. S. (2007). Ecotourism: which school of thought should prevail? In J. E. S. Higham (ed.), *Critical Issues in Ecotourism: Understanding a Complex Tourism Phenomenon*. Oxford: Elsevier, pp. 428–434.

Honey, M. and Stewart, M. (2002). Introduction. In: M. Honey (ed.), *Ecotourism and Certification: Setting Standards in Practice*. Covelo, CA: Island Press, pp.1–29.

ICZM Malaysia. (1998). *Sabah Coastal Zone Profile*. Town and Regional Planning Department, Sabah. Kota Kinabalu. (ICZM: Integrated Coastal Zone Management)

Indonesia Travel Plan. (2011). *Easy Going Sumatra: Swinging with the Apes*. www.indonesiatravelplan.co.uk/indonesia-tour.htm (accessed Dec. 18, 2011).

Irwin, A. (2001). Wild at heart. *New Scientist*, 169 (2280): 26.

IUCN. (2011). *The IUCN Red List of Threatened Species. Version 2011.2.* www.iucnredlist. org (accessed April 15, 2012).

Ives, J. D. and Messerli, B. (1989). *The Himalayan Dilemma: Reconciling Development and Conservation.* New York: Routledge.

Kaplan, G. and Rogers, L. (1994). *Orang-Utans in Borneo.* Armidale, NSW, Australia: University of New England Press.

Kaplan, G. and Rogers, L. (2000). *The Orangutans: Their Evolution, Behavior, and Future.* Cambridge, MA: Perseus Publishing.

Kemp, B. (2003). The old man of the forest. *Dreamscapes*, July–Aug., 17–29.

Kilbourn, A. M., Karesh, W. B., Bosi, E. J., *et al.* (1998). Disease evaluation of free-ranging orangutans (*Pongo pygmaeus pygmaeus*) in Sabah, Malaysia. In: C. K. Baer (ed.), *Proceedings of the American Association of Zoo Veterinarians and American Association of Wildlife Veterinarians Joint Conference.* Philadelphia, USA: American Association of Zoo Veterinarians Media, pp. 417–421.

Kosasih, E., *et al.* (1977). *Viral hepatitis B exposure rate among captive orangutans, Pongo pygmaeus.* Dept. of Clinical Pathology, Faculty of Medicine, University of North Sumatra and Provincial Top Referral Hospital, Medan.

Kuze, N., Dellatore, D., Banes, G. L., Pratje, P., Tajima, T., and Russon, A. E. (2011). Factors affecting reproduction in rehabilitant female orangutans: Young age at first birth and short inter-birth interval. *Primates*, 53 (2): 181–192.

Lardoux-Gilloux, I. (1994). Rehabilitation centers: their struggle, their future. In: J. J. Ogden, L. A. Perkins, and L. Sheehan (eds.), *Proceedings of the International Conference on "Orangutans: The Neglected Ape."* San Diego: Zoological Society of San Diego.

Leiman, A. and Ghaffer, N. (1996). Use, misuse, and abuse of the orang-utan – exploitation as a threat or the only real salvation? In: V.J. Taylor and N. Dunstone (eds.), *The Exploitation of Mammal Populations.* London: Chapman & Hall, pp. 345–357.

Lindberg, K. (1998). Economic aspects of ecotourism. In: K. Lindberg, M. Epler Wood, and D. Engeldrum (eds.), *Ecotourism: A Guide for Planners and Managers*, vol. 2. North Bennington, VT: Ecotourism Society, pp. 87–117.

Litchfield, C. (2001). Responsible tourism with great apes in Uganda. In: S. McCool and R. N. Moisey (eds.), *Tourism, Recreation and Sustainability: Linking Culture and the Environment.* Oxon, UK: CABI Publishing, pp. 105–132.

Low, T. W. (2004). *Can Ecotourism Help Protect Orang-utans?* Sustainable Tourism Development, APU University, Cambridge/Chelmsford, UK: Unpublished BSc (Hons) thesis.

Macfie, E. J. and Williamson, E. A. (2010). *Best Practice Guidelines for Great Ape Tourism.* Gland, Switzerland: IUCN/SSC Primate specialist Group (PSG).

MacKinnon, J. R. (1977). Rehabilitation and orangutan conservation. *New Scientist*, 74: 697–699.

McCarthy, J. (1999). *Nature based tourism. Case study: Gunung Leuser, Indonesia.* http:// wwwscience.murdoch.edu.au/teach/n279/n279content/casestudies/g-leuser/nbtcasestudy. html (accessed Sept. 12, 2005).

Muehlenbein, M. P., Ancrenaz, M., Sakong, R., *et al.* (2012). Ape conservation physiology: Fecal glucocorticoid responses in wild *Pongo pygmaeus morio* following human visitation. *PLoS ONE*, 7(3): e33357.

Muehlenbein, M. P., Martinez, L. A., Lemke, A. A., *et al.* (2010). Unhealthy travelers present challenges to sustainable primate ecotourism. *Travel Medicine and Infectious Disease*, 8: 169–175.

Orams, M. B. (1995). Towards a more desirable form of ecotourism. *Tourism Management*, 16(1): 3–8.

Orams, M. B. (1997). The effectiveness of environmental education: Can we turn tourists into "greenies"? *Progress in Tourism and Hospitality Research*, 3: 295–306.

OrangHutan Tours. (2012). *Samboja Orangutan Tour*. www.oranghutantours.com/samboja_ orangutan_tours.htm (accessed April 28, 2012).

Orangutan Information Centre. (2011). *Human Orangutan Conflict Monitoring and Mitigation in North Sumatra*. Unpublished final grant report to BOS Canada, Toronto, Canada.

Pan Eco Foundation. (2000). *Sumatran Orangutan Conservation Programme*. Project brochure. Medan, Sumatra.

Payne, J. and Andau, P. (1989). *Orang-Utan: Malaysia's Mascot*. Kuala Lumpur: Berita Publishing Sdn. Bhd.

Peterson, D. (1993). *The Deluge and the Ark: A Journey into Primate Worlds*. New York: Avalon.

Rasa Ria Resort. (2012). Nature. www.shangri-la.com/kotakinabalu/rasariaresort/sports-recreation/nature/ (accessed April 15, 2012).

Responsible Travel. (2012). *Orangutan Conservation Vacation*. www.responsiblevacation. com/vacation/1289/orangutan-conservation-vacation (accessed June 5, 2012).

Reuters. (2005). Thai police crack case of the missing orang-utan. *Reuters News Service*, Aug. 13., 2005

Rijksen, H. D. (1978). *A Field Study of Sumatran Orang Utans (Pongo pygmaeus abelii, Lesson 1872), Ecology, Behavior and Conservation*. Wageningen, the Netherlands, Mededlingen Landbouwhogeschool: H. Veenman and Zonen B.V.

Rijksen, H. D. (1982). How to save the mysterious 'man of the forest'? In: L. E. M. de Boer (ed.), *The Orang Utan: Its Biology and Conservation*. The Hague: Dr. W. Junk Publishers, pp. 317–341.

Rijksen, H. D. (1997). *Orang Utan Viewing Centre in Sumatra: Recommendations for Improving of the Bohorok Facility*. Unpublished report commissioned by the Director General, PHPA, Indonesia.

Rijksen, H. D. and Meijaard, E. (1999). *Our Vanishing Relative: The Status of Wild Orang-Utans at the Close of the Twentieth Century*. Kluwer Academic Publishers.

Rijksen, H. D. and Rijksen-Graatsma, A. G. (1975). Orangutan rescue work in North Sumatra. *Oryx*, 13: 63–73.

Rosen, N. and Byers, O. (eds.). (2002). *Orangutan Conservation and Reintroduction Workshop: Final Report*. Apple Valley, MN: IUCN/SSC Conservation Breeding Specialist Group.

Ross, S. and Wall, G. (1999). Evaluating ecotourism: The case of North Sulawesi, Indonesia. *Tourism Management*, 20: 673–682.

Russell, C. L. (1995). The social construction of orangutans: An ecotourist experience. *Society and Animals*, 3(2): 151–70.

Russell, C. L. (2004). Establishing guidelines for primate-focused ecotourism: Lessons from whalewatching. Paper presented at the International Primatological Society Congress, Torino, Italy, Aug.

Russell, C. L. and Ankenman, M. J. (1996). Orangutans as photographic collectibles: Ecotourism and the commodification of nature. *Tourism Recreation Research*, 21(1): 71–78.

Russon, A. E. (2009). Orangutan rehabilitation and reintroduction. In: S.A. Wich, S. S. Utami Atmoko, T. Mitra Setia, *et al.* (eds.), *Orangutans: Geographic Variation in Behavioral Ecology and Conservation*. Oxford University Press, pp. 327–350.

Russon, A. E. and Susilo, A. (1999). The effects of the 1997–98 droughts and fires on oran-
gutans in Sungai Wain Protection Forest, E. Kalimantan, Indonesia. In: H. Suhartoyo and
T. Toma (eds.), *Impacts of Fire and Human Activities on Forest Ecosystems in the Tropics:
Proceedings of the Third International Symposium on Asian Tropical Forest Management*.
Samarinda, Indonesia: Tropical Forest Research Center, Mulawarman University and
Japan International Cooperation Agency, pp. 348–372.

Sabah Tourism Board. (2012). *Sepilok orangutan sanctuary*. www.sabahtourism.com/en/des-
tination/32/ (accessed April 15, 2012).

Sajuthi, D., Karesh, W., McManamon, R., *et al.* (1991). Medical aspects in orangutan
reintroduction. *Proceedings of the International Conference on "Conservation of the Great
Apes in the New World Order of the Environment"*, Indonesia, 1991.

Sarawak Forestry Corporation. (2012). Sarawak National Parks and Reserves: Semenggoh
Wildlife Centre. www.sarawakforestry.com/htm/snp-nr-semenggoh.html (accessed April
15, 2012).

Sarawak Forestry Department. (2002). *Visitors Statistics to the Wildlife Centre, Wildlife
Rehabilitation Centre and Nature Reserve*. www.forestry.sarawak.gov.my/forweb/np/about/
ff/nrvis.htm (accessed Sept. 20, 2005).

Sepilok Orangutan Appeal. (2003). *Update on murdered orangutans*. www.usenet.com/news-
groups/talk.politics.animals/msg04322.html (accessed May 8, 2005).

Sherman, P. B. and Dixon, J. A. (1991). The economics of nature tourism: Determining if
it pays. In: T. Whelan (ed.), *Nature Tourism: Managing for the Environment*. Covelo, CA:
Island Press, pp. 89–131.

Singleton, I. and Aprianto, S. (2001). *The semi-wild orangutan population at Bukit Lawang:
A valuable 'ekowisata' resource and their requirements*. Unpublished paper presented at the
Workshop on Eco-tourism Development at Bukit Lawang, Medan, Indonesia, April.

Sirakaya, E. (1997). Attitudinal compliance with ecotourism guidelines. *Annals of Tourism
Research*, 24(1): 919–950.

Sirakaya, E. and Uysal, M. (1997). Can sanctions and rewards explain conformance behaviour of tour operators with ecotourism guidelines? *Journal of Sustainable Tourism*, **5**:
322–332.

Smits, W. T. M., Heriyanto, and Ramono, W. S. (1995). A new method for rehabilitation
of orangutans in Indonesia: A first overview. In: R. D. Nadler, B. M. F. Galdikas, L. K.
Sheeran, and N. Rosen (eds.), *The Neglected Ape*. New York: Plenum, pp. 69–77.

Sorenson, H. (2004). *Sabah Ecotourism*. http://borneoecotours.com/news/details.
asp?newsid=57 (accessed Sept. 22, 2005).

Steele, P. (1995). Ecotourism: An economic analysis. *Journal of Sustainable Tourism*, 3(1):
29–44.

Sugardjito, J. and van Schaik, C. P. (1991). Current population status, threats, and conservation measures. *Proceedings of the International Conference on "Conservation of the Great
Apes in the New World Order of the Environment"*, Indonesia, 1991.

Sumatran Orangutan Conservation Programme. (2005). www.sumatranorangutan.org
(accessed Sept. 20, 2005).

Sumatran Orangutan Society. (2000). *SOS Newsletter*, **4**, February update.

TIES. (2005). *TIES Global Ecotourism Fact Sheet*, updated in 2006. Washington, DC.
(TIES – The International Ecotourism Society).

Tilson, R. L., Seal, U. S., Soemarna, K., *et al.* (eds.). (1993). *Orangutan population and
habitat viability analysis report of the Captive Breeding Specialist Group/Species Survival*

Commission of the IUCN. Unpublished report to PHPA, based on workshop held in Medan, Sumatra, Indonesia, Jan. 1993.

Wagner, L. (2004). *Where To See Orangutans In South East Asia. Part 1: Orangutan Rehabilitation Centres*. www.asiahotelbookings.net/travel/tale/gen/orangutans1.htm (accessed May 19, 2004).

Wallis, J. and Lee, D. R. (1999). Primate conservation: The prevention of disease transmission. *International Journal of Primatology*, 20(6): 803–826.

Walpole, M. J. and Goodwin, H. J. (2000). Local economic impacts of dragon tourism in Indonesia. *Annals of Tourism Research*, 27 (3): 559–576.

Warren, K. S. (2001). *Orang-utan Conservation – Epidemiological Aspects of Health Management and Population Genetics*. School of Veterinary and Biomedical Sciences, Murdoch University, Perth, Australia: Unpublished PhD Dissertation.

Weaver, D. (2002). The evolving concept of ecotourism and its potential impacts. *International Journal of Sustainable Development*, **5** (3): 251–264.

Wong, J. (2005). *Orangutan provoked by visitors?* http://thestar.com.my/news/story.asp?file=/2005/6/5/nation/11138820andsec=nation (accessed June 7, 2005).

Woodford, M. H., Butynski, T. M., and Karesh, W. B. (2002). Habituating the great apes: The disease risks. *Oryx*, 36 (2): 153–160.

Yeager, C. P. (1997). Orangutan rehabilitation in Tanjung Puting National Park, Indonesia. *Conservation Biology*, 11 (3): 802–805.

Yeager, C. P. (ed.) (1999). *Orangutan Action Plan*. WWF-Indonesia, PHPA (Indonesia).

6 The impact of tourism on the behavior of rehabilitated orangutans (*Pongo abelii*) in Bukit Lawang, North Sumatra, Indonesia

David F. Dellatore, Corri D. Waitt, and Ivona Foitovà

Introduction

Combining tourism and orangutan (*Pongo spp.*) rehabilitation, originally thought to work symbiotically, is no longer recommended (Beck *et al.*, 2007; Macfie & Williamson, 2010; Rijksen & Meijaard, 1999; Rosen & Byers, 2002; Russon & Susilo, this volume). The effects of unregulated tourism at orangutan rehabilitation programs are somewhat predictable (e.g. higher rates of disease and mortality resulting from increased contact with humans, and inadequate foraging skills resulting from higher dependence on provisioning), but they have not yet been quantified and studied systematically. Tourism's effects on wildlife are recognized as being understudied (Berman & Li, 2002; Grossberg *et al.*, 2003; Kruger, 2005). This study aimed to investigate these effects via systematic focal observations of 13 rehabilitant and wild Sumatran orangutans (*Pongo abelii*) ranging in areas open to tourists around Bukit Lawang, North Sumatra. Results are discussed in terms of their implications for how tourism operations need to be restructured to better manage and protect this critically endangered species.

Conservation situation, rehabilitation, and tourism

With the rapid, large-scale degradation, fragmentation, and transformation of wild orangutan habitat since the 1970s, increasing numbers of orangutans are being displaced from the forest and captured for the illegal wildlife trade (Rijksen & Meijaard, 1999). It has been illegal in Indonesia to capture and keep wild orangutans since 1924 (Rijksen & Meijaard, 1999) so when captive orangutans are discovered they are confiscated by governmental authorities and placed in rehabilitation and/or reintroduction programs. These programs often differ in their methods and specific goals, but all aim to return orangutans rescued from captivity to free forest

Primate Tourism: A Tool for Conservation?, ed. Anne E. Russon and Janette Wallis. Published by Cambridge University Press. © Cambridge University Press 2014.

life and to contribute to orangutan conservation by, for example, helping fight the illegal wildlife trade (Rijksen & Meijaard, 1999).

Orangutan rehabilitation began in the 1960s when orangutans were thought to be nearly extinct in the wild (Harrison, 1962). At that time, experts believed rehabilitants could contribute to orangutan conservation by reinforcing depleted wild populations, tourism would help these projects and orangutan conservation, and the awareness that tourism fosters could serve as a major tool to generate global conservation momentum and convince government officials to take stronger action to protect biodiversity (Rijksen, 1978; Rijksen & Meijaard, 1999). For these reasons, tourism was introduced at many early orangutan rehabilitation projects. The Bukit Lawang orangutan rehabilitation program was among those that welcomed and encouraged tourism with the ex-captive orangutans it was rehabilitating.

Enabling ex-captive orangutans to readapt to free life in the forest usually requires substantial behavioral and cognitive restructuring, and withdrawing from human dependence is now recognized as critical to success (Grundmann, 2006; Rijksen & Meijaard, 1999; Russon, 2002, 2009). Any activities that maintain contact with humans, tourism included, foster rather than reduce human dependence (Russon, 2001). This understanding has resulted in newer orangutan reintroduction programs banning tourism with rehabilitant orangutans (Smits *et al.*, 1995). Bukit Lawang is among the sites that maintain tourism operations centered on viewing ex-captive orangutans released into the area's forest (Beck *et al.*, 2007).

Orangutan rehabilitation and tourism at Bukit Lawang

Bukit Lawang (BL, also called the Bohorok Center), located in the western part of Gunung Leuser National Park (GLNP), hosted an official orangutan rehabilitation project from 1973 to 1991 and continued releases until 2001 (Ian Singleton, pers. comm.). The BL rehabilitation project was established, in addition to one that operated at Ketambe from 1971 to 1975, due to increasing interest in rehabilitation and conservation at that time (Rijksen, 1978). It was terminated in 2001 when the Sumatran Orangutan Conservation Program established alternative orangutan rehabilitation facilities. Although no longer a rehabilitation site, the BL area still supports several rehabilitant orangutans and their surviving progeny. National park rangers provide two scheduled supplementary feedings of bananas and milk daily for rehabilitants and, on occasion, for wild orangutans. Feedings occur at a raised wooden platform on which rangers distribute provisions by hand, directly, to each orangutan that arrives.

Orangutan tourism started at BL soon after the rehabilitation project's inception and the project was actively attracting tourists by 1977 (Frey, 1978; Rijksen & Meijaard, 1999). BL orangutan tourism came to include direct interaction, allowing Western "volunteers" access to orangutans in the name of helping the rehabilitation process; it was partly for this reason that the project was terminated (Cochrane, 1998; McCarthy, 1999). The BL area also still hosts an orangutan tourism industry, based on orangutans that range in the area, which still allows both passive viewing and direct interaction. Official GLNP office figures show a total of 310 970 visitors (228 045 foreign) from 1985 to October 2009. The highest annual total, 21 577 foreign and 9561 domestic visitors, was reached in 1995. These totals do not include unregistered visitors, who may more than double the official figures (Rijksen & Meijaard, 1999).

Although it is now forbidden to touch, feed, or disturb the orangutans, infractions still occur in the forest for tourist enjoyment. Many visitor guides bring rucksacks full of fruit into the forest that they or tourists give to orangutans. Some guides guarantee orangutan sightings to visitors, and visitors can pressure guides to break the rules. Most guides provide "scouts" who travel ahead of each tour group, calling and luring orangutans to the trail system until the tourist group arrives. Further, there is now mobile phone reception throughout much of the region, making long-range scouting and networking possible. A small proportion of the tourists who the first author (DFD) surveyed informally in the course of this study reported being unhappy with such practices but many more reported encounters with orangutans, with most expressing their satisfaction in holding and/or feeding them. This pattern is further reinforced by the many travel blogs and videos posted online that document these kinds of encounters.

This study assessed the effects of current orangutan tourism on BL's rehabilitated and wild orangutans, as one facet of a larger study assessing the behavioral health of the area's orangutan population.

Methods

Study site

The study took place in Bukit Lawang within the GLNP, North Sumatra, Indonesia (see Figure 6.1). A 2 km² study area was established (03°32.770'N, 098°07.000'E) in an area of mixed lowland dipterocarp forest, defined by the trail system in place.

Orangutans

The forest supports both a wild and a rehabilitated ex-captive orangutan population and their offspring. Data were collected on 13 habituated orangutans selected on the basis of their home ranges including the trail system: seven rehabilitated adult females with dependent offspring, two wild adult females with dependent offspring, three rehabilitated adolescent females, and one wild juvenile male (see Table 6.1).

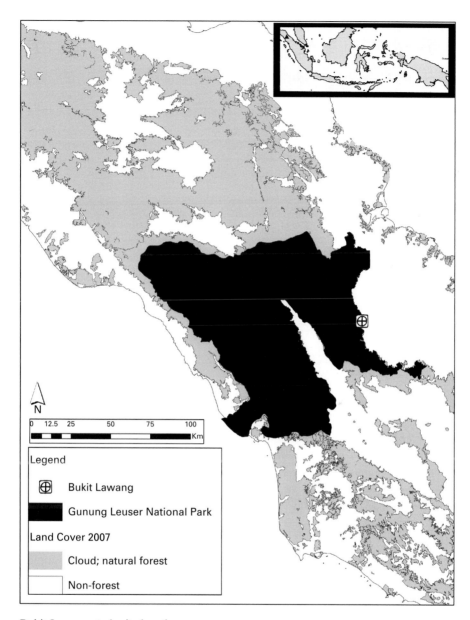

Figure 6.1   Bukit Lawang study site location.

Orangutan behavioral observations

The data reported represent 796 hours of observations collected from May 22 to July 31, 2007, based on focal animal sampling with continuous recording within full-day (nest to nest) follows (Altmann, 1974; Martin & Bateson, 1993). Once a follow was initiated, it was continued for three to five consecutive days.

Table 6.1 Orangutan subjects

Orangutan	Offspring	Sex/Age (years)	Background
Borjong		F / ~ 20+	Wild
→	Infant	M / ~ 1	Born to wild mother
Edita		F / 24	Rehabilitated
→	Sepi	F / 8	Born to rehabilitant mother
Jecky		F / 11	Rehabilitated
Lucky		F / ~ 20+	Wild
→	Infant	M / ~ 1	Born to wild mother
→	Damar	M / ~ 7	Born to wild mother
Mina		F / 28	Rehabilitated
→	Juni	F / 8	Born to rehabilitant mother
Pesek		F / 38+	Rehabilitated
→	Infant	M / 1	Born to rehabilitant mother
→	April	F / 10	Born to rehabilitant mother
Ratna		F / 20	Rehabilitated
Sandra		F / 18	Rehabilitated
Suma		F / 31	Rehabilitated

Behavioral observations consisted of recording all the focal individual's activities: foraging, resting, traveling, human-oriented behavior (behavior resulting from human activity), and other (i.e. social, nesting, play alone). Feeding on natural forest foods and on human provisions were coded as "foraging" and "'human (subset human feeding)," respectively. All of the above activities were treated as behavioral states and recorded after the focal orangutan had engaged in the activity for at least 15 seconds. Social behaviors, time spent on the ground, and feeding were recorded regardless of duration. Estimates of the orangutan's vertical height (m) and daily travel path were also recorded throughout the day.

Range use was plotted by taking a GPS waypoint of the orangutan's location every 30 minutes during follows. The data represented here are from 742 GPS points taken from all subjects' follows over the course of the study.

Observations concerning tourism

The impact of tourism on orangutan behavior was coded by recording the orangutan behaviors elicited when tourists were in the forest and within visible or audible range of the focal orangutan (i.e. within 50 m). Behaviors were identified as elicited by tourists if the orangutan ceased other behaviors (foraging, resting, etc.) and oriented and/or moved toward humans (Hsu *et al.*, 2009). The number of tourists (excluding guides, normally one or two guides per group) encountered daily and tourist guide behaviors (feeding and/or calling orangutans, etc.) were also recorded during follows.

Analyses

Behavioral data

Daily rates of time dedicated to all behavior categories were calculated (mean, standard deviation) for each orangutan to construct activity budgets. The Wilcoxon signed ranks test was used to determine whether there were significant differences in activity when tourists were present versus absent (significance level set at 0.05). Non-parametric statistics were used because these data were non-normally distributed.

Habitat use

GPS points representing orangutan locations during follows were analyzed using the Kernel Density Estimation (KDE) tool within the Spatial Analysis extension for ArcGIS ArcMap©, to determine areas of high use/concentration in the forest. KDE is a nonparametric, probabilistic method that calculates home range boundaries based on utilization distribution (UD) data (Silverman, 1986; Worton, 1989). UD, the relative amount of time an individual spends in any one place, provides a model for intensity and overlap of range use (Seaman & Powell, 1996). The density given is an estimate of the amount of time spent in a given area, in this case as the number of GPS points recorded in each location. KDE information was then overlaid onto a trail map to show usage patterns. Daily path length was calculated by totaling estimates of the distance traveled between adjacent waypoints for a complete day.

Results

Activity budgets

To handle complexities in comparing activity budgets, data from only one age and sex class, adult females, were analyzed. Data from the sole independent male were dropped. Compared with wild adult female Sumatran orangutans, these BL adult female orangutans had very different activity budgets: they devoted less time daily to natural foraging (overall mean differences > 20%) and more to rest and other activities (see Table 6.2).

Comparisons with other rehabilitant populations are hard to interpret because the activity budgets reported confound age, sex, species, rehabilitation methods, rehabilitation habitat, and duration of forest life. We limited comparisons to the Bukit Tiga Puluh rehabilitant population because it represents the same species and does not experience tourism (see Table 6.3). This comparison similarly suggests that time spent daily on natural foraging was relatively low and on "other" behavior was relatively high in BL orangutans. This difference can be explained by human-related activities, which represented 9% of their activity budget.

Table 6.2 Activity budgets from rehabilitated vs. wild adult female Sumatran orangutans (adapted from Morrogh-Bernard *et al.* 2009)

Study site	Feed (%)	Rest (%)	Travel (%)	Other (%)
Rehabilitated (this study)				
Bukit Lawang	34.6	34.3	19.4	11.8
Wild SuaqBalimbing	54.9	25.9	16.9	2.3

Table 6.3 Cumulative mean activity rates: tourists absent (abs) vs. present (pres) (*N* = 13 orangutans)

	Forage		Rest		Travel		Play		Social	
Tourists	Abs	Pres	Abs	Pres	Abs	Pres	Abs	Pres	Abs	Pres
Mean rate	0.34	0	0.32	0.41	0.17	0	0.01	0	0.02	0
STD	0.08	0	0.09	0.22	0.06	0	0.02	0.02	0.04	0

Travel and habitat use

BL adult female orangutans' mean daily travel distance was 647 ± 300 m SD, which is low relative to wild Sumatran populations (for sexually active females and mothers respectively, 1077 ± 368 m SD and 833 ± 306 m SD in Suaq Balimbing; 722 ± 293 m SD and 675 ± 282 m SD in Ketambe; Singleton *et al.*, 2009). Notably, they restricted their travel to a small area, much of it heavily influenced by humans. As the KDE analysis shows (see Figure 6.2), the BL females did not travel outside the main tourism areas; the furthest waypoint recorded was only approximately 0.5 km from the nearest trail. The two areas of highest-intensity use, the feeding platform and the area known as "damar tree," were areas where tourist groups often congregated. These two findings imply tendencies to remain near trails and heavily visited tourist areas, where provisions were often offered. Although no physical barriers were in place, a range restriction was in effect, as the cumulative range use heavily coincided with areas of high tourist use.

The orangutans were actively and consistently lured to the trails. From tourist groups, guides commonly made a type of whooping call, which often drew orangutans to the trails or tourists because food provisioning typically followed. Researchers could hear these calls most days as far as 300 m away, and assumed that if these calls were audible to researchers, they were also audible to the focal orangutan. A total of 2237 calls from the trails were recorded, with a mean of 72 ± 79 SD calls per day and a one-day maximum of 324.

These orangutans still used arboreal pathways, at a mean height of 10.2 ± 3.87 m SD. However, they were also regularly lured down from the trees with the promise of feeding. If such luring behavior continues, they may at some point start losing

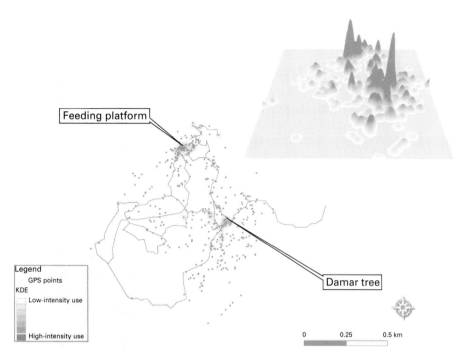

Feeding platform

Damar tree

Legend
GPS points
KDE
Low-intensity use

High-intensity use

0   0.25   0.5 km

Figure 6.2   Kernel Density Estimation (KDE) overlaid onto study site trail map (Inlay shows a three-dimensional projection of KDE analysis).

arboreal habits. Of this study's total observation time, the orangutans spent a cumulative daily mean of 5 ± 8% SD of their time on the ground. No other Sumatran sites provide comparable data, so we could not determine whether this value is relatively high or low. It is notable that in BL, 41% of the time spent on the ground was due directly to tourist presence in the forest: orangutans descended either to approach groups (often on the trail system itself) or to eat provisioned foods. The remaining time on the ground was spent resting, playing, or eating (rattan shoots or ant/termite nests), behaviors rarely if ever seen in wild Sumatran orangutans.

Tourist presence and impact

A total of 1131 tourists were encountered in the forest while following orangutans. Daily averages were 31 ± 20 SD tourists and 6 ± 3 SD tourist groups (maximum 84 and 16, respectively). Tourist group size averaged 5 ± 2 SD (range 1–31 tourists plus 1–4 guides/assistants). A typical tourist group viewing an orangutan is shown in Figure 6.3.

Human activity consumed 8 ± 9% SD of the cumulative activity budget; individual values are shown in Figure 6.4. The highest values were from Mina, Pesek, and Damar, who spent as much as 30% of their waking time with humans. Wilcoxon signed ranks analyses revealed that the BL population spent significantly less time on three types of natural behavior (foraging, traveling, other–social) when tourists

Figure 6.3   A tourist group in the forest (the orangutan, Damar, is in the top right corner).

were present versus absent (see Table 6.4). In short, when tourists were present, virtually all natural orangutan activity stopped. The orangutans did not forage, but instead spent the majority of time watching tourist groups, seemingly waiting to be provisioned. Of their time with tourists, however, a daily mean of only $3 \pm 4\%$ SD was spent being fed by them. This watching behavior consumed more time than actually being provisioned. It follows that the orangutans would be engaging in other behaviors, such as foraging, if they were not stopping to watch and wait for tourist groups.

Tourists also affected social behavior, partly because they distorted ranging and activity budgets and partly because they provided provisions. Tourist-related provisioning may have adverse effects through its influence on sociality, by artificially extending "food-rich" conditions and by concentrating feeding around specific locations (feeding platforms and the damar tree).

Provision-related aggression has been seen in these BL orangutans, in the form of higher-ranking individuals chasing lower-ranking ones when tourists offered illegal provisions. Dominance rank was determined from asymmetrical outcomes of interactions and displacements among BL orangutans over the course of this study (Utami *et al.*, 1997). Mina and Lucky were seen chasing away and/or biting their own offspring on four occasions in conflicts over tourist-offered provisions. Such competition was never recorded when feeding on natural forest foods or at the

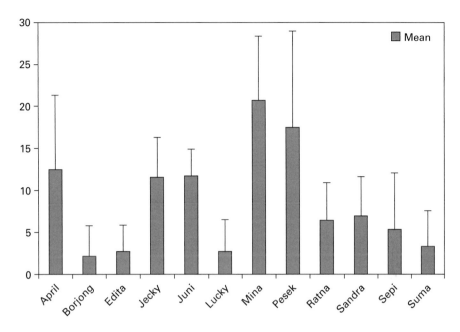

Figure 6.4   Mean individual daily percentage of time spent on human-related activities (± standard deviation).

Table 6.4 Difference in orangutans' mean activity rates between tourists present vs. absent periods

Wilcoxon signed ranks test	Forage	Rest	Travel	Other: play	Other: social
Z	−3.186	−1.329	−3.184	−1.89	−2.41
Asymp. Sig. (2-tailed)	0.001	0.184	0.001	0.059	0.016

feeding platform, where provisioning was controlled and distributed more evenly. It only occurred during illegal tourist provisioning events where provisions were limited to whatever quantity the guide or tourist was offering.

Finally, there is a history of orangutans attacking tourists, guides, and national park rangers in BL. Mina, with the highest mean time engaged with humans, was well-known for attacking tourist groups in the forest in pursuit of food; six confirmed incidents occurred during the course of this study (reportedly, with injuries to guides and tourists requiring surgery) (Darma Pinem, pers. comm.). These aggressive interactions have also led to humans attacking the orangutans. On one occasion during this study, DFD witnessed a guide striking Edita with a thick branch while Edita was trying to remove a rucksack full of fruit from a guide's back. The sack of food had presumably been brought in, illegally, to feed orangutans. Mina has sustained machete wounds to her forehead in the past, and more recently to her hand, due to her reputation for attacking tourist groups.

Table 6.5 Activity budgets from populations of Sumatran rehabilitant orangutans, with tourism (Bukit Lawang) and without tourism (Bukit TigaPuluh: Peter Pratje, pers. comm)

Population	n	Years since release	Age classes[a]	Sex	Feed (%)	Rest (%)	Travel (%)	Other[b] (%)
Bukit TigaPuluh	15	<1–3	J–AL	?	41.6	37.8	16.8	3.8
Bukit Lawang (this study)	12	>6	AL–AD	F	34.6	34.3	19.4	11.8[b]

Notes:
[a] Age classes: J, juvenile; AL, adolescent; S, subadult; AD, adult.
[b] Includes human-related and social activity.

Discussion

This study provides evidence from BL for one of four main types of unsustainability in tourism operations that Kruger (2005) identified: direct effects on wildlife behavior and health. This study also found that the influences of orangutan-focused tourism on orangutan behavior were products of tourists' and their guides' inappropriate behaviors. These findings concur with others in showing that changes in wildlife behavior are affected by tourists' behavior, not simply their numbers or presence (Burns & Howard, 2003; Grossberg et al., 2003; Treves & Brandon, 2005). One important reason for these changes at BL is that provisioning has created dependence in the orangutan population and may allow for serious disease transmission potential through fostering close contact with humans (Berman et al., 2007; Macfie & Williams, this volume; Wallis & Meuhlenbein, this volume).

Activity budgets and diet

With food availability considered the single most important factor in determining animal activity budgets (Orams, 2002), the disruption that tourist provisioning causes in BL orangutans' foraging patterns has serious implications. The tourism situation at BL includes year-round provisioning and tourist-offered foods that could result in a diet consistently high in energy-rich foods, so BL orangutans may enjoy a better positive balance between energy intake and output than wild orangutans (e.g. Knott, 1998, Wich et al., 2006). Possible consequences include shortened interbirth intervals (Kuze et al., 2012) and physiological disorders such as pre-diabetic syndrome (Sapolsky, this volume), both of which could represent serious threats to orangutan well-being.

This is not only a concern from the point of view of the orangutans' nutrition, but also of maintaining natural behavioral patterns. Foraging depends heavily on learning, including where to go and what to eat. When individuals forage less frequently, they learn less and become less efficient (Orams, 2002). Consequently, the offspring of these rehabilitants may also be affected. In addition to undermining foraging expertise, tourism also distorts activity budgets: BL orangutans often sat and waited to be provisioned by tourist groups, regardless of forest food availability. This time spent monitoring humans, not in vigilance but seemingly waiting for provisioning, uses time that could be used instead for foraging (Treves, 2000). One mother–daughter pair, Pesek and April, showed this pattern: both spent high rates of time with humans. Pesek was known to cross the river that borders the national park regularly and enter a nearby restaurant, where staff fed her to entertain the restaurant's patrons. April, already an independent adolescent, also devoted much time to human-related activities. Further, Pesek had a dependent infant who accompanied her on these human activities, so if he survives, he may adopt similar practices. Thus the provisioning and tourist interaction may result in a self-propagating cycle of human dependence.

Together, these factors could increase the risks of disease transmission, through increasing proximity to conspecifics and to humans, and in inciting aggression over tourist-offered provisions (Table 6.5).

Range use

Using food to attract wildlife is popular because it increases the likelihood of wildlife encounters (Orams, 2002). Encounters are particularly important in BL, where nearly the entire tourism industry is built around orangutans that are normally semi-solitary and elusive. BL orangutans had become so accustomed to provisioning that waiting for humans affected their ranging and diet. KDE analysis demonstrated very restricted ranging. Given that the BL orangutans in this study did not travel far from the trail system, where provisioning often occurred, even when no tourists were in the area, it could also be argued that their ranges and range use were restricted beyond times when tourists were present.

Range restriction and dietary shifts, as related specifically to provisioning, have been linked to increased intraspecific aggression in some other primates (Berman *et al.*, 2007, this volume). Provisioned foods, typically available in one small area at certain times, are more spatially and temporally concentrated than foods found naturally in the forest and may be higher in quality. These factors can lead to increased proximity and aggressive competition for high-value resources along with range restriction (Hill, 1999; Hsu *et al.*, 2009). In BL, illegal tourist provisioning clearly caused these problems. Official provisioning did not incite aggression or competition or create incentives for orangutans to remain near trails and tourists. Higher-quality provisions can also induce dietary distortions that undermine health (e.g. pre-diabetic syndrome: see Sapolsky, this volume).

With orangutans being the most arboreal of the great apes, the ability to travel effectively in an arboreal, three-dimensional environment has been identified as a crucial point of re-adaptation to forest life. Effective arboreal travel enables orangutans to move safely through the canopy, locate their most important food sources, and avoid terrestrial predators. The latter are especially important in Sumatra, where tigers present major terrestrial threats (Povinelli & Cant, 1995;

van Noordwijk *et al.*, 2009). Thus Sumatran orangutans, unlike Bornean orangutans, are almost exclusively arboreal (Rijksen, 1978; Thorpe & Crompton, 2009). However, the BL orangutans in this study were often on the ground, not because arboreal travel was impossible, but to walk along the trail system, build and rest in ground nests, or play with dirt. As in other documented cases, it could be that supplementary feeding encourages propensities to descend (Riedler, 2007). BL, with its official set-time platform feedings and randomly timed illegal feedings on the trail system throughout the day, could well multiply this effect on the orangutans' behavior. The three BL orangutans in this study with zero time recorded on the ground, Lucky, Damar, and Borjong, were all wild-born individuals who had never been involved in rehabilitation. They had become habituated enough to accept provisioning, but only from tourist groups along the trail system and never at the feeding platform. Their having being lured this far already suggests coming to the ground is a future likelihood.

Mortality

Influences of BL tourism on orangutans may also include risks of disease transmission, because of the human proximity that tourism creates. This study did not seek evidence on disease in the BL orangutans subjected to tourism. However, orangutan infant mortality rates at BL (59%) (Kuze *et al.*, 2012) are 8.42 and 3.47 times higher respectively than they are in wild Sumatran populations at Suaq Balimbing (7%) and Ketambe (17%) (van Noordwijk & van Schaik, 2005; Wich *et al.*, 2004).

The likelihood that such high infant mortality rates at BL are purely due to natural causes seems low, given orangutan mortality rates from other sites. Further, three infant orangutans died within a one-year period during the time this study was conducted (Darma Pinem, pers. comm.), and two cases of filial cannibalism were observed by rehabilitant females that frequented tourist areas (Dellatore *et al.*, 2009). With this many infant deaths, the documented infant mortality rate, and the cannibalism, the BL orangutan population can be said to be operating sub-optimally and any possible barriers to their adaptation to forest life, poorly managed tourism included, should be reduced or removed.

Human proximity to nonhuman primates, especially great apes, creates a considerable risk of disease exchange due to close phylogenetic relationships (Adams *et al*,. 2001; Kilbourn *et al.*, 2003; Wallis & Lee, 1999; Woodford *et al.*, 2002). Tourism operators often discount disease concerns, citing the lack of solid evidence of negative impact on wildlife (Berman *et al.*, 2007; Lerche, 1993; Orams, 2002; Wallis & Lee, 1999; Woodford *et al.*, 2002). However, there are many documented cases indicating a high probability of disease transmission between humans and primates (Köndgen *et al.*, 2007; Pusey *et al.*, 2008). Previous studies in nonhuman primates have shown that infant mortality rates can be an indicator of the impact of tourism (Berman *et al*,. 2007). Thus, with regular tourist–orangutan interactions like that pictured in Figure 6.5, there is serious cause for concern about tourists transmitting diseases to orangutans at BL. This is especially true for infants under one year old, because they are highly vulnerable.

Figure 6.5   Western tourists feeding Pesek (who is accompanied by her one-year-old infant).

Recommendations for tourism

Bukit Lawang shows signs of having fallen into the tourism life-cycle of growing demand leading to increased development, which ultimately works to destroy what it purportedly tries to protect (Adams & Infield, 2003; Burns & Howard, 2003; Cochrane, 1998; Hillery *et al.*, 2001; Tremblay, 2001). Based solely on the fact that BL visitors are accosted by aggressive orangutans, its tourism could be considered unsustainable (Burns & Howard, 2003). The orangutans themselves are also threatened by tourism: not only does tourism put them at risk of anthropozoonoses, it creates risks of injury when orangutans compete for human-offered foods or humans intervene with "problem-causing" orangutans.

This is not to say that all orangutan tourism should cease in BL. The local economy has centered on orangutan tourism for decades and grown in scale with increasing visitor numbers (Panut Hadisiswoyo, pers. comm.). The people living in and around BL deserve to benefit from orangutan tourism because, in accepting it, they are probably foregoing valuable resources by not developing the area in other ways. The protections afforded to orangutans are typically seen as limiting humans, because it is often more profitable (at least in the short term) to convert forestland for development such as agriculture (e.g. farms, oil palm, or rubber tree plantations), mines, or logging (Adams & Infield, 2003; ITTO, 2005; Norton-Griffiths, 1995). Surrounding BL and across Indonesia an estimated 6.2 million hectares of land are now planted with oil palm plantations (Sheil *et al.*, 2009).

National parks and other protected areas are not immune to clearing and agricultural development, even though it is illegal. A series of GLNP encroachments have occurred and still do across much of northern Sumatra and approximately 22 000 ha of the park now need restoration in just one subdistrict (Panut Hadisiswoyo, pers. comm.). Further, in 2010 a local person illegally cleared and planted oil palms in the BL area, within the GLNP. Although this illegal development was promptly stopped, other illegal activities within the national park are not controlled (Wich & Utami Atmoko, 2010; Wich *et al.*, 2011). Without orangutan tourism in Bukit Lawang, illegal development could well increase and further degrade the area's forest both inside and outside the GLNP. Thus the future of orangutan conservation in the BL area depends importantly on support from people living nearby. Stopping tourism would remove much of the local community's income and thereby probably their support, so it would probably further endanger BL area orangutans by reducing incentives to resist forest conversion projects, regardless of their legality.

The problem may be better handled from the perspective of stakeholder conflict, in that tourism industry stakeholders are seeking greater access to the orangutans (Burns & Howard, 2003) while wildlife conservationists seek to limit and control that access (de la Torre *et al.*, 2000; Duchesne *et al.*, 2000). Tourist guides and the GLNP authority seem hesitant to limit what paying tourists can see and experience. Yet with the negative effects on orangutan behavior and the high risk of injury and disease transmission to both orangutans and humans, it is imperative that measures be taken to limit levels of interaction between orangutans and humans. In this light, one suggestion for reducing tourism pressures on the area's orangutans is shifting BL away from being simply an orangutan viewing center and toward becoming a visitors' educational gateway into the GLNP (Singleton & Aprianto, 2001). A more holistic tourist experience is possible, as many other sights and species in the forest could be included that were commonly ignored by guides and not pointed out to visitors during this study. Altering this pattern could better serve the tourists, the local community, the forest, and the orangutans by broadening the area's tourism activities and its economic dependency beyond orangutan tours. This could serve orangutan conservation by raising awareness of the interdependencies within the forest and increasing understanding of animal behavior in natural environments.

The value expected from such tourism is well summed up by van Schaik, who states, "Although I am well aware of the negative sides of ecotourism, they pale into insignificance compared to the threats that are now faced by the wild orangutan" and who considers that "the best way to turn an initially indifferent person into an ardent conservationist is to give him or her the privilege to follow a habituated wild primate" (van Schaik, 2001, p. 34). The caveat is that, while these experiences may foster conservation attitudes, attitudes do not necessarily or easily translate into the changes in behavior needed to enhance conservation.

Improving health management should also have high priority in BL orangutan tourism, given the high potential for disease transmission under current conditions. There are three health risk management philosophies for ape conservation – non-intervention (processes are allowed to flow unimpeded), precautionary (protection from exposure is made at all costs, including complete exclusion of humans from the area), and risk-based (potential threats are addressed and dealt with in advance) (Travis *et al.*, 2008). In BL, nonintervention is currently in place, with no effective limit to the amount of contact between humans and orangutans. Some have already asked that the precautionary approach be applied and orangutan tourism be banned with rehabilitants (Rosen & Byers, 2002). The chances are that this would result in massive problems in BL, with a strong backlash directed at agencies held responsible for this action and perhaps at the orangutans themselves. Thus a risk-based strategy would be best, which would include: hazard identification, risk assessment, risk management, and risk communication (Travis *et al.*, 2006).

This study has contributed to hazard identification and risk assessment, but further work is needed on these and the remaining components. Concerning risk management and communication, for instance, education programs are needed for guides, visitors, and local residents. Tourists are seemingly learning very little about orangutans and their conservation situation, or care little about either, as evidenced by their wanting to be close to and to feed orangutans. Orangutan infants, the most susceptible to disease, are often sought out and subjected to close contact; this also occurs at other sites, as many tourists prefer seeing babies (Russell, 1995). Many of the tour operators appeared to think only in the short term, because if they continue to contact and feed orangutans this will have further negative effects. The fact that there have been three infant deaths in the past year but regular human contact with both tourists and guides continues shows that the potential for disease transmission is not known, not understood, and/or deliberately disregarded. Although systematic data were not collected on education in this study, comments that DFD overheard from guides to tourists seldom concerned conservation and often included incorrect information about orangutan physiology and behavior. Studies in Tanjung Puting have yielded similar results (Russell, 1995).

Since tourists must be accompanied by a guide to enter the GLNP, properly educating, training, and supervising guides is an important way of improving their influences on tourists and tourist behavior (Grossberg *et al.*, 2003). Without properly educating guides about health protocols and the reasons for them, there is no chance that they will regulate their own or tourists' behaviors appropriately (Hsu *et al.*, 2009).

Well-designed education initiatives could serve risk communication, which can be an effective method to help ensure the long-term commensalism of humans and apes (Jones-Engel *et al.*, 2006). DFD, now working with the Indonesian NGO, the Orangutan Information Centre (OIC), initiated the Gunung Leuser Ecotourism Development Program (GLEDP) in conjunction with GLNP authorities and the local guide association (HPI). The GLEDP is an initiative aiming to provide active guide training courses as well as passive visitor education in the form of signboards that display the OIC's newly developed and GLNP adopted park guidelines, an orangutan-themed visitor center, and an instructional film screened to visitors before they enter the park. The OIC has also published and distributed a comprehensive guidebook and pocket guidebook to the GLNP, which, in addition to including information on BL and orangutan conservation, details the importance of adhering to visitor guidelines (OIC, 2009).

The OIC is also working with the GLNP authority and HPI on matters such as restricting tourist group size (maximum of five guests with two guides), banning visitors taking bags into the forest, and banning any group lunch/picnic activities in the forest. The OIC is also working to adjust the cost of the GLNP permit, as the current price per day, 20 000 IDR (~$2.50 US), is considered much too low. Visitors in Uganda are charged a government minimum of $30 US to visit chimpanzees and $500 US to visit gorillas, under strict rules (Uganda Wildlife Authority, 2011) and may have to book their visit years in advance (Nielsen & Spenceley, 2010). In BL and other national parks where orangutans range, visitors can simply drop in, pay an extremely low fee, and visit orangutans in their natural habitat. This we feel reduces the value of the experience, and may be encouraging visitors to act less responsibly than they might otherwise. We have suggested to the park authority that they (initially) raise admittance to 100 000 IDR (~$12.50 US) per visitor. This issue is complicated, however, because Indonesian national park permit fees are regulated by the national government, so proposed changes must be approved in Jakarta.

The involvement of GLNP authorities is clearly essential, and a large part of the solution for problems in BL depends on their having a stronger field presence. They alone have the authority to enforce park regulations and regulate behaviors directly. That they have been involved in the implementation of this new initiative is promising, but the fate of BL orangutans still depends on whether all stakeholders can begin to work in a more conservation-oriented fashion to help protect the local orangutan population and the economy that has come to depend so highly on visiting these orangutans.

Conclusion

The major findings of this study are as follows: BL orangutans' ranging is being restricted, their activity budgets significantly distorted, and their health potentially undermined by the tourism operations in place. The orangutans remaining in the BL area, rehabilitants included, are still members of a critically endangered species living within a protected national park. With so few remaining free-ranging populations, every individual counts. Therefore, continuing current practices that turn them away from forest-oriented activities and potentially threaten their health undermines rather than contributes to conservation. Strategies have been proposed that could improve the situation for these orangutans and steps have been taken to put some of these into place. Important tasks remaining include risk management and risk communication, both of which are currently being facilitated by the GLEDP. Further, the national park authority must become more active in patrolling the forest and enforcing newly established guidelines, which are to be in line with the new IUCN best practices for great ape tourism (Macfie & Williamson, 2010, this volume).

Acknowledgments

We thank the Indonesian Institute of Sciences (LIPI), the Indonesian Ministry of Forestry (PHKA), and the Gunung Leuser National Park Authority, for permitting this research to take place. We also thank the Orang Utan Republik Education Initiative LP Jenkins Memorial Fellowship for providing financial support. This research was conducted in partial fulfillment of the requirements for the degree of Master of Science in Primate Conservation at Oxford Brookes University, UK. It was approved scientifically and ethically by Indonesia's Institute of Sciences (LIPI) and PHKA, and the GLNP Authority. As such it complied with the legal requirements of research in Indonesia, the country in which it was conducted.

References

Adams, H. R., Sleeman, J. M., Rwego, I., and New, J. C. (2001). Self-reported medical history survey of humans as a measure of health risk to the chimpanzees (*Pan troglodytes schweinfurthii*) of Kibale National Park, Uganda. *Oryx*, 35: 308–312.

Adams, W. M. and Infield, M. (2003). Who is on the gorilla's payroll? Claims on tourist revenue from a Ugandan national park. *World Development*, 31: 177–190.

Altmann, J. (1974). Observational study of behavior: sampling methods. *Behaviour*, 49: 227–267.

Beck, B., Walkup, K., Rodrigues, M., *et al.* (2007). *Best Practice Guidelines for the Re-introduction of Great Apes*. Gland, Switzerland: SSC Primate Specialist Group of the World Conservation Union.

Berman, C. M. and Li, J. H. (2002). Impact of translocation, provisioning and range restriction on a group of *Macaca thibetana*. *International Journal of Primatology*, 23: 383–397.

Berman, C. M., Li, J. H., Ogawa, H., *et al.*. (2007). Primate tourism, range restriction and infant risk among *Macaca thibetana* at Mt Huangshan, China. *International Journal of Primatology*, 28: 1121–1141.

Burns, G. L. and Howard, P. (2003). When wildlife tourism goes wrong: a case study of stakeholder and management issues regarding dingoes on Fraser Island, Australia. *Tourism Management*, 24: 699–712.

Cochrane, J. (1998). *Organisation of Ecotourism in the Leuser Ecosystem*. Medan, Indonesia: Report to the Leuser Management Unit.

de la Torre, S., Snowdon, C. T., and Bejarano, M. (2000). Effects of human activities on wild pygmy marmosets in Ecuadorian Amazonia. *Biological Conservation*, 94: 153–163.

Dellatore, D. F., Waitt, C. D. and Foitovà, I. (2009). Two cases of mother-infant cannibalism in orangutans. *Primates*, 50: 277–281. DOI 10.1007/s10329–009–0142–5.

Duchesne, M., Cote, S. D., and Barrette, C. (2000). Responses of woodland caribou to winter ecotourism in the Charlevoix biosphere reserve, Canada. *Biological Conservation*, 96: 311–117.

Frey, R. (1978). Management of orangutans. In: J. McNeely, D. S. Rabor, and E. A. Sumardja (eds.), *Wildlife Management in Southeast Asia*. Bogor, Indonesia: BIOTROP Special Publication 8, pp. 199–215.

Grossberg, R., Treves, A., and Naughton-Treves, L. (2003). The incidental ecotourist: measuring visitor impacts on endangered howler monkeys at a Belizean archaeological site. *Environmental Conservation*, 30: 40–51.

Grundmann, E. (2006). Back to the wild: will reintroduction and rehabilitation help the long-term conservation of orang-utans in Indonesia? *Social Science Information*, 45: 265–284.

Harrison, B. (1962). The immediate problem of the orang-utan. *Malayan Nature Journal*, 16: 4–5.

Hill, D. A. (1999). Effects of provisioning on the social behaviour of Japanese and rhesus macaques: implications for socioecology. *Primates*, 40: 187–198.

Hillery, M., Nancarrow, B., Griffin, G., and Syme, G. (2001). Tourist perception of environmental impact. *Annals of Tourism Research*, 28: 853–867.

Hsu, M. J., Kao, C., and Agoramoorthy, G. (2009). Interactions between visitors and formosan Macaques (*Macaca cyclopis*) at Shou-Shan Nature Park, Taiwan. *American Journal of Primatology*, 71: 214–222.

ITTO. (2005). *Annual Review and Assessment of the World Timber Situation*. Yokohama, Japan: International Tropical Timber Organization.

Jones-Engel, L., Engel, G. A., Schillaci, M. A. *et al.* (2006). Considering human–primate transmission of measles virus through the prism of risk analysis. *American Journal of Primatology*, 68: 868–879.

Kilbourn, A. M., Karesh, W. B., Wolfe, N. D., *et al.* (2003). Health evaluation of free-ranging and semi-captive orangutans (*Pongo pygmaeus pygmaeus*) in Sabah, Malaysia. *Journal of Wildlife Diseases*, 39: 73–83.

Knott, C. D. (1998). Changes in orangutan caloric intake, energy balance, and ketones in response to fluctuating fruit availability. *International Journal of Primatology*, 19: 1061–1079.

Köndgen, S., Kühl, H., N'Goran, P. K., *et al.* (2007). Pandemic human viruses cause decline of endangered great apes. *Current Biology*, 18: 1–5.

Kruger, O. (2005). The role of ecotourism in conservation: panacea or Pandora's box? *Biodiversity and Conservation*, 14: 579–600.

Kuze, N., Dellatore, D., Banes, G. L., *et al.* (2012) Factors affecting reproduction in rehabilitant female orangutans: young age at first birth and short inter-birth interval. *Primates*, 53: 181–192. DOI 10.1007/s10329–011–0285-z.

Lerche, N. W. (1993). Emerging viral diseases of nonhuman primates in the wild. In: M. E. Fowler (ed.), *Zoo and Wild Animal Medicine*. Philadelphia: Saunders, pp. 340–344.

Macfie, E. J. and Williamson, E. A. (2010). *Best Practice Guidelines for Great Ape Tourism*. Gland, Switzerland: IUCN/SSC Primate Specialist Group (PSG).

Martin, P. and Bateson, P. (1993). *Measuring Behaviour*. New York, NY: Cambridge University Press.

McCarthy, J. (1999). *Nature Based Tourism Case Study: Gunung Leuser, Indonesia*. Perth: Murdoch University.

Morrogh-Bernard, H. C., Husson, S. J., Knott, C. D., *et al.* (2009). Orangutan activity budgets and diet: A comparison between species, populations and habitats. In: S. A Wich, S. S. Utami Atmoko, T. Mitra Setia, and C. P. van Schaik (eds.), *Orangutans: Geographic Variation in Behavioral Ecology and Conservation*. New York: Oxford University Press, pp. 119–134.

Nielsen, H. and Spenceley, A. (2010). *The Success of Tourism in Rwanda – Gorillas and More*. Background paper for the African success stories study. Washington, DC: World Bank and the Netherlands Development Organization.

Norton-Griffiths, M. (1995). Economic incentives to develop the rangelands of the Serengeti: implications for wildlife conservation. In: A. R. E. Sinclair and P. Arcese (eds.), *Serengeti II: Dynamics, Management, and Conservation of an Ecosystem*. University of Chicago Press, pp. 588–604.

OIC. (2009). *Guidebook to the Gunung Leuser National Park*. Medan, Indonesia: Yayasan Orangutan Sumatera Lestari – Orangutan Information Centre.

Orams, M. B. (2002). Feeding wildlife as a tourism attraction: a review of issues and impacts. *Tourism Management*, 23: 281–293.

Povinelli, D. J. and Cant, H. G. H. (1995). Arboreal clambering and the evolution of self-conception. *Quarterly Review of Biology*, 70: 393–421.

Pusey, A. E., Wilson, M. L., and Collins, D. A. (2008). Human impacts, disease risk, and population dynamics in the chimpanzees of Gombe National Park, Tanzania. *American Journal of Primatology*, 70: 738–744.

Riedler, B. (2007). *Activity patterns, habitat use, and foraging strategies of juvenile Sumatra orangutans (Pongo abelii) during the adaptation process to forest life*. Vienna, Austria: University of Vienna: Unpublished Diploma thesis.

Rijksen, H. D. (1978). *A Field Study on Sumatran orang utans (Pongo pygmaeus abelii)*. Wageningen, The Netherlands: Veenman H, Zonen BV.

Rijksen, H. D. and Meijaard, E. (1999). *Our Vanishing Relative: the Status of Wild Orangutans at the Close of the Twentieth Century*. Dordrecht: Kluwer Academic Publishing.

Rosen, N. and Byers, O. (2002). *Orangutan Conservation and Reintroduction Workshop: Final Report*. Apple Valley, MN, USA: IUCN/SSC Conservation Breeding Specialist Group.

Russell, C. L. (1995). The social construction of orangutans: an ecotourist experience. *Society & Animal*, 3: 151–170.

Russon, A. E. (2001). Rehabilitating orangutans (*Pongo pygmaeus*): behavioral competence. *Laboratory Primate Newsletter*, 40: 8–9.

Russon, A. E. (2002). Return of the native: cognition and site-specific expertise in orangutan rehabilitation. *International Journal of Primatology*, 23: 461–478.

Russon, A. E. (2009). Orangutan rehabilitation and reintroduction: Successes, failures, and role in conservation. In: S. A. Wich, S. S. Utami Atmoko, T. Mitra Setia, and C. P. van Schaik (eds.), *Orangutans: Geographic Variation in Behavioral Ecology and Conservation*. New York: Oxford University Press, pp. 327–350.

Seaman, D. E, and Powell, A. (1996). An evaluation of the accuracy of kernel density estimators for home range analysis. *Ecology*, 77: 2075–2085.

Sheil, D., Casson, A., Meijaard, E., *et al.* (2009). *The Impacts and Opportunities of Oil Palm in Southeast Asia: What do we know and what do we need to know?* Occasional paper No. 51. Bogor, Indonesia: CIFOR.

Silverman, B. W. (1986). *Density Estimation for Statistics and Data Analysis*. London, UK: Chapman & Hall.

Singleton, I. and Aprianto, S. (2001). *The Semi-Wild Orangutan Population at Bukit Lawang; a Valuable 'Ekowisata' Resource and their Requirements*. Unpublished paper presented at a "Workshop on eco-tourism development at Bukit Lawang" held in Medan, April 2001. Medan, Indonesia: PanEco, Yayasan Ekosistem Lestari.

Singleton, I., Knott, C. D., Morrogh-Bernard, H. C., *et al.* (2009). Ranging behavior of orangutan females and social organization. In: S. A. Wich, S. S. Utami Atmoko, T. Mitra Setia, and C. P. van Schaik (eds.), *Orangutans: Geographic Variation in Behavioral Ecology and Conservation*. New York: Oxford University Press, pp. 205–214.

Smits, W. T. M., Heriyanto, and Ramono, W. S. (1995). A new method for rehabilitation of orangutans in Indonesia: A first overview. In: R. D. Nadler, B. M. F. Galdikas, L. K. Sheeran LK, and N. Rosen (eds.), *The Neglected Ape*. New York: Plenum Press, pp. 69–77.

Thorpe, K. S. and Crompton, R. H. (2009). Orangutan positional behavior: Interspecific variation and ecological correlates. In: S. A. Wich, S. S. Utami Atmoko, T. Mitra Setia and C. P. van Schaik (eds.), *Orangutans: Geographic Variation in Behavioral Ecology and Conservation*. New York: Oxford University Press, pp. 33–48.

Travis, D. A., Hungerford, L., Engel, G. A., and Jones-Engel, L. (2006). Disease risk analysis: a tool for primate conservation planning and decision making. *American Journal of Primatology*, 68: 855–867.

Travis, D. A., Lonsdorf, E. V., Mlengeya, T., and Raphael, J. (2008) A science-based approach to managing disease risks for ape conservation. *American Journal of Primatology*, 70: 1–6.

Tremblay, P. (2001). Wildlife tourism consumption: consumptive or non-consumptive? *International Journal of Tourism Research*, 3: 81–86.

Treves, A. (2000). Theory and method in studies of vigilance and aggregation. *Animal Behaviour*, 60: 711–722.

Treves, A. and Brandon, K. (2005). Tourism impacts on the behavior of black howling monkeys (*Alouatta pigra*) at Lamanai, Belize. In: J. Paterson and J. Wallis (eds.), *Commensalism and Conflict: The Human-primate Interface*. OK: American Society of Primatologists, pp. 146–167.

Uganda Wildlife Authority. (2011). *UWA Conservation Fees/Tariffs July 2011 – June 2013*. Kampala, Uganda: Uganda Wildlife Authority.

Utami, S. S, Wich, S. A., Sterck, E. H. M., and van Hooff, J. A. R. A. M. (1997). Food competition between wild orangutans in large fig trees. *International Journal of Primatology*, 18: 909–927.

van Noordwijk, M. A., Sauren, S. E. B., Nuzuar, *et al.* (2009). Development of independence: Sumatran and Bornean orangutans compared. In: S. A. Wich, S. S. Utami Atmoko, T. Mitra Setia, and C. P. van Schaik (eds.), *Orangutans: Geographic Variation in Behavioral Ecology and Conservation*. New York: Oxford University Press, pp. 189–204.

van Noordwijk, M. A. and van Schaik, C. P. (2005). Development of ecological competence in Sumatran orangutans. *American Journal of Physical Anthropology*, 127: 79–94.

van Schaik, C. P. (2001). Securing a future for the wild orangutan. In *The Apes: Challenges for the 21st Century Conference Proceedings*. Brookfield: Brookfield Zoo, pp. 29–35.

Wallis, J. and Lee, D. R. (1999). Primate conservation: the prevention of disease transmission. *International Journal of Primatology*, 20: 803–826.

Wich, S. A., Riswan, J. J., *et al.* (2011). *Orangutans and the Economics of Sustainable Forest Management in Sumatra*. Trykkeri AS, Norway: UNEP, GRASP, PanEco, YEL, ICRAF, GRID-Arendal.

Wich, S. A. and Utami-Atmoko, S. S. (2010). *Report on Continuing Encroachment in and around the Ketambe Research Area and Gunung Leuser National Park*. Medan, Indonesia: Yayasan Ekosistem Lestari, Sumatran Orangutan Conservation Program, PanEco Foundation.

Wich, S. A., Utami Atmoko, S. S, Mitra Setia, T., *et al.* (2004). Life history of wild Sumatran orangutans (*Pongo abelii*). *Journal of Human Evolution*, 47: 385–398.

Wich, S. A., Utami-Atmoko, S. S., Mitra Setia, T., *et al.* (2006). Dietary and energetic responses of *Pongo abelii* to fruit availability fluctuations. *International Journal of Primatology*, 27: 1535–1550.

Woodford, M. H., Butynski, T. M., and Karesh, W. B. (2002). Habituating the great apes: the disease risks. *Oryx*, 36: 153–160.

Worton, B. J. (1989). Kernel methods for estimating the utilization distribution in home-range studies. *Ecology*, 70: 164–168.

Part III

African primates

7 Lemurs and tourism in Ranomafana National Park, Madagascar: economic boom and other consequences

Patricia C. Wright, Benjamin Andriamihaja, Stephen J. King, Jenna Guerriero, and Josephine Hubbard

Introduction

Because Madagascar has been isolated from other continents for more than 150 million years (Jernvall & Wright, 1998; Kremen *et al.*, 2008), it is characterized today by an extraordinary biodiversity that is unique to the island. Over the past 1000 years, 90% of Madagascar's natural habitat has been destroyed, likely due primarily to human impact (Green & Sussman, 1990; Perez *et al.*, 2005). Today many of its endemic species of plants and animals are threatened or endangered due to a combination of habitat destruction and hunting (Banks *et al.*, 2007; Mayor *et al.*, 2004; Wright *et al.*, 2008). In very recent years tourism may have become an additional factor affecting Malagasy biodiversity. Madagascar's diverse ecosystems, including spiny desert, subtropical dry forest, and rainforests, are all now contained in protected areas visited by tourists (Garbutt, 2009; Mittermeier *et al.*, 2010).

Before the national park system was organized, tourism in Madagascar was primarily limited to beach resorts. A 15-year Environmental Action Plan was started in 1990 through which the national protected area system was established and the Association Nationale pour la Gestion des Aires Protégées (ANGAP) was organized to manage it (Wright & Andriamihaja, 2002). In 1990 there were two national parks in Madagascar, but by 2008, 18 had been established (Figure 7.1). Today the majority of ecologically minded tourists visit three of those parks: Mantadia, Isalo, and Ranomafana. In 2008, ANGAP was renamed Madagascar National Parks (MNP).

Ranomafana National Park (RNP), the fourth national park in Madagascar, was established in 1991 and contains 43 601 ha of continuous rainforest ranging in elevation from 600 to 1450 m (Wright, 1992; Wright & Andriamihaja, 2002). The Ranomafana forest had previously been partially protected by its natural steepness and unsuitability for farming. The park provides an economic advantage to local communities by protecting the watershed for the hydroelectric power plant

Primate Tourism: A Tool for Conservation?, ed. Anne E. Russon and Janette Wallis. Published by Cambridge University Press. © Cambridge University Press 2014.

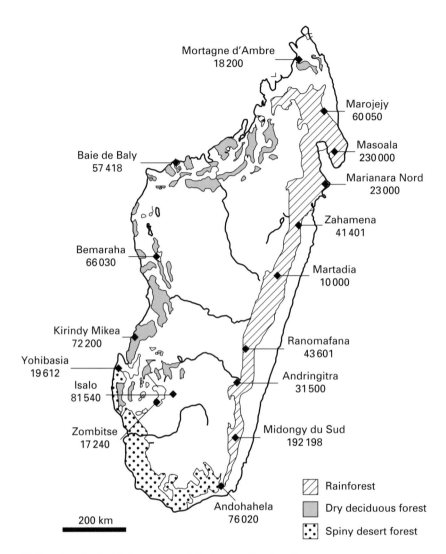

Mortagne d'Ambre
18 200

Marojejy
60 050

Masoala
230 000

Baie de Baly
57 418

Marianara Nord
23 000

Zahamena
41 401

Bemaraha
66 030

Martadia
10 000

Kirindy Mikea
72 200

Ranomafana
43 601

Yohibasia
19 612

Isalo
81 540

Andringitra
31 500

Zombitse
17 240

Midongy du Sud
192 198

Rainforest

Dry deciduous forest

200 km

Andohahela
76 020

Spiny desert forest

Figure 7.1   National parks in Madagascar and their areas (hectares).

that produces all the electricity for the regions of Fianarantsoa, Ambositra, and Antsirabe. The creation of the park, together with improved infrastructure such as roads (2007) and modern hotels (1999–2012), has also inspired nature tourism (DeFries *et al.*, 2009). In a system known as the Droit Entrée aux Aires Protegées (DEAP), entrance fees to the park are divided between the national park system and conservation projects proposed by the local communities (Peters, 1998).

There are many places in Madagascar where lemurs flee for fear of hunters or are already hunted to local extinction, but at RNP a tourist can observe six to eight species of habituated lemur groups in a morning walk in the rainforest. Due to the unique biodiversity of Madagascar, RNP has been a focus of long-term scientific research initiatives (Lovejoy, 2006; Wright & Andriamihaja, 2002; Wright, *et al.*, 2012). Lemur researchers have played an important role in habituating the lemurs, establishing the park trail system, and encouraging tourism through raising international public awareness. Scientists have provided training for tourist guides to raise the quality of the guide service to a professional level (Wright & Andriamihaja, 2002, 2003), and tourist guides often collaborate with trained research technicians to guide their clients efficiently to see lemurs up close. Since 2004, 58 local tourist guides have been trained, tested, and certified by the MNP Service. In 2003, a modern international research center, Centre ValBio (CVB), was established adjacent to the RNP (Wright, 2004). In 2012, there were more than 85 local residents employed at CVB, assisting over 100 researchers each year from all over the world. In 2012, selected tourists began to visit CVB and tour operators asked for scientific information to share with their clients. Illustrated field guides describing local lemurs, fruits, and trees are being developed by CVB staff and will be available for tourists. Because of the high lemur diversity, the fact that there are at least four critically endangered lemurs and the accessibility of Ranomafana wildlife in general, the park has become a popular site for the production of educational films (BBC, Great Britain; NHK, Japan; National Geographic, USA), television documentaries, and, in 2014, a three-dimensional film on lemurs by IMAX. The primary value of such films, along with books and magazine articles about Madagascar, comes from the increased national and international public awareness they generate which, in turn, stimulates commitment to protect Ranomafana National Park and its wildlife. These productions also generate additional income for MNP from filming fees.

Scientific tourism that often includes volunteers who work with scientists to help collect data and spread awareness of conservation issues has also developed at Ranomafana. The Earthwatch Institute is a good example of this system. From 1996 to 2008, more than 20 Earthwatch groups came to Ranomafana to study lemurs. Not only have they helped with the research, but they often return to Ranomafana, spearheading projects in education, health, and reforestation.

Here we discuss the effects of the dramatic increase in tourism brought about by RNP's natural beauty, its rare primates, its recently improved infrastructure, and the interest generated by international researchers. We have a special opportunity at RNP to examine data collected on the three rarest lemur species before and after tourism: *Propithecus edwardsi*, the Milne-Edwards' sifaka; *Prolemur simus*, the greater bamboo lemur; and *Hapalemur aureus*, the golden bamboo lemur. These three species are seen by nearly every visitor to the park. A goal in this paper is to examine population trends for these lemur species during 6 years without tourism and 15 years with tourism. For one species, *P. edwardsi*, we can also compare body masses of adults across this period. If the lemurs are stressed by the presence of tourists, we hypothesize that tourism may have negative effects on the population, as indicated by decreased population size, group size, and body mass; if they are not stressed, we predict that there would be no decrease through time, and no difference between years when tourists were absent and present. We also calculate the economic gain to the region around the RNP since tourism began. Finally, we weigh the costs and benefits of tourism to lemurs versus to human residents and make suggestions for improving the benefits to lemur conservation with tourism.

Methods

Lemur data

Study site
The lemur data derive from a 22-year study of *P. edwardsi* and a 16-year study of one group of *P. simus* in the submontane rainforest of RNP in southeastern Madagascar (Figure 7.1). On average, RNP receives 3000 mm of rain annually (Wright, 2006). The climate varies seasonally, with both rainfall and temperatures being predictably higher during the months of December to March than during the remainder of the year (Hemingway, 1996; Overdorff, 1993). Wright (1995) has previously described the study site in detail.

Study subjects
Individuals included in our study were members of four social groups of *P. edwardsi* and one group of *P. simus* within the Talatakely trail system (elevation from 900 m to1100 m), an area that was selectively logged between 1986 and 1989 and that is relatively homogeneous in habitat quality (Arrigo-Nelson, 2006; Morelli *et al.*, 2009; Tan, 1999; Tecot, 2008; Wright, 1998). *H. aureus*, a new species discovered in 1986 (Meier *et al.*, 1987, Tan, 1999); *P. simus*, a species rediscovered in 1987 (Tan, 1999, 2006; Wright *et al.* 1987); and *P. edwardsi* (Wright, 1995; Wright *et al.*, 2012) live sympatrically in Talatakely (Tan, 1999; Wright *et al.*, 2008) and are targeted species for examining behavioral differences with and without tourism.

The Milne-Edwards' sifaka, *P. edwardsi*, is an anatomical folivore (Hill, 1953) that includes a variety of leaves, fruits, and flowers in its diet (Arrigo-Nelson, 2006; Hemingway, 1995, 1998). Over the course of this study, social groups ranged in size from three to nine individuals, typically with one or two breeding females per group (Pochron & Wright, 2003; Wright, 1995). On average, females produced one offspring every two years (Morelli *et al.*, 2009). The population dynamics of these sifakas have been reported elsewhere (Pochron *et al.*, 2004; Wright *et al.*, 2012). Group size numbers are taken for the month of May for each year.

Study individuals have been successfully captured with anesthetic darts in the field, measured, and released on a nearly annual basis since 1987 using an established protocol (Glander *et al.*, 1992; Wright, 1995), under veterinary supervision, and with the approval of the Stony Brook University IACUC (Institutional Animal Care and Use Committee) and Madagascar's CAFF/CORE (Madagascar Animal Care Committee). While captured, individuals were fitted with color-coded tags

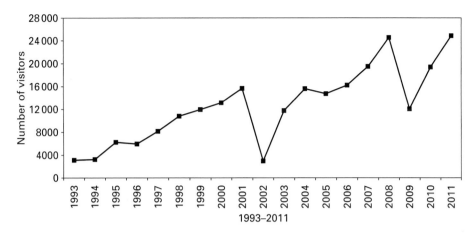

Figure 7.2   Numbers of international tourists visiting Ranomafana National Park, 1993–2011. Note the drop in tourist numbers in 2002 and 2009, both years with political turmoil.

and collars to facilitate their identification in the field. Our data include morphometric data collected from 56 different individuals of all ages and both sexes. Most individuals were sampled more than once, for a total of 210 captures (King *et al.*, 2011). Measurements were taken by trained researchers under controlled conditions in the CVB laboratory at RNP while the animals were under sedation following capture. Body mass was measured with a portable spring scale and recorded to the nearest 0.1 kg following the guidelines of Smith and Jungers (1997).

Prolemur simus is a bamboo specialist with 95% of its diet coming from different parts of one species of bamboo (Tan, 1999). At RNP we have studied one group for 16 years, capturing and collaring individuals as described above for *P. edwardsi*; this group's size has fluctuated from two to eleven individuals. There have been two breeding females in the group, each producing one offspring each year. The population dynamics of this group have been reported elsewhere (Norosoarinaivo *et al.*, 2009; Tan, 1999; Wright *et al.*, 2008, 2012).

Hapalemur aureus is a bamboo dietary specialist that lives in monogamous groups that range from two to six individuals (Meier *et al.*, 1987; Tan, 1999). Seven *H. aureus* groups have been followed in Talatakely since 1994 (Grassi, 2001; Guerriero *et al.*, 2009; Norosoarinaivo, 2000; Tan, 1999; Wright *et al.*, 2012). The data reported in this paper are observations of four groups, with and without tourist observers, and include height in trees and activity patterns (resting, traveling, feeding, social behavior).

Tourism data

Our tourism data include the number of tourists visiting RNP each year since the Park's inception (Figure 7.2) (MNP records). Estimates of the number of tourists at RNP before the park was created are from PCW records.

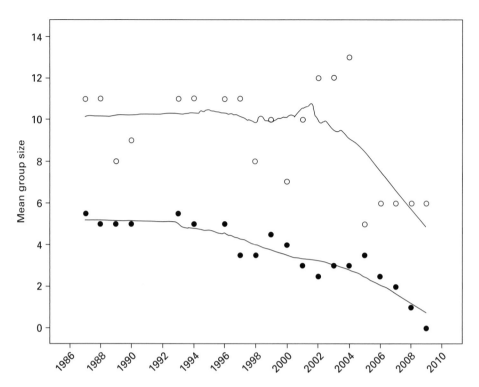

Figure 7.3   Changes in mean group size since 1986. Open circles, *Prolemur simus* (*n* = 1 group); filled circles, *Propithecus edwardsi* (*n* = 2 groups). Lines represent loess regressions.

Results

Lemur populations, body mass, group size, and behavior

Our analyses of the lemur population size, body mass, and group size of *P. edwardsi* show that all three have decreased since 1987, and the most precipitous drop was in the years 2005–2010. For our study population of *Prolemur simus*, the population size and group size have both decreased. In contrast, the populations and group sizes of *H. aureus* have increased since 1987.

The mean membership during May of two groups of *P. edwardsi* decreased by half from 1986 to 2009 (Figure 7.3). Figure 7.4 shows group size changes for four *P. edwardsi* groups. From 1996 to 2006, Group I maintained three to five individuals, in 2007 it consisted of a pair, in 2009 a new male appeared, and by 2012 the group's size had increased to four. Group II had nine individuals in 1986 through 1991 and seven individuals until 1996. With one exception, Group II included three to four individuals from 1996 to 2005, dropped to two individuals in 2006, and disappeared in 2008. Group III included two to four members from 1996 to 2006 and then disappeared after the death of the adult female by predation by *Cryptoprocta ferox* in January 2007. Group IV, which had a robust size ranging from five to nine individuals from 1998 to 2006, had reduced to a single five-year-old natal female by

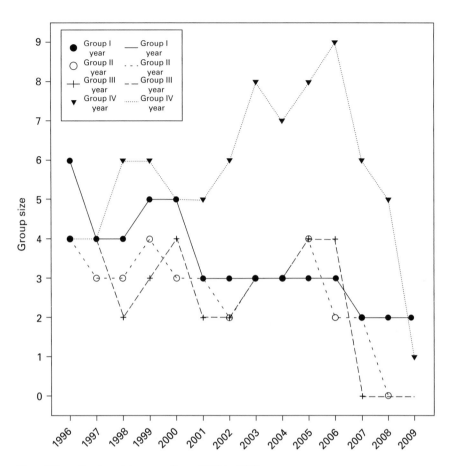

Figure 7.4   Size of four *P. edwardsi* groups from 1996 to 2009.

May, 2009. A new male arrived in 2010 with an infant born in 2011. The *P. simus* group size consistently varied from seven to thirteen individuals from 1986 to 2004 (Tan, 2006; Wright *et al.*, 2008) (Figure 7.3). In 2004, the two adult males in the group disappeared and with no adult males, there was no group increase during 2004–2006 (Wright *et al.*, 2008). Even after an adult male joined the group, infants born in 2007 and 2009 died before reaching six months of age. Infant survival until one year of age in lemurs is 50% (Wright *et al.*, 2012). The group size in May 2005 was five individuals (Norosoaraivo *et al.*, 2009) and had dropped to two individuals by 2010 (CVB records).

Propithecus edwardsi population size decreased by half in 2007 (Figure 7.5). *P. simus*, with the same home range as the four *P. edwardsi* groups, also decreased in population size by approximately half after 2005.

The mean body mass of adult *P. edwardsi* individuals from these four groups decreased by 7% over the 21-year period, from 5.7 kg in 1987 to 5.3 kg in 2008 (Arrigo-Nelson, 2006; King *et al.*, 2011; Wright, unpublished data). Further, female sifakas ranging in pristine forest, which tourists rarely visit, had statistically

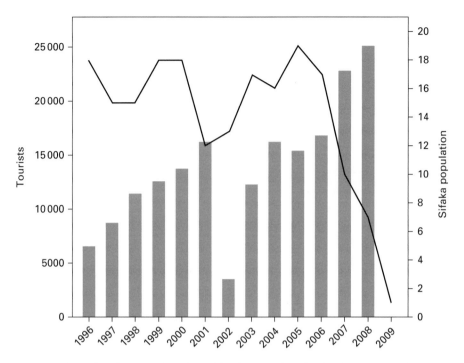

Figure 7.5   Number of tourists (bars) and total combined population size for four groups of *Propithecus edwardsi* (line) from 1996 to 2009.

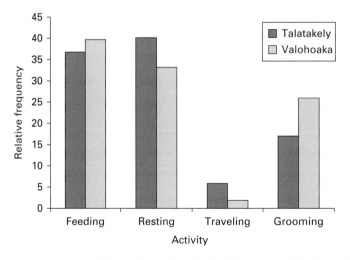

Figure 7.6   Activity budget of *Propithecus edwardsi* in Talatakely vs. Valohoaka The figure shows relative frequencies of major activities between the disturbed forest of Talatakely and the undisturbed forest of Valohoaka. A chi-squared test showed a significant difference between Talatakely and Valohoaka activity budgets ($\chi^2 = 61.14$, $df = 8$, $p < 0.05$), with more feeding and grooming in the undisturbed site (from Hubbard, 2011).

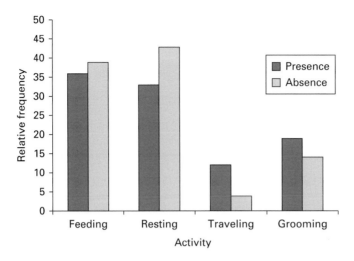

Figure 7.7 Activity budgets of *Propithecus edwardsi* with tourists present vs. absent. The figure shows relative frequencies of activities when tourists were absent vs. present within the disturbed forest of Talatakely. A chi-squared test showed a significant difference in activity budgets within Talatakely in the presence vs. the absence of tourists ($\chi^2 = 17.27$, $df = 3$, $p < 0.05$). Individuals fed less and traveled and groomed more with tourists present (from Hubbard, 2011).

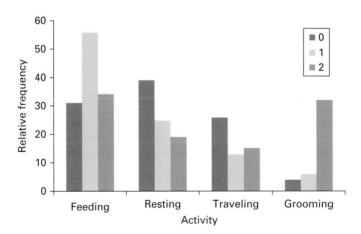

Figure 7.8 Disturbance and activity budget in *P. edwardsi*. The figure shows relative frequencies of major activities in the disturbed forest of Talatakely when tourists were present, at different levels of disturbance $0 = 1$ or 2 tourists, $1 = 2$ to 6 tourists, $2 =$ more than 7 tourists. A chi-squared test showed a significant difference between activity budgets over these levels of disturbance ($\chi^2 = 20.83$, $df = 6$, $p < 0.05$). When there were large groups of tourists the lemurs traveled five times more than when there were few or no tourists observing them (from Hubbard, 2011).

Figure 7.9   Tourists observing the critically endangered golden bamboo lemur (*Hapalemuraureus*), inset, in Ranomafana National Park. (© NRowe-alltheworldsprimates.org.)

significant higher adult body mass than those ranging in disturbed forest visited heavily by tourists (Arrigo-Nelson, 2008).

In addition to the decline in number of *P. edwardsi* individuals, a study in 2010 found that the presence of tourists may also alter their natural behavior. By measuring the disturbance level of the tourists we found that the more disturbing behaviors such as flash photography or shaking trees may enforce these changes in behavior. We first compared the activity budgets between Talatakely and Valohoaka. This showed that in the tourist-disturbed area of Talatakely, individuals were feeding and grooming less while traveling and resting more (Figure 7.6). Upon closer examination of the data within Talatakely, it was found that in the presence of tourists individuals fed and rested less and traveled and groomed more (Figure 7.7). The data was analyzed further using only the data from the tourists present in Talatakely, which further showed that as the disturbance level of the tourists increased, resting

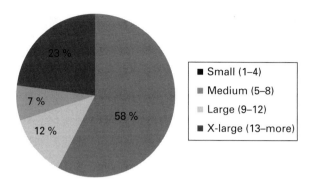

Figure 7.10   Group size of tourists watching *Prolemur simus.*

and grooming decreased while traveling greatly increased (Figure 7.8) (Hubbard, 2011, unpublished data).

For the monogamous *H. aureus*, the group size has fluctuated from two to six individuals consistently, even with the increase of tourism at RNP. We do not have consistent data on body mass over time for this species. On the other hand, we found that the presence of tourists may alter the natural behavior of *H. aureus*. In 2008, we followed individual bamboo lemurs both in the presence of tourists and in their absence (Guerriero *et al.*, 2009). On average, there were 13.2 tourists observing bamboo lemurs at any one time, with a maximum of 55 tourists beneath one group. When tourists were present, female *Hapalemur* moved almost 2 m higher in the canopy on average (Figure 7.9). The height of males was unaffected by the presence of tourist groups. We found a tendency for adults of both sexes to feed less, rest less, and travel more during tourist visits (Guerriero *et al.*, 2009).

In October 2011, which is the high tourist season, we followed the one group of *P. simus* consisting of only two group members for ten days within the Ranomafana National Park, recording numbers of tourists. RNP is the only location where tourists can see this critically endangered species, and because they are often low and highly visible, they are an important tourist attraction. Figure 7.10 shows the group size of tourists observing this critically endangered species. More than one-third of the tourist observers were in a grouph of over 20 individuals. Perhaps this is a factor that resulted in the lemur group size now being reduced to two individuals.

Tourists in Madagascar

Interest in Madagascar's biodiversity, spearheaded by publicity generated by researchers and conservationists, has created a flourishing tourist industry. Within the last decade, tourism has become an important contributor to the GNP in Madagascar and 65% of the tourists who visit Madagascar visit the protected areas (Madagascar Bureau of Tourism Records, 2008). Tourism generated 400 million US dollars nationwide in 2008, and Madagascar "had enjoyed average economic

Table 7.1 Ecotourism development in the peripheral zone of Ranomafana National Park (total numbers)

Year	Shops selling artisanal products	Camp sites	Restaurants	Guides	Trackers	Apprentices to guides and trackers	Hotels
2003	10	3	13	45			10
2004	10	3	13	45			10
2005	10	3	15	48			11
2006	10	3	15	46			11
2007	10	4	21	48	15		11
2008	10	4	21	50	15	22	13
2009	15	4	21	51	27		14
2010	16	4	21	50	30	14	16
2011	16	4	20	51	30	14	16

growth of 5% a year" due primarily to "a booming tourist industry" (*The Economist,* Feb. 7, 2009, p. 42). Table 7.1 shows RNP area ecotourism development for 2003–2011 and Figure 7.2 shows the total number of tourists to RNP, 1996–2011 (data from the official MNP database). In 17 years the number of tourists visiting RNP has increased from a few thousand to more than 24 000.

In RNP alone, tourism has increased nearly forty-fold since the park was created (Figure 7.2). An annual increase in tourists of around 20% has been steady from 1999 to 2012, except for 2002 and 2009 when there was political turmoil. This increase has been challenging. In 2006, tour groups began arriving in modern buses with 20–30 tourists per bus. The high season for tourism begins in July and continues until December and during this six-month period in 2008, over 24 000 tourists viewed RNP lemurs (Figure 7.2). Nearly all the tourists at RNP visited our study groups of endangered species of lemurs. Most tours are conducted in the morning hours only, with nearly 170 tourists per day on average viewing the lemurs within a five-hour block. Lemurs appear tolerant of human observation, having been habituated for research early in their lives or since their birth, although subtle changes in their behavior during tourist visits suggest they may be less tolerant than they first appeared.

Revenue to the region generated by the RNP

Before the RNP was established there were fewer than 20 tourists per year who visited the area (PCW, pers. comm.). For the first ten years after the creation of the RNP, access was a problem for tourism because of poor roads, lack of comfortable hotels, limited communication infrastructure, and no marketing. Tourist numbers rose slowly at first. Beginning in 1999, hotels maintaining international standards opened (e.g. Domaine Nature in 1999, Centrest in 2001, Setam Lodge in 2006). Each of these hotels offers approximately 26 clean and spacious bungalows. These

Table 7.2 Distribution of funds generated in 2011 by ecotourism
and research for the region of Ranomafana National Park

Distribution	US dollars
Hotels	83 270
Restaurants	248 500
DEAP (RNP fees)	150 000
Handicraft sales	132 700
Health	25 000
Education	50 000
Medicinal plants	5000
Food CVB	50 000
Salaries	
CVB staff	171 000
Tourist guides	132 500
Hotel staff	15 000
Total	~1 812 400

hotels are primarily owned and managed by local Malagasy. Previously established local hotels, such the Manja, have expanded and upgraded to a higher standard. Each hotel has a restaurant with excellent food, and the Domaine Nature has a local band which plays traditional Malagasy music on the weekends. Currently in 2013, there are 16 hotels located in the area around RNP.

As a result of the rapid increase in RNP tourism, DEAP funds from park entrance fees have increased substantially every year, tripling from 2004 to 2008. In 2008 nearly $150 000 US was distributed among 58 villages in the Ranomafana region. These funds are managed by the local people and supervised by the national park authorities. Each year the villages located around the national park propose projects, such as building wells for clean water; maintenance of schools; buying seeds; building small dams and rice storage facilities; buying a boat for villages to take crops to market; starting the infrastructure for establishing a market day; establishing a local library; developing animal husbandry, such as chickens and pigs; building bee boxes for honey; and buying baby fish for fish farming. Each village requests one project per year and more than half of these requests are awarded by a local committee composed of representatives from each village. Those that use the money effectively are rewarded by success in future competitions. This economic return is linked directly to tourism, so that in the years without tourism due to political turmoil (2002 and 2009), there was a decided decrease in income generated in local villages. The total minimum revenue generated by tourism and research for the Ranomafana region including hotel, restaurant, handicraft, park fees (DEAP), tourist guide fees, and research station technician salaries in 2011 was approximately $1 800 000 US (Table 7.2). This represents about 65% of the local income of the region (Mayor's office statistics).

Discussion

Lemur population reduction

First of all, Ranomafana National Park has two critically endangered, one endangered, and four threatened lemur species that are easily visible to tourists (Schwitzer *et al.* 2014). A fourth critically endangered species, *Varecia variagatus*, the black and white lemur, can be viewed after a two-hour walk. Over the past decade the populations of the most viewed lemurs in Ranomafana National Park have fluctuated, some staying steady and other species diminishing (Tecot *et al.*, 2013, Wright *et al.*, 2012). *Propithecus edwardsi* and *P. simus* first decreased in group size (Figure 7.3), a sign of impending extinction seen in other primate groups (Boinski & Sirot, 1997). Although confounded by differences in selective logging, the finding that adult sifakas' body mass was higher in pristine forest, areas that tourists rarely visit, than in the disturbed forest visited heavily by tourists was also a red flag (Arrigo-Nelson, 2008).

Many factors could have contributed to this decrease in lemurs, however, and tourism is only one among them. King *et al.* (2005) suggested that increased dry seasons, which have been seen over the last decade, might make milk production difficult for older mothers, resulting in early infant deaths. Wright (2006) and Dunham *et al.* (2011) expanded on this argument, suggesting that more severe cyclones, paired with increasingly longer dry seasons associated with global climate change, could be a factor in lemur population decreases. Deaths by fossa (*C. ferox*) predation have increased in Ranomafana National Park, perhaps because peripheral zone forests have been burned and the hunting area of fossa has been reduced, causing more frequent return times to groups (Gerber *et al.*, 2010; Irwin *et al.*, 2009; Karpanty *et al.*, 2008; Wright, 1998). This predation has contributed to the decrease in our *P. edwardsi* population; raptor and/or fossa predation could be impacting the *P. simus* populations (Karpanty, 2006; Karpanty & Grella, 2001; Karpanty & Wright, 2007). By comparing *Propithecus* predator alarm call frequency from 1987–1993 (low tourism) and from 2000–2007 (high tourism), we see a reduction by 80% in alarm calls (Karpanty, 2007). Has tourism caused the visited groups to be less aware of predator danger? Data do not indicate a reduction in predators (Gerber *et al.*, 2010; Karpanty 2006). Another possibility, indicated by the weight discrepancy between *P. edwardsi* populations in selectively logged versus undisturbed habitats, is not related to tourism, but rather that there is an "extinction debt" (Tilman *et al.*, 1994) caused by exploitation for timber in the period 1986–1989 which resulted in the loss of 25 m-diameter tree canopies. Prior to the timber exploitation, large trees produced large amounts of fruit for the larger lemur groups (Arrigo-Nelson, 2008). It is possible that the combination of all these stresses is exacerbated by increased tourism, resulting in the precipitous decline of lemurs.

Tourism can affect primate populations in many ways, and it is important that we try to understand the impact. There is some indication that high numbers of tourists increase stress on the lemurs, and groups of nearly 20 to 30 tourists observing one group of four to six lemurs probably has a significant impact. In a 2008 study, we found that golden bamboo lemurs moved farther up in the trees with increased tourism (Guerriero *et al.*, 2009). In a 2010 study, we also found that *P. edwardsi* individuals in the tourist disturbed area of Talatakely were on average farther from their nearest neighbor than in the pristine forest of Valohoaka (Hubbard, 2011, unpublished data).

More individual lemurs exit our tourist groups than immigrate into the groups (Morelli *et al.*, 2009). It is possible that immigration of unhabituated lemurs into groups visited by tourists may be impeded because naïve lemurs choose to avoid immigration into tourist areas. Emigration from the tourist groups may be increased by lemurs deciding to find a less visited group. Additionally, disease could be spread by tourists to lemurs, especially if close viewing is sought and allowed (e.g. Goldsmith, 2000; Fujita *et al.*, 2009; Wallis & Lee, 1999; in this volume see chapters by Goldsmith, Muehlenbein & Wallis, Williamson & Macfie), or stress from tourists could lower the lemurs' immune response (Wright *et al.*, 2009). We have indications that *P. simus* with constant contact with tourists ($n = 2$) have contracted human nematodes, while *P. simus* groups not visited by tourists ($n = 5$) have no intestinal parasites (Tuomas Aivelo, pers. com.).

Nature tourism, especially ecotourism, has been suggested to be an economic incentive that helps pay for conservation (Aylward *et al.*, 1996; Bookbinder *et al.*, 1998; Gossling, 1999; Honey, 2008; Horwich, 1998; Horwich & Jones, 1998; Maille & Mendelssohn, 1993; Menkhaus & Lober, 1996; Wilkie & Carpenter, 1999). After the initial enthusiasm and rising numbers of visitors, however, there is now increased awareness that nature tourism may not be the "silver bullet" needed to save biodiversity (Berman *et al.*, 2007; Davenport *et al.*, 2002; Obua, 1997; Walker, 1997; Wall, 1997; Wilkie & Carpenter, 1999; Yu *et al.*, 1997). Previously documented drawbacks of nature tourism include transfer of disease from humans to great apes (Goldsmith, 2000, this volume), stress on the animals visited (Berman *et al.*, 2007; Butynski & Kalina, 1998; Ruesto, *et al.*, 2009), and damage to the habitat (Higham, 2007; Hunter & Green, 1995; Obua, 1997; Walpole *et al.*, 2003; PCW personal observations).

One of the major accomplishments of the conservation effort of the past 15 years in Madagascar is the establishment of a functioning MNP system with trained personnel, modern infrastructure, and a network of over 20 parks. Spatial analyses in and around protected areas demonstrate clearly that park status indeed has been a deterrent to serious loss of forest cover and biodiversity (Lovejoy, 2006; MNP records). Part of this success can be attributed to the fact that parks of small size are capable of containing and maintaining rich biodiversity, making it relatively easy to generate adequate funding for neighboring communities and develop tourism that satisfies local needs. The people living near the parks are poor, and the money generated by parks can make a big difference to their standard of living (Wright & Andriamihaja, 2002). In addition, the local people have strong traditions and respect for laws, unlike the frontier situation prevailing in many places in the Americas, Asia, and Africa (Obua, 1997; Ruesto *et al.*, 2009; Wallace & Pierce, 1996; Wilkie & Carpenter, 1999; Yu *et al.*, 1997).

One reason for opening new parks in Madagascar is increased nature tourism. Furthermore, once opened, these parks can be partially maintained through the tourist revenue they provide (Maille & Mendelsohn, 1993; Peters, 1998; Wright & Andriamihaja, 2002). With ~$1.8 million US generated annually from tourism in the region of Ranomafana, the economic advantage is substantial and life-changing for the local residents. In addition to direct revenue, ecosystem services provided by the national parks including clean water, electricity, pure air, reduced erosion, and carbon sequestration have made a positive economic impact. In fact, most of the headwaters of Madagascar's river systems are sustained from the forested water-sheds of National Parks. If these disappear, irrigation for agriculture, hydroelectric power, and drinking water for humans and livestock will vanish.

Population size, body weight, and group size of lemurs

Our results show that the population size, body weight, and group size decreased as tourism increased in *P. edwardsi*, an endangered species. In the critically endangered *P. simus,* the size of the one group visited by tourists also showed a decline (Wright *et al.*, 2008, Figure 7.3). We also found that tourism altered the natural behavior of *H. aureus* and *P. edwardsi* (Hubbard, 2009, and Figures 7.7, 7.8, 7.10).

The decrease of lemur populations corresponds to the years when tourist num-bers were over 20 000. However, this association could be coincidental and there may be other factors, such as climate change (King, *et al.*, 2005; Wright, 2006), predation by fossa (Irwin *et al.*, 2009; Morelli *et al.*, 2009; Wright, 1998; Wright & Karpanty, 2006), increase in parasite loads (Wright *et al.*, 2009), or a delayed response to habitat alteration (Arrigo-Nelson, 2006; DeFries *et al.*, 2009) that could be contributing to the lemurs' diminishing populations. Our most unsettling evi-dence that tourism may play a primary role in lemur decline is the case of sifaka group IV, which had never been visited by tourists until 2007, after two of the four sifaka groups normally visited by tourists disappeared. In 2006 there were ten individuals in Group IV, and by May 2009 only one individual remained in that range. It should be noted that these trends are reversible and three years later the population seems to be recovering, with Group I increasing to four individuals and Group IV to three individuals. Signs of resilience in these populations suggest that a change of tourism rule enforcement might make a difference.

Linking tourists and researchers

From the very beginning RNP was closely linked to research scientists with long-term commitments to its biodiversity and to the local human population. Research fees go directly to local salaries. Research knowledge, such as the long-term sys-tem of ecological and peripheral village monitoring, is used in park management planning. Research news also attracts the international and national media, which promotes public awareness and good public relations as well as more tourists to generate more funds for the park and its neighbors.

Village residents directly benefit from the park by receiving revenues generated by tourists, who are attracted by wildlife – especially wildlife that is easily seen because of habituation by researchers. The benefits of the park to village residents are evident, as an increasing number of them own bicycles and cars, send their children to high school, roof their houses with tin, and begin having fewer children and decreased childhood mortality.

However, decreased populations in the most popular lemur species is also documented. We are seeking better ways to increase awareness of lemur conservation by tourists, as well as better ways to decrease the negative impact of tourism on lemur populations at the viewing site.

Problems and solutions

Tourism has increased steadily at RNP (Figure 7.2), and in 2012 there were up to 55 tourists viewing one group of lemurs at one time. This human crowding diminishes the wildlife experience for the tourists themselves in addition to exacerbating the problems that it may be causing for the lemurs. We make the following suggestions to improve RNP tourism for the lemurs as well as for the tourists. These are suggestions to be shared with tour operators, guidebook authors, and park managers.

1) Tour operators might explain to tourists that large groups of tourists may disturb lemurs, and that tourists may have to wait their turn to see the rarest species, a system proven successful with gorillas and other sensitive species (Honey, 2008).
2) A six-person limit for watching the lemurs is now set by park authorities. However, monitoring the numbers of tourists watching the lemurs and enforcement of these limits is lacking. We propose that the rules be enforced.
3) A booking system could be established so that tourists do not have to wait to see the animals, but know their time slot.
4) More trails might be opened inside the park to spread out the impact of tourists. In response to tourist crowding, the MNP opened six new tourist circuits in 2010 and tourists have begun to take advantage of these new routes. We will monitor the effects of these additional trails on the behavior and populations of the lemurs.
5) With the opening of a new building at Centre ValBio in 2012, there is a lecture hall where researchers can convey knowledge about lemurs directly to tourists, enriching their experience during scheduled evening or lunchtime lectures. The main focus of these lectures will be on the specialness of the lemurs and how the tourists can help to protect them.
6) Authors might adjust tourist guide books to include April–June as excellent months for lemur tourism (Bradt, 2012; Filou & Stiles, 2012). By extending the tourist season over nine months instead of six, the experience will improve for

both lemurs and tourists. September to November (spring) is the most stressful time for lemurs because it is both the dry season and the season when infants are born (Tecot, 2008), and spreading out visitors over more months would reduce this added stress.

7) The National Office of the Environment's (ONE) system of gathering Environmental Impact Statements for each park is advantageous for monitoring and evaluating changing dynamics. Incorporating monitoring of the relationship between the size of tourist groups and the behavioral reactions of the lemurs visited should be included in the annual ONE Environmental Impact Statement.

8) International donors should continue their funds to support the Madagascar National Parks. Although in Madagascar there were discussions of the possibility that parks could earn enough income to become self-sufficient, the truth is that even in wealthy countries like the United States or Costa Rica, parks must be subsidized (Aylward et al., 1996; Boza, 1993; Honey, 2008).

9) Direct links should be made between saving the lemurs and tourist contributions to wildlife protection. Brochures and videos with this link to saving these endangered species, as well as direct conservation discussions by guides and researchers, should be effected as soon as possible. Donations of time and funds are good places for tourists to start to save these primates.

It is clear that tourism may help the economic condition of the villages surrounding parks, but tourism alone is not the "silver bullet" that will provide funds to protect the remaining wilderness (Vedder & Webber, 1990; Yu et al., 1997). In particular, the RNP case offers an example in which improving human economic conditions attracts more people to move to the park area (Brooks et al., 2009), and may be aggravating rather than relieving pressures on the areas visited and their primate residents. Local economies can be negatively impacted by political problems, because tourists are warned to stay away, and efforts should be made to protect against these periodic economic fluctuations. However, with careful management, nature tourism, now one of the top three income generators of Madagascar, can be an advantage to the lemurs as well as the people of Madagascar. We clearly need to improve on ways that nature tourism can be linked directly to wildlife conservation.

Acknowledgments

We acknowledge the government of Madagascar and the CAFF/CORE oversight committee for authorization to do work in Madagascar. We are grateful to the Director General of MNP, Guy Suzon Ramangason, Ranomafana park manager, Rakotoarijaona Mamy, and the RNP research and tourism coordinator, Rakotonirina Josiane. We appreciate the help with logistics and infrastructure from MICET, and Centre ValBio, including Anna Feistner, Pascal Rabeson, and Florent

Ravoavy. Thanks to our research assistants who have helped collect these data including Georges René Randrianirina, Remi Rakotosoa, the late Georges Rakotonirina, Raymond Ratsimbazafy, Laurent Randrianasolo, and Telo Albert. Particular thanks to the many donors who have supported the work: National Science Foundation, USA, Wenner-Gren Foundation, Earthwatch Institute, Stony Brook University Study Abroad Program and the University of Helsinki Center for excellence grant. All work reported here was approved by IACUC Stony Brook University and complied with the laws of Madagascar. Toni Lyn Morelli and Summer Arrigo-Nelson assisted with darting season data collection. We are grateful to Stacey Tecot and Eileen Larney for excellent comments on the manuscript. Thanks to Erin Achilles for her assistance with the editing, figures, tables, and references. Many thanks to Anne E. Russon and Janette Wallis for inviting us to join them in producing this volume.

References

Arrigo-Nelson, S. (2006). *The impact of habitat disturbance on the feeding ecology of the Milne-Edwards' Sifaka (Propithecus edwardsi) in Ranomafana National Park, Madagascar [dissertation]*. Stony Brook, NY: Stony Brook University.

Arrigo-Nelson, S. (2008). Ecological and social correlates of ectoparasite infections in Milne-Edwards' sifakas (*Propithecus edwardsi*), in Ranomafana National Park, Madagascar. *American Journal of Physical Anthropology*, 46: 61.

Aylward, B., Allen, K., Echeverria, J., and Tosi, J. (1996). Sustainable ecotourism in Costa Rica: The Monteverde Cloud Forest Preserve. *Biodiversity and Conservation*, 5: 315–343.

Banks, M. A., Ellis, A., Wright, P. C. (2007). Global population of a critically endangered lemur, *Propithecus perrieri. Animal Conservation*, 10: 254–262.

Berman, C., Ionica, C., Li, J. H., Ogawa, H., and Yin, H. (2007). Primate tourism, range restriction and infant risk among *Macaca thibetana* at Mount Huagshan, China. *International Journal of Primatology*, 28: 1123–1141.

Boinski, S. and Sirot, L. (1997). Uncertain conservation status of squirrel monkeys in Costa Rica, *Saimiri oerstedi oerstedi* and *Sairmiri oerstedi citrinellis. Folia Primatologica*, 68: 181–193.

Bookbinder, M. P., Dinerstein, E., Rijal, A., Cauley, H., and Rajouria, A. (1998). Ecotourism's support of biodiversity conservation. *Conservation Biology*, 12: 1399–1404.

Boza, M. A. (1993). Conservation in action: past, present, and future of the national park system of Costa Rica. *Conservation Biology*, 7: 239–247.

Bradt, H. (2012). *Bradt Guide: Madagascar*. Old Saybrook, CT: The Globe Pequot Press, Inc.

Brooks, C. P., Holmes, C., Kramer, K., Barnett, B., and Keitt, T. H. (2009). The role of demography and markets in determining deforestation rates near Ranomafana National Park, Madagascar. *PLoS ONE*, 4: e783.

Butynski, T. M. and Kalina, J. (1998). Gorilla tourism: A critical look. In: E. J. Milner-Gulland (ed.). *Conservation of Biological Resources*. Oxford: Blackwell Science. pp. 294–313 and 359–369.

Davenport, L., Brockelman, W., Wright, P., Ruf, K., and Del Valle, F. (2002). Ecotourism tools for parks. In: J. Terborgh, C. van Schaik, M. Rao, and L. Davenport (eds.),

Making Parks Work: Strategies for Preserving Tropical Nature. California: Island Press. pp. 279–306.

DeFries, R., Rovero, F., Wright, P., *et al.* (2009). From plot to landscape scale: linking tropical biodiversity measurements across spatial scales. *Frontiers in Ecology and the Environment*, 8: 153–160.

Dunham, A. E., Erhart, E. M., and Wright, P. C. (2011). Global climate cycles and cyclones: consequences for rainfall patterns and lemur reproduction in southeastern Madagascar. *Global Change Biology*, 17: 219–227.

Filou, E. and Stiles, P. (2012). *Lonely Planet: Madagascar*. Australia: Lonely Planet Press.

Fujita, S., Ogasawara A., and Kageyama, T. (2009). Prevalence of *Clostridium perfringens* in intestinal microflora of non-human primates. In: C. Chapman and M. Huffman (eds.), *Primate Parasite Ecology*. Cambridge University Press. pp. 271–281.

Garbutt, N. (2009). *Mammals of Madagascar*. New Haven, CT: Yale University Press.

Gerber, B. D., Karpanty, S. M, Crawford, C., Kotschwar, M., and Randrianantenaina, J. (2010). An assessment of carnivore relative abundance and density in the eastern rain forests of Madagascar using remotely-triggered camera traps. *Oryx* 44: 219–222.

Glander, K. E., Wright, P. C., Daniels, P. S., and Merenlender, A. M. (1992). Morphometrics and testicle size of rain forest lemur species from southeastern Madagascar. *Journal of Human Evolution*, 22: 1–17.

Goldsmith, M. L. (2000). Effects of ecotourism on the behavioral ecology of Bwindi gorillas, Uganda: preliminary results. *American Journal of Physical Anthropology*, 30: 161.

Gossling, S. (1999). Ecotourism: a means to safeguard biodiversity and ecosystem functions? *Ecological Economics*, 29: 303–320.

Grassi, C. (2001). *The behavioral ecology of Hapalemur griseus griseus: the influences of microhabitat and population density of this small-bodied prosimian folivore* [dissertation]. Austin, TX: The University of Texas at Austin.

Green, G. M. and Sussman, R. W. (1990). Deforestation history of the eastern rain forests of Madagascar from satellite images. *Science*, 248: 212–215.

Guerriero, J., Larney, E., King, S., and Wright, P. C. (2009). *The influence of tourism on height and activity budget of golden bamboo lemurs (Hapalemur aureus) in Ranomafana National Park, Madagascar*. Undergraduate Research and Creative Activities (URECA), Stony Brook University, Stony Brook, NY [abstract].

Hemingway, C. A. (1995). *Feeding and reproductive strategies of the Milne-Edwards' Sifaka, Propithecus diadema edwardsi* [dissertation]. Durham, NC: Duke University.

Hemingway, C. A. (1996). Morphology and phenology of seeds and whole fruit eaten by Milne-Edwards sifaka, *Propithecus diadema edwardsi*, in Ranomafana National Park. *International Journal of Primatology*, 17: 637–659.

Hemingway, C. A. (1998). Selectivity and variability in the diet of Milne-Edwards' sifakas (*Propithecus diadema edwardsi*): implications for folivory and seed-eating. *International Journal of Primatology*, 19: 355–377.

Higham, J. (ed.) (2007). *Critical Issues in Ecotourism: Understanding a Complex Tourism Phenomenon*. Oxford: Elsevier.

Hill, W. C. O. (1953). *Primates: Comparative Anatomy and Taxonomy I-Strepsirhini*. Edinburgh: University Press.

Honey, M. (2008). *Ecotourism and Sustainable Development: Who Owns Paradise?* Covelo, CA: Island Press.

Horwich, R. H. (1998). Effective solutions for howler conservation. *International Journal of Primatology*, 19: 579–598.

Horwich, R. and Jones, M. (1998). Wildlife conservation crosses international borders: profile of a community conservation organization. *Journal of Wildlife Rehabilitation*, 1: 29.

Hubbard, J. (2009). *Tourism and its affect on lemurs in Ranomafana National Park.* Unpublished SBU Study Abroad research paper, pp. 1–25.

Hubbard, J. (2011). *Tourists, Friend or Foe: The Effects of Tourism on Propithecus edwardsi in Ranomafana National Park, Madagascar.* Undergraduate Research Paper for Study Abroad.

Hunter, C. and Green, H. (1995). *Tourism and the Environment: A Sustainable Relationship?* London: Routledge

Irwin, M. T., Raharison, J. L., and Wright, P. C. (2009). Spatial and temporal variability in predation on rainforest primates: do forest fragmentation and predation act synergistically? *Animal Conservation*, 12: 220–230.

Jernvall, J. and Wright, P. C. (1998). Diversity components of impending primate extinctions. *Proceedings of the National Academy of Sciences of the United States of America*, 95: 11279–11283.

Karpanty, S. M. (2006). Direct and indirect impacts of raptor predation on lemurs in Southeastern Madagascar. *International Journal of Primatology*, 27: 239–261.

Karpanty, S. M. (2007). Behavioral responses of *Propithecus edwardsi* to an experimental multiple-predator community in Ranomafana National Park, Madagascar. *American Journal of Physical Anthropology*, 132: 140.

Karpanty, S. M., Crawford, C., Gerber, B., and Kotshwar, M. (2008). Using camera-traps and traditional knowledge to estimate population parameters and movements of carnivores in southeastern Madagascar: tools toward better understanding lemur predation risk from a predator's perspective. *American Journal of Physical Anthropology*, 135: 128.

Karpanty, S. M. and Grella, R. (2001). Lemur responses to diurnal raptor vocalizations in the eastern rainforests of Madagascar. *Folia Primatologica*, 72: 100–103.

Karpanty, S. M. and Wright, P. C. (2007). Predation on lemurs in the rainforest of Madagascar by multiple predator species: observations and experiments. In S. L. Gursky and K. A. I. Nekaris (eds.), *Primate Anti-predator Strategies*. New York: Springer Press, pp. 75–97.

King, S. J., Arrigo-Nelson, S. J., Pochron, S. T., *et al.* (2005). Dental senescence in a long-lived primate links infant survival to rainfall. *Proceedings of the National Academy of Sciences of the United States of America*, 102: 16579–16583.

King, S. J., Morelli, T. L., Arrigo-Nelson, S., *et al.* (2011). Morphometrics and pattern of growth in wild sifakas (*Propithecus edwardsi*) at Ranomafana National Park, Madagascar. *American Journal of Primatology*, 73: 155–172.

Kremen, C., Cameron, A., Moilanen, A., *et al.* (2008). Aligning conservation priorities across taxa in Madagascar with high-resolution planning tools. *Science*, 320: 222–226.

Lovejoy, T. E. (2006). Protected areas: a prism for a changing world. *Trends in Ecology and Evolution*, 21: 329–333.

Madagascar Bureau of Tourism Records (2008). Public record requested from the Ministry of Tourism, Antananarivo, Madagascar, 2012.

Maille, P. and Mendelsohn, R. (1993). Valuing ecotourism in Madagascar. *Journal of Environmental Management*, 38: 213–218.

Mayor, M. I., Sommer, J. A., Houck, M. L., *et al.* (2004). Specific status of Madagascar's endangered sifakas. *International Journal of Primatology*, 25: 875–900.

Meier, B., Albignac, R., Peyrieras, A., Rumpler, Y., and Wright, P. (1987). A new species of *Hapalemur* (Primates) from southeast Madagascar. *Folia Primatologica*, 48: 211–215.

Menkhaus, S., and Lober, D. J. (1996). International ecotourism and the valuation of tropical rainforests in Costa Rica. *Journal of Environmental Management*, 47: 1–10.

Mittermeier, R.A., Louis, E. E. Jr., Richardson, M., *et al.* (2010). *Lemurs of Madagascar*. Washington, DC: Conservation International.

Morelli, T. L., King, S., Pochron, S. T., Wright, P C. (2009). The rules of disengagement: takeovers, infanticide and dispersal in a rainforest lemur, *Propithecus edwardsi, Behaviour*, 146(4–5): 499–523.

Norosoarinaivo, J. A. (2000). Contribution à l'étude du comportement chez *Hapalemur aureus,* du stade enfant jusqu'au stade juvenile dans le Parc National de Ranomafana [DEA thesis]. Antananarivo, Madagascar: University of Antananarivo.

Norosoarinaivo, J. A., Tan, C., Rabetafka, L., and Rakotondravony (2009). Impact du tourisme sur Prolemur simus à Talatakely, dans le Parc National de Ranomafana. *Lemur News*, 14: 43–46.

Obua, J. (1997). The potential, development, and ecological impact of ecotourism in Kibale National Park, Uganda. *Journal of Environmental Management*, 50: 27–38.

Overdorff, D. J. (1993). Similarities, differences, and seasonal patterns in the diets of *Eulemur rubriventer* and *E. fulvus rufus* in the Ranomafana National Park, Madagascar. *International Journal of Primatology*, 14: 721–753.

Perez, V. R., Godfrey, L. R., Nowak-Kemp, M., Burney, D. A., Ratsimbazafy, J., and Vasey, N. (2005). Evidence of early butchery of giant lemurs in Madagascar. *Journal of Human Evolution*, 49: 722–742.

Peters, J. (1998). Sharing national park entrance fees: forging new partnerships in Madagascar. *Society and Natural Resources*, 11: 517–530.

Pochron, S. T., Tucker, W. T., Wright, P. C. (2004). Demography, life history and social structure in *Propithecus diadema edwardsi* from 1986–2000 in Ranomafana National Park, Madagascar. *American Journal of Physical Anthropology*, 125: 61–72.

Pochron, S. T. and Wright, P. C. (2003). Variability in adult group compositions of a prosimian primate. *Behavioral Ecology and Sociobiology*, 54: 285–293.

Ruesto, L. A., Sheeran, L. K., Matheson, M. D., and Li, J. H. (2009). Tourist behavior and decibel levels correlate with threat frequency in Tibetan macaques (*Macaca thibetana*) at Mt. Huangshan, China. *Primate Conservation*, 24: 145–151.

Schwitzer, C., Mittermeier, R. A., Johnson, S. E., *et al.* (2014). Averting lemur extinctions amidst Madagascar's political crisis. *Science*, 343: 842–843.

Smith, R. J. and Jungers, W. L. (1997). Body mass in comparative primatology. *Journal of Human Evolution*, 32: 523–559.

Tan, C. L. (1999). Group composition, home range size, and diet of three sympatric bamboo lemur species (Genus Hapalemur) in Ranomafana National Park, Madagascar. *International Journal of Primatology*, 20: 547–566.

Tan, C. L. (2006). Behavior and ecology of gentle lemurs (Genus *Hapalemur*). In: L. Gould and M. L. Sauther (eds.), *Lemurs: Ecology and Adaptation*. New York: Springer Press, pp. 369–382.

Tecot, S. R. (2008). *Seasonality and predictability: the hormonal and behavioral responses of the red-bellied lemur, Eulemur rubriventer, in Southesastern Madagascar*. PhD Dissertation, University of Texas, Austin.

Tecot, S.R., Gerber, B.D., King, S.J., Verdolin, J.L., and Wright, P.C. (2013). Risky business: sex differences in mortality and dispersal in a polygynous, monomorphic lemur. *Behavioral Ecology*, 24 (4): 987–996. Published online Feb. 28, 2013: doi:10.1093/beheco/art008.

The Economist (2009). Turmoil in Madagascar: I'm the King of the ....Oops. *The Economist*, February 7, 2009, p.4.

Tilman, D., May, R. M., Lehman, C. L., and Nowak, M. A. (1994). Habitat destruction and the extinction debt. *Nature*, 371: 65–66.

Vedder, A. and Webber, A. W. (1990). The mountain gorilla project. In: A. Kiss (ed.). *Living with Wildlife: Wildlife Resource Management with Local Participation*. Washington, DC: World Bank Technical Publication 130: 83–90.

Wallace, G. N, and Pierce, S. M. (1996). An evaluation of ecotourism in Amazonas, Brazil. *Annals of Tourism Research*, 23: 843–873.

Walker, S. L. (1997). Perceived impacts of ecotourism development. *Annals of Tourism Research*, 24: 743–745.

Wall, G. (1997). Is ecotourism sustainable? *Journal of Environmental Management*, 21: 483–491.

Wallis, J, and Lee, D. R. (1999). Primate conservation: the prevention of disease transmission. *International Journal of Primatology*, 20: 803–826.

Walpole, M., Karanja, G., Sitati, N., *et al.* (2003). *Wildlife and People: Conflict and Conservation in Masai Mara, Kenya*. London: IIED Wildlife and Development Series, IIED.

Wilkie, D. S. and Carpenter, J. F. (1999). Can nature tourism help finance protected areas in the Congo Basin? *Oryx*, 33: 332–338.

Wright, P. C. (1992). Primate ecology, rainforest conservation, and economic development: building a national park in Madagascar. *Evolutionary Anthropology*, 1: 25–33.

Wright, P. C. (1995). Demography and life history of free-ranging *Propithecus diadema edwardsi* in Ranomafana National Park. *International Journal of Primatology*, 16: 835–854.

Wright, P. C. (1998). Impact of predation risk on the behaviour of *Propithecus diadema edwardsi* in the rain forest of Madagascar. *Behaviour*, 135: 483–512.

Wright, P. C. (2004). Centre ValBio: long-term research commitment in Madagascar. *Evolutionary Anthropology*, 13: 1–2.

Wright, P. C. (2006). Considering climate change effects in lemur ecology and conservation. In: L. Gould and M. L. Sauther (eds.), *Lemurs: Ecology and Adaptation*. New York: Springer Press, pp. 385–401.

Wright, P. C. and Andriamihaja, B. A. (2002). Making a rain forest national park work in Madagascar: Ranomafana National Park and its long-term research commitment. In: J. Terborgh, C. van Schaik, M. Rao, and L. Davenport (eds.), *Making Parks Work: Strategies for Preserving Tropical Nature*. California: Island Press, pp. 112–136.

Wright, P. C. and Andriamihaja, B. A. (2003). The conservation value of long-term research: A case study from Parc National de Ranomafana. In: S. Goodman and J. Benstead (eds.), *Natural History of Madagascar*. University of Chicago Press.

Wright, P. C., Arrigo-Nelson, S. J., Hogg, K. L., *et al.* (2009). Habitat disturbance and seasonal fluctuations of lemur parasites in the rain forest of Ranomafana National Park, Madagascar. In: C. Chapman and M. Huffman (eds.), *Primate Parasite Ecology*. Cambridge University Press, pp. 311–330.

Wright, P. C., Daniels, P. S., Meyers, D. M., Overdorff, D. J., and Rabesoa, J. (1987). A census and study of *Hapalemur* and *Propithecus* in southeastern Madagascar. *Primate Conservation*, 8: 84–88.

Wright, P. C., Erhart, E. M., Tecot, S., *et al.* (2012). Long-term lemur research at Centre Valbio, Ranomafana National Park, Madagascar. In: P. M. Kappeler and D. P. Watts (eds.), *Long-term Field Studies of Primates*. Berlin and Heidelberg: Springer, pp. 67–100.

Wright, P. C., Johnson S. E., Irwin, M. T., *et al.* (2007). The crisis of the critically endangered greater bamboo lemur (*Prolemur simus*). *Primate Conservation*, 23: 11–22.

Wright, P. C. and S. Karpanty (2006). Predation on lemurs in the rainforest of Madagascar by multiple predator species: Observations and experiments. In: S. L. Gursky and K. A. I. Nekaris (eds.), *Primate Anti-predator Strategies*. New York: Springer Press, pp. 75–97.

Wright, P. C., King, S. J., Baden, A., and Jernvall, J. (2008). Aging in wild female lemurs: sustained fertility with increased infant mortality. In: S. Atsalis and S. Margulis (eds.), *Reproductive Aging in Primates*. Switzerland: Karger Press, pp. 17–28.

Yu, D. W., Hendrickson, T., and Castillo, A. (1997). Ecotourism and conservation in Amazonian Peru: short-term and long-term challenges. *Environmental Conservation*, 24: 130–138.

8 Some pathogenic consequences of tourism for nonhuman primates

Robert M. Sapolsky

Introduction

As should be obvious to anyone who has ever encountered primatologists and their obsessions, nonhuman primates can hold a strong emotional and intellectual sway. One appeal of nonhuman primates is their sheer variability. They occupy rainforest, desert, mountains, and grasslands. They can live in habitats ranging from the snows of Japan to the heat of Ethiopia. Their social structures include cooperative breeders, such as marmosets, and markedly uncooperative ones, such as savanna baboons, pair-bonded gibbons, and polygamous and polyandrous bonobos. There are also solitary orangutans and hamadryas who live in a complex, multi-tier system of small stable harems of half a dozen or so individuals that can temporarily merge into collections of hundreds of individuals.

Another realm of variability in primates is in their diet. At one extreme are species such as the mountain gorilla, whose diet is sufficiently narrow and specialized as to play a role in its endangerment. And at the other are opportunistic omnivores that flexibly exploit a wide array of food resources. Among the most extreme examples of the latter are savanna baboons. These species eat grass blades and corms, parts of trees, shrubs and tubers, insects, and they both hunt and scavenge meat. This has allowed baboons to successfully occupy the grasslands, forests, and arid highlands of Africa.

"Baboons are opportunistic omnivores" is a phrase that might be used to describe their dietary flexibility by, say, a physiological ecologist. But this dietary versatility can just as readily be described in a way far more relevant to those interested in conservation and wildlife management: "Baboons like to eat the same things as humans." And, almost inevitably, this is not a good thing for either species.

The most pressing manifestation of this problem is crop-raiding by baboons, in which a troop of animals can devastate a maize field in astonishingly short amounts of time. This can obviously have a hugely adverse impact on farmers. Because of crop-raiding, baboons are among the most hated of species among African agriculturalists. Conservation-minded efforts have been made to combat this problem with, for example, electric fences or conditioned food aversion (e.g. Forthman,

Primate Tourism: A Tool for Conservation?, ed. Anne E. Russon and Janette Wallis. Published by Cambridge University Press. © Cambridge University Press 2014.

Strum, & Muchemi, 2005; and see Strum & Manzolillo Nightingale, this volume). Far more often and, from the farmer's perspective, far more successful, animals are speared, shot, or poisoned (pers. obs.).

The versatility of baboons also generates problems when the humans encountered are tourists and the people working in the tourist industry. The most visible conflict that arises is a variant of the problem for a farmer, namely baboons raiding employees' quarters at lodges and camps or tourists' buffet lunches on the veranda. Far too often, the response is the same as that of farmers to baboon crop-raiding, namely the killing of baboons.

But in some ways, a more insidious problem arises when baboons can access the garbage dumps of tourist lodges and camps, and take to foraging on the refuse. In this chapter, I review the surprising array of adverse consequences of such garbage eating for populations of baboons in East Africa. These findings were derived from the fieldwork of a small number of researchers (including the author), where physiological data were obtained from baboons tranquilized for examination. The study sites were Amboseli National Reserve and Masai Mara National Reserve, both in Kenya, and Awash National Park, Ethiopia. The subjects were olive baboons (*Papio anubis*) and yellow baboons (*Papio cynocephalus*).

Background: Savanna baboon social behavior

Savanna baboons live in large, multi-male and multi-female troops ranging from 20 to 100 animals. In troops living under natural conditions, animals typically sleep in a grove of large trees or on cliff faces to avoid predators, and forage in the open savanna during the day.

The social system of savanna baboons is well understood (cf Cheney & Seyfarth, 2008; Melnick & Pearl, 1987). Some of its features include:

a) A despotic male hierarchy. Male rank shifts over time. Attaining a high rank is mostly a function of winning critical fights, but maintaining it is more a function of social intelligence, "political" skill, and impulse control.

b) High rates of male–male aggression and of aggressive displacement onto innocent bystanders; in this atmosphere, socially subordinate males show frequent indices of stress-related disease (Sapolsky, 2005).

c) Male reproductive success that is heavily, but not exclusively, a function of rank. Subordinate males must employ alternative strategies (typically in the form of establishing affiliative relationships with potential mates) to increase their own reproductive success.

d) Males typically leave their natal troop around the time of puberty and subsequently join another troop, where they slowly build social affiliations and climb up in the hierarchy.

e) Females who spend their entire lives in the same troop, and whose dominance rank is static and hereditary. Thus, a female's first daughter establishes a rank one below hers, the next daughter one below that, and so on.

f) High rates of female–female affiliative behaviors, particularly among relatives, low rates of affiliation between adult females and males, and virtually no affiliation between adult males.

In the context of garbage eating around tourist facilities, these characteristics are important because they help determine which individuals are most affected by garbage eating.

When baboons eat like Westernized humans: metabolic consequences

Troops of savanna baboons typically forage for 3–4 hours and travel around 10 km a day (Altmann & Altmann, 1970). For a quadruped weighing 15–40 kg as an adult, this represents considerable energy expenditure. Commensurate with that, studies with isotope-labeled water show that an adult female baboon requires approximately 3000 calories/day (Altmann *et al.*, 1993).

Probably the most striking features of the transition to eating garbage is that baboons then walk only about one-third of the typical distance each day, spend less than half the typical time obtaining food, and decrease their energy expenditure by more than 16% (Altmann & Muruthi, 1988). Importantly, amid this marked shift, the natural and garbage diets provide approximately the same number of calories (Altmann & Muruthi, 1988).

The most striking consequence of this transition to a sedentary lifestyle of subsisting on garbage is obesity. In one such troop, female weight and body mass index (BMI) increased approximately 50% and body fat stores rose from 2% to 23% (Altmann *et al.*, 1993). The picture in males was more complex and informative, in that there were 20% and 15% increases in body weight and BMI, respectively, along with a great deal more variability than among females. As noted above, most (but not all) male baboons leave their natal troop at puberty, transferring into their adult troop. Closer examination of the data showed that the most dramatic obesity was observed among natal males (i.e. those who had access to garbage their entire lives), and the least among transfer males (whose exposure only began at the time of their adolescent transfer into the troop). This is not only likely to reflect chronicity of exposure to a garbage diet, but the effects of early nutrition on life-long metabolic programming (Barker & Hales, 2001) and perhaps preferences established in early learning as well.

Because a garbage diet does not involve an increase in calories consumed, the obesity was attributable to the decrease in activity. Thus, there was an interesting failure of some, but not all aspects of metabolic regulation. Specifically, while baboons' bodies did not regulate to the extent of decreasing calorie intake to match the lower energy expenditures, nonetheless, there was sufficient regulation to prevent an *increase* in food intake, despite a surplus of food and time to consume it. This dissociation reflects the complex regulation of appetite, in which feedback signals arise from gastrointestinal peptides that can access the brain, stomach distention, and circulating levels of metabolic hormones and nutrients (reviewed in Schwartz *et al.*, 2000; Strubbe & van Dijk, 2002).

There are also metabolic consequences of a garbage diet that are less obvious than the obesity, as revealed by studies of two "garbage" troops (Kemnitz *et al.*, 2002). One consequence concerned circulating levels of cholesterol. As background, elevated cholesterol levels (hypercholesteremia) were long considered to increase the risk of cardiovascular disease independently by causing the formation of atherosclerotic plaques (reviewed in Ferns, 2008). More recent work shows that hypercholesteremia is more likely to accelerate plaque growth, rather than initiate it, because inflammatory damage to blood vessels is a prerequisite for elevated cholesterol levels to be pathogenic (reviewed in Ferns, 2008).

Among the males of garbage-eating troops, there were elevated levels of total cholesterol, VLDL (very low-density lipoprotein) cholesterol, apolipoprotein B (VLDL's transport protein in the circulation), and HDL (high-density lipoprotein) cholesterol. VLDL cholesterol is typically characterized as "bad" cholesterol, insofar as it is the type most readily deposited in atherosclerotic plaques. HDL cholesterol is considered "good," as it is in the process of being transported from plaques to the liver, where it is degraded (Kemnitz *et al.*, 2002). The rise in levels of both VLDL and HDL cholesterol initially seems puzzling. However, the net result of these increases was a decrease in the "good/bad" HDL/VLDL ratio, ultimately the most meaningful indicator of the pathogenic potential of cholesterol levels. Veterinary surveys of baboon populations some decades ago revealed that facets of cardiovascular disease occur even among wild populations (McGill *et al.*, 1960); thus, there could well be adverse health consequences because of the changes in cholesterol profiles in male garbage eaters.

Another set of changes concerned circulating levels of insulin and glucose in garbage-eating baboons. As background, the amount of glucose in the bloodstream is a function of the amount that enters the circulation (from diet and/or from mobilization from storage sites during, for example, stress or starvation) and the amount removed from the circulation (where it enters target cells for immediate use for energy or, when there is a surplus, for long-term storage). Insulin is the hormone most responsible for the entry of glucose into target cells; its secretion can be stimulated by proximal signals of eating (such as stomach distension), by the anticipation of eating (for example, the bodies of people who eat dinner at the same time each evening are typically conditioned to secrete insulin shortly before that time) and, most importantly, by increased circulating levels of glucose (Konturek *et al.*, 2003).

Among garbage-eating males, there were two- and three-fold increases in insulin levels in the two troops, and a tripling among females in both troops (Kemnitz *et al.*, 2002). In approximately half the baboons, elevated levels of insulin were accompanied by normal glucose levels. This profile indicates a body that can still effectively regulate glucose, amid its requiring more endocrine "work" to do so (i.e. the hyperinsulinemia). However, with sufficiently high insulin levels, target cells decrease their sensitivity to an insulin signal (i.e. become "insulin-resistant") and their capacity to regulate glucose becomes impaired, leading to elevated glucose in the circulation. This is the profile seen in adult-onset diabetes, where there are elevated levels of both insulin and glucose (Banks *et al.*, 2003). Such hyperglycemia is dangerous, in that the excess glucose can, among other adverse consequences, damage blood vessels, kidneys, and eyes. Approximately half of the garbage-eating baboons had progressed to this profile of adult-onset diabetes.

The once-obscure adult-onset diabetes (also known as insulin-resistant, or Type 2 diabetes) has become one of the leading causes of death in the Western world, not just because of the increased number of calories consumed combined with decreased amounts of exercise, but also because of another feature, processed, Westernized food. A characteristic of such food is that it is denser in calories and digested more rapidly than a more natural diet. This produces a large and rapid rise in circulating glucose levels after a meal, which then produces an atypical spike of insulin secretion. Such a spike is more likely to produce insulin resistance than the same amount secreted in low, steady levels (Qi *et al.*, 2008). Thus, the high-density caloric nature of the garbage diet likely played a role in the hyperinsulinemia of the baboons.

Finally, the subgroup with elevated insulin and glucose levels also had elevated levels of the hormone leptin (Banks *et al.*, 2003). This hormone is secreted by fat cells in response to signals of nutrient availability; logically, it acts in the brain to decrease appetite, and decreases levels of insulin and glucose (Friedman & Halaas, 1998). For a baboon, elevated levels of leptin along with the elevated levels of glucose and insulin indicate "leptin resistance" (i.e. a decrease in the efficacy of leptin). As such, the hormone not only fails to regulate insulin and glucose levels, but also fails to curb appetite, as evidenced by such baboons consuming calories in excess of their energetic needs.

The trio of obesity, elevated levels of insulin, and elevated levels of cholesterol constitutes the profile of "metabolic syndrome" (also known as Syndrome X), which is a major predisposing factor for both diabetes and cardiovascular disease. The approximate 50% of baboons with this profile did not differ from other baboons without it in body weight, age, dominance rank, or amount of garbage consumed, suggesting that their uniquely pathogenic profile reflects the unmasking of some susceptibility factor (Banks *et al.*, 2003).

When baboons eat like Westernized humans: infectious disease consequences

The transition to a garbage diet does not just carry adverse metabolic consequences, but immunological ones as well. This reflects the abnormally close proximity of baboons to each other when clustered around a garbage pit, which facilitates animal-to-animal pathogen transfer. Moreover, food in the refuse could well carry human pathogens that are novel to the immune system of baboons (particularly given the fact that these could readily be pathogens derived from far-flung tourist populations, rather than local humans).

Three sets of findings suggest that there are such infectious disease consequences. First, when compared with baboons eating natural diets, garbage eaters were found to have a higher incidence of periodontal disease. The authors of this study interpreted this as likely to have arisen from both the high sugar content of the refuse and the exposure to food contamination from oral bacteria of humans (Phillips-Conroy *et al.*, 1993). As a second infectious disease example, garbage-eating baboons were found to have higher levels than other baboons of antibiotic-resistant enteric bacteria; this was particularly worrisome, given evidence of baboon-to-baboon bacterial transfer (Rolland *et al.*, 1985). A final example concerns an outbreak of tuberculosis in a garbage-eating baboon troop due to consumption of infected meat at the garbage dump of one lodge. In humans, tuberculosis infection can be asymptomatic for long periods and, once manifest, can emerge relatively slowly. Among numerous nonhuman primate species, however, including baboons, the disease is far more virulent than in humans, often producing death within weeks. In this case, as a result, essentially all baboons who were sufficiently aggressive to gain access to the contaminated meat in the garbage dump died (Sapolsky & Else, 1987; Tarara *et al.*, 1985).

Discussion

The interest in animal behavior that fuels wildlife tourism is understandable, particularly so (in the view of this primatologist) when it concerns nonhuman primates. There are many reasons for this appeal, a strong one being the sense of familiarity and similarity with these close primate relatives. When similarity concerns potentially near-identical diets, there is the potential for adverse health consequences for the nonhuman primates. These include the infectious disease risk of high population density that has been the scourge of the West for centuries, and the metabolic diseases of excess that have become the scourge of the modern West.

The prescriptions that one might suggest for humans are not new – that tourist camps and lodges need to minimize their impact upon their surroundings, including being more vigilant in dealing with refuse, and that such responsible behavior be required by tourists in making their choices as to where to stay. The prescription that one might suggest to baboons in these settings is not new either – to make a point of steering clear of that most destructive primate species. When it comes to humans, lunches, even if discarded, are rarely free.

Acknowledgments

Studies by the author described herein were made possible by grants from the H.F. Guggenheim Foundation, the Templeton Foundation, and the Leakey Foundation, and were approved by the Office of the President, Republic of Kenya. We thank the Office of the President, Republic of Kenya, for permission to carry out this work. All procedures were approved by the APPLAC Committee of the Division of Laboratory Medicine, Stanford University.

References

Altmann, J. and Altmann, S. (1970). *Baboon Ecology: African Field Research*. University of Chicago Press.

Altmann, J. and Muruthi, P. (1988). Differences in daily life between semiprovisioned and wild-feeding baboons. *American Journal of Primatology*, 15: 213–221.

Altmann, J., Schoeller, D., Altmann, S., *et al.* (1993). Body size and fatness of free-living baboons reflect food availability and activity level. *American Journal of Primatology*, 30: 149–56.

Banks, W., Altmann, J., Sapolsky, R., *et al.* (2003). Serum leptin levels as a marker for a Syndrome X-like condition in wild baboons. *Journal of Clinical Endocrinology and Metabolism*, 88: 1234–1240.

Barker, D. and Hales, C. (2001). The thrifty phenotype hypothesis. *British Medical Bulletin*, 60: 5–20.

Cheney, D. and Seyfarth, R. (2008). *Baboon Metaphysics: The Evolution of a Social Mind*. University of Chicago Press.

Ferns, G. (2008). New and emerging risk factors for CVD. *Proceedings of the Nutrition Society*, 67: 223–231.

Forthman, D. L., Strum, S. C., and Muchemi, G. M. (2005). Applied conditioned taste aversion and the management and conservation of crop-raiding primates. In: J. D. Paterson and J. Wallis (eds.), *Commensalism and Conflict: the Human-Primate Interface*. Oklahoma: The American Society of Primatologists, pp. 421–443.

Friedman, J. and Halaas, H. J. (1998). Leptin and the regulation of body weight in mammals. *Nature*, 395: 763–770.

Kemnitz, J. W., Sapolsky, R. M., Altmann, J., *et al.* (2002). Effects of food availability on serum insulin and lipid concentrations in free-ranging baboons. *American Journal of Primatology*, 57: 13–19.

Konturek, S. J., Zabielski, R., Konturek, J. W., and Czarnecki, J. (2003). Neuroendocrinology of the pancreas; role of brain-gut axis in pancreatic secretion. *European Journal of Pharmacology*, 48: 1–14.

McGill, H., Strong, J., Holman, R., and Werthessen, N. (1960). Arterial lesions in the Kenya baboon. *Circulation Research*, 8: 670–679.

Melnick, D. and Pearl, M. (1987). Cercopithecines in multimale groups: Genetic diversity and population structure. In B. Smuts, R. Cheney, R. Seyfarth, *et al.* (eds.), *Primate Societies*. University of Chicago Press, pp. 121–134.

Phillips-Conroy, J., Hildebolt, C., Altmann, J., *et al.* (1993). Periodontal health in free-ranging baboons of Ethiopia and Kenya. *American Journal of Physical Anthropology*, 90: 359–371.

Qi, L., Hu, F. B., and Hu, G. (2008). Genes, environment, and interactions in prevention of type 2 diabetes: a focus on physical activity and lifestyle changes. *Current Molecular Medicine*, 8: 519–532.

Rolland, R., Hausfater, G., Marshall, B., and Levy, S. (1985). Antibiotic-resistant bacteria in wild primates: increased prevalence in baboons feeding on human refuse. *Applied Environmental Microbiology*, 49: 791–794.

Sapolsky, R. (2005). The influence of social hierarchy on primate health. *Science*, 308: 648–652.

Sapolsky, R. and Else, J. (1987). Bovine tuberculosis in a wild baboon population: Epidemiological aspects. *Journal of Medical Primatology*, 16: 229–235.

Schwartz, M., Woods, S., Porte, D., *et al.* (2000). Central nervous system control of food intake. *Nature*, 404: 661–671.

Strubbe, J. H. and van Dijk, G. (2002). The temporal organization of ingestive behaviour and its interaction with regulation of energy balance. *Neuroscience and Biobehavioral Reviews*, 26: 485–498.

Tarara, R., Suleman, M., Sapolsky, R., *et al.* (1985). Tuberculosis in wild baboon (*Papio cynocephalus*) in Kenya. *Journal of Wildlife Diseases*, 21: 137–140.

9 Baboon ecotourism in the larger context

Shirley C. Strum and Deborah L. Manzolillo Nightingale

Introduction

Biodiversity conservation is a growth industry. Controversies abound over approaches and philosophies and about whose rights and whose ethics should apply. In this chapter we address some of these issues based on 38 of our 42 years of experience on the ground, conserving baboons in the face of threats resulting from change in land use, rapid human population growth, escalating rates of poverty, sedentarization of people, and human–baboon conflict – over crops and livestock. One of our strategies has been to promote primate-based tourism. "Walking with Baboons," our ecotourism product, is distinctly different from other well-known types of primate tourism because it takes place in the arid savannah, in contrast to the more usual forest-based primate tours. The project is located in Kenya which has a well-developed tourism industry with a history of ecotourism among Maasai livestock pastoralists who traditionally were not hostile to baboons.

The baboon ecotourism venture came late in the history of the long-term baboon research. Moreover, it has unique aspects that may not be replicable in other primate settings. Nonetheless, we have been successful in building conservation momentum within the neighboring community. Unintentionally, we expanded and intensified our interactions with it not just around primate-focused tourism but in areas of mutual interest and concern. The process built inter-individual and inter-community bonds which increased the community's motivation to conserve. We discovered that activities apparently tangential to conservation can make important contributions. This implies that, to succeed as a conservation technique, primate ecotourism needs to be embedded in the larger local social and cultural contexts. This chapter presents the larger context for "Walking with Baboons" as well as the specifics.

What is ecotourism?

In terms of the number of people involved, income generated, and employment capacity, travel and tourism is the world's largest industry at over $4 trillion US

Primate Tourism: A Tool for Conservation?, ed. Anne E. Russon and Janette Wallis. Published by Cambridge University Press. © Cambridge University Press 2014.

in economic benefits annually. "Wildlife" tourism is one of the highest earners of foreign exchange in several countries, including Kenya (Honey, 1999/2008; Pennisi *et al.*, 2004; Sindiga, 1999). Mass tourism, however, has brought serious environmental and socio-cultural impacts that can overshadow the economic benefits. The definition of ecotourism and its role in conservation is still hotly debated (for discussion see Higham, 2007; Higham & Carr, 2003; Honey, 2002, 1999/2008). Historically, the terminology has evolved as the industry diversified from general tourism into specialties: wildlife tourism became nature tourism, and much of it now carries the label of ecotourism.

Views on ecotourism are highly polarized. At the positive end, it is seen as a sustainable alternative to mass tourism (Honey, 1999/2008; Pennisi *et al.*, 2004; but see Honey, 2002), as a means for people in peripheral areas to make an economic transition (Hall & Boyd, 2005), or as a way to communicate a message about conservation (Beaumont, 2001; Hill *et al.*, 2007; Orams, 1997). At the negative end, the focus is on its failings: environmental degradation, disruption of wildlife behavior, and insufficient income generation (Boo, 1991; Bramwell & Lane, 1993; Wheeler, 1993, 2005; see this book). Without some consensus on the definition of ecotourism (Higham, 2007), criteria are liable to range from those that are so strict as to disqualify most projects (Boyd & Butler, 1996) to broad criteria that include most nature tourism as ecotourism (Ballantine & Eagles, 1994).

In this chapter we define ecotourism as it has been formulated by The International Ecotourism Society (TIES). Originally it was "responsible travel to natural areas that conserves the environment and improves the well-being of local people" (The Ecotourism Society Newsletter 1, 1991 as quoted in Honey, 1999/2008, p. 6). Since then the definition has evolved to become "travel to fragile, pristine and usually protected areas that strives to be low impact and (usually) small scale. It helps educate the traveler; provides funds for conservation; directly benefits the economic development and political empowerment of local communities; and fosters respect for different cultures and for human rights" (Honey, 1999/2008, p. 25). Different countries may emphasize different aspects of ecotourism definitions; in some cases, for instance, the emphasis is on maximizing ecological sustainability (Ceballos-Lascurain, 2000; Seema *et al.*, 2006). But avoiding adverse environmental impacts does not guarantee that local communities will actually see economic benefits from tourism revenues. In Kenya, achieving benefits for the community has become a key component in ecotourism (Honey, 1999/2008; Koch *et al.*, 2002). This involves local participation in ownership of facilities and of products. While local participation is a noble ideal, it is often the most difficult to achieve (Honey, 1999/2008; Watkin, 2003).

Baboon ecotourism in context

Long-term field sites are exposed to the impacts of environmental and socio-economic changes over time. Local communities, resident wildlife, and those of us carrying out research must cope with these changes. In response, field workers, including primatologists, get involved in community development projects hoping that they will promote conservation as well.

For over three decades of research on baboons (*Papio anubis*), our study troops have lived in high altitude savanna in Kenya. Our Uaso Ngiro Baboon Project (begun in 1970 as the Gilgil Baboon Project) started working with communities in 1981, long before most primate research projects realized the need to involve local communities in conservation efforts (Community Based Conservation or CBC, Barrow & Murphree, 2001). Our approach has changed over the past 28 years as we gained experience. This chapter is not a scholarly review of ecotourism or even of community-based conservation. We offer, instead, a view "from the trenches" situated in the Kenya context to provide a framework for primate-focused tourism and conservation.

Why baboon ecotourism is different from other primate tourism ventures

Kenya: The Big Five

The earliest conservation practices revolved around maintaining food sources for local humans. These efforts were a response to human needs, and not an altruistic concern for wild fauna and flora (Western & Wright, 1994). In the eighteenth and nineteenth centuries, a "fortress conservation" model emerged based on a preservationist view of separating people from wildlife (Nash, 1967; Shetler, 2007). In colonies such as Kenya, the preservationist view, combined with the need to generate income, encouraged the removal of ownership of and access to wildlife and natural resources by local communities. In effect, traditional peoples were alienated from the lands they had occupied for centuries.

Tourism is a conservation tool, but hardly a new one. Tourism provided the main impetus for the creation of national parks in the 1800s (Honey, 1999; Watkin, 2003). There has been a close parallel between conservation and tourism development in Kenya. In the years since 1945, a number of major parks and reserves were created in the rangelands, those areas suitable for grazing wild and domestic herbivores. Consequently, Maasai pastoralists had direct contact with wildlife conservation activities and tourism on a large scale, but at the same time were alienated from their traditional lands and had their lifestyle demeaned (Southgate, 2006).

The abundance and variety of wildlife in these areas ensured that Kenya became a major tourist destination for those wishing to see (and not hunt) The Big Five: Lions, Elephants, Rhino, Buffalo, and Leopard. As Kenya's tourism industry grew, so did the country's population and the need to provide land and employment for its citizens. Habitat loss, the conversion of land to agriculture, and the spread of urban areas meant that these highly valuable tourism species were increasingly confined to protected areas and the remaining pastoral grazing lands. Mass tourism in National Parks and Reserves resulted in overcrowding and damage to park areas. The Big Five were Big Money Earners, but they were a menace to local people, threatening lives and livelihoods. Discerning tourists became interested in other forms of tourism in Kenya and walking, sports, cultural, and historical tourism emerged.

In Kenya, therefore, primate tourism grew out of the need to diversify from large charismatic species and the minibus experience. In a country where dramatic game species viewing is commonplace, the experience of walking through the bush in the company of wild (yet habituated) baboons is considered exotic.

Kenya was one of the first countries in Africa where community conservation initiatives were actively promoted (Ayoo, 2007; Barrow *et al.*, 2000; Murphree, 2000; Western, 1994). The strategy at the heart of CBC is that the people who suffer the costs of biodiversity conservation should share its benefits (Gibson, 1999; Hutton *et al.*, 2005; Western & Wright, 1994). CBC was pioneered in the early 1970s around Amboseli National Park (Honey, 1999/2008; Western, 1994). Kenya's 1975 Wildlife Policy proposed that wildlife should pay its way outside parks, a major premise of the Wildlife Act passed in 1977 (Western, 1994). The Amboseli program provided local Maasai pastoralists with opportunities for revenue from the park, hunting concessions (before the 1977 ban), and livestock husbandry adjacent to the park. A comprehensive and long-term ecological monitoring program was put in place that continues to this day. Amboseli has been the model for other CBC initiatives in Kenya and elsewhere.

CBC also tries to ensure that benefits from conservation activities are sustainable and distributed equitably and transparently (Barrow & Murphree, 2001). One aim of community involvement in Kenya was to reverse problems caused by state control over resources, especially in the establishment of protected areas (Ayoo, 2007; Barrow *et al.*, 2000; Murombedzi, 1998). Because wildlife was a key pillar of the tourism industry, and because of the presence of wildlife on community lands adjacent to national parks, community-based tourism enterprises were seen as a viable form of conservation (Ayoo, 2007; Honey, 1999/2008; Koch *et al.*, 2002; Sindiga, 1999). Although the last two decades have not seen the establishment of any major new national parks in Kenya, local communities and private ranchers in Kenya have recently and voluntarily set aside over 2000 km². This is significant because approximately two-thirds of all wildlife in Kenya is found outside parks (Manzolillo Nightingale & Western, 2006). The empirical data generated by ecological monitoring indicate that the community approach has succeeded; wildlife numbers remain high in Amboseli and other areas with community involvement but are declining elsewhere both inside and outside protected areas (Western & Manzolillo Nightingale, 2004; Western *et al.*, 2009).

Pastoralists and wildlife: helping people cope with environmental change

Historically, the Maasai have coexisted with wildlife in Kenya's rangelands. Rangelands cover three-quarters of Kenya's land surface and include almost all of its large herbivore population (Western *et al.*, 2009); it is no accident that they hold most of Kenya's National Parks. Livestock is a key resource for Maasai pastoralists in the rangelands and livestock biomass eclipses that of wild ungulates in most of these arid and semi-arid areas (Muchiru *et al.*, 2008). Maasai track seasonal rains and grass production which allows them to maximize herd productivity. They mirror the seasonal migrations of wildlife in the savannas (Western & Manzolillo Nightingale, 2004). The Maasai saw wildlife as "second cattle," relied on for food or used for barter at times when cattle herds had been decimated by drought and disease (Western, 1982; Western & Manzolillo Nightingale, 2004). Historical events, including land use change and subdivision, have confined both Maasai and wildlife to increasingly smaller areas, cut off from dry season resources and key drought refuges. No longer able to subsist on livestock, the Maasai have become increasingly vulnerable to droughts and other stressors that also negatively affect wildlife. Evidence suggests that maintaining space for pastoral livelihoods along with setting up community sanctuaries, conservancies, and wildlife-based ecotourism projects can stem the loss of wildlife and biodiversity in the rangelands (Waithaka, 2004; Western & Manzolillo Nightingale, 2004; Western et al., 2009).

The stage is now set

History of the Baboon Project

Our research on baboons in the rangelands began in 1970 on Kekopey Ranch near Gilgil in Kenya's Rift Valley (Harding, 1976; Strum, 1975). The main study group, the Pumphouse Gang (PHG), has been under continuous observation since 1970. Two to three other troops have also been simultaneously studied during this period. Kekopey Ranch, 45 000 acres (18 210 ha) of high-altitude savanna habitat with a large complement of wildlife, was originally part of the best grazing land that the Maasai relinquished during a 1904 Agreement with the British (Hughes, 2006). With rainfall of just under 700 mm a year, it is too dry for rain-fed agriculture but its excellent grazing made it a refuge for wildlife (Blankenship & Qvortrup, 1974).

A major change in land use, in 1979, catapulted the Gilgil Baboon Project into community work. The land was sold to a cooperative society and smallhold farms appeared in 1980 despite the area's unsuitability for agriculture. By 1984 less than 25% of the area was under cultivation. Nevertheless, conflict began when baboons started raiding the newly established farms. In response, we studied the development of crop-raiding and tested various means of control (Forthman, 1986; Strum, 1994b). In 1980, we also began working with the local people hoping to reduce the skyrocketing anthropogenic mortality of baboons. The newcomers were not actually a "community": they did not have traditional ties to this kind of landscape or to one another. Farmers from elsewhere migrated to Kekopey because of a growing shortage of land in their home areas – a scenario repeated elsewhere in Kenya and Africa.

As conflict between the farmers, baboons, and other wildlife escalated, the Baboon Project began to apply the new CBC model: providing employment (local Kenyans became research assistants), opportunities for alternative incomes (a Wool-craft project for spinning wool, dying it with local natural dyes, and making carpets for

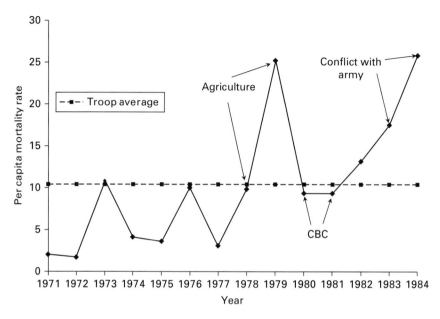

Figure 9.1   Per capita mortality rate from the start of research until translocation from Kekopey (1970–1984). The dotted line represents the average per capita mortality rate for the entire period. Key events are highlighted (change in land use, initiation of community-based conservation (CBC), and conflict with the army).

sale), and education (building the first local primary school). Education was a long-term strategy that should *eventually* provide options off the land, the ultimate reso-lution of conflict. In the meantime, the first two activities promoted a change in attitudes and actions: the farmers became less willing to kill the baboons.

The Baboon Project's long-term perspective provides insights about the primate–human interactions which are the context for ecotourism. Land use change set off a chain of behavioral responses by the baboons, the people, and the Baboon Project. Baboon mortality increased sharply after settlement (Figure 9.1; years 1978 and 1979). In 1981, the study troop split into two groups (PHG: Pumphouse and WBY: Wabaya) because of the opportunity to raid (Strum, 2010). The baboons showed great adaptability. Initially both raiders and nonraiders succumbed to humans. As the raiding lifestyle developed, raiders became more aware of the dangers and could avoid them while the nonraiders did not (Figure 9.2). We also started working with the community. This resulted in a decline in mortality rate (Figure 9.1, 1980–1982) and in human-related deaths (Figure 9.2). The baboons then started visiting a nearby army camp and mortality increased again (1984). The army was not part of the "community." Instead, the army gave an ultimatum: solve the baboon problem or they would be shot. This was the motivation for translocation.

In 1984, the Baboon Project translocated three of the study groups from Kekopey, two to the Eastern Laikipia Plateau 200 km away near the remote police outpost of Il Polei and one to the Western Laikipia (Strum, 1987/2001; Strum & Southwick, 1986). The main translocation site was 15 000 acres (6070 ha) of private land. Like

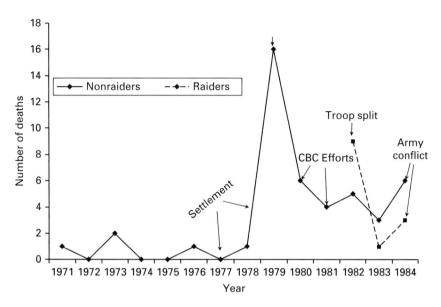

Figure 9.2 The number of human-related deaths during the period before translocation for the two main study groups, one who remained nonraiders and one who split off to raid. Key events are indicated.

Kekopey, it contained more wildlife than livestock. However, the release site bordered land communally owned by Maasai pastoralists.

By 1986, the translocated troops were ranging on Maasai communal land. The baboons' new home was also savanna, but dryer than Kekopey (mean annual rainfall 350–400 mm) and sparsely populated by pastoralists whose livestock included a mix of cattle, sheep, and goats. The lack of rain made it impossible to grow crops. The area was not subdivided into smallholdings, was not fenced, and was used only seasonally.

The Baboon Project (renamed the Uaso Ngiro Baboon Project, UNBP) started community work once again using the CBC model of employment and education. As one of the study troops slowly shifted its home range, our community involvement had to expand to new areas. Eventually, UNBP was employing ten Kenyan research assistants and supporting six budding primary schools in the area. For most of a decade, relations between the research project, the baboons, other wildlife, and the local communities were cordial. A conservation committee made up of local leaders and wildlife agency representatives handled the few incidents of baboon predation on small livestock.

Human movement patterns changed dramatically during and after the El Niño event of 1998. The surplus of forage that year disrupted livestock migration and contributed to a complex process of rapid sedentarization that led to increases in human–wildlife conflict as baboons increasingly preyed on young goats or sheep. In addition, ongoing land degradation of these areas included invasion by prickly pear

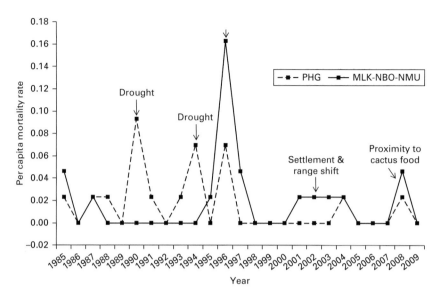

Figure 9.3   Per capita morality rate from anthropogenic sources after translocation (1985–2009). Mortality peaks are marked with associated contextual events.

(*Opuntia stricta*). Because baboons eat the opuntia fruits and disperse the seeds, the perception of some in the community was that baboons were responsible for the rapid invasion of the cactus.

Until that time, our community-based conservation efforts (both at Kekopey and at Il Polei) had been successful. On balance, local people felt that the benefits they received outweighed the costs of having baboons (and other wildlife) around. Since 1998, more frequent conflict has tipped this balance, specifically for those who had personally lost a goat or sheep. While the translocation solved the crop-raiding conflict, the long-term mortality data serve to illustrate how finely tuned community relations need to be to guarantee the future of just this one species in one location. The spikes in mortality after translocation are due to conflict with humans in non-agricultural contexts (Figure 9.3). Although the translocated raider baboon group (Malaika, MLK) was no longer "raiding," the two translocated troops (PHG and MLK) exploited opportunities differently; there is little synchrony between the groups in anthropomorphic mortality. The motivation for killing the baboons changed as the ecological, political, and demographic context changed (see Figure 9.3). Each time, UNBP has had to adopt new tactics to mitigate this problem. This is where baboon ecotourism entered the picture.

Why baboon walks?

Walpole and Thouless (2005) identify a common problem for the commercial viability of community-based ecotourism. Wildlife is at low density outside protected areas; in Kenya, it is also very wary of people because of past displacement by livestock or hunting by poachers (Waithaka, 2004). In an effort to deal with this situation, UNBP created "Walking with Baboons" for Kenya's first community "eco-lodge" at Il Ngwesi. Constructed in 1996 with donor funds, the lodge was owned and managed by the local community (Honey,1999/2008; Walpole & Thouless, 2005).

It took time for elephant, zebra, lion, and other wildlife to repopulate (and they did) but from the start there were many baboon troops. The lodge agreed that "Walking with Baboons" could be a unique attraction. Since 2004, UNBP has introduced "Walking with Baboons" to two other communities. The histories of these baboon walks illustrate how external factors like local politics and the vulnerability of tourism to national and international events can influence the success of community-based ecotourism.

Twala Cultural Manyatta

Quite remarkably, ecotourism at Il Polei was a bottom-up, not a top-down project. Four women's groups originally initiated ecotourism adjacent to UNBP's research site on the Eastern Laikipia Plateau. Il Polei is situated between the Mukogodo Forest, the Loldaiga Hills, and the Uaso Ngiro River. The topography is hilly arid savanna with dramatic granitic rock outcrops (kopjes) and a view of Mt. Kenya. The women built the Twala Cultural Manyatta ("manyatta" is the Maasai word for a ceremonial settlement) in 1997 as a place to sell beadwork to tourists but also as a site to preserve what they recognized as their own fast-disappearing traditions. Cultural "manyattas" have proliferated as part of the "cultural" aspect of ecotourism in Kenya. The most visited ones border the main national parks and reserves. Ideally, visitors see traditional ways of life, ceremonies, customs, and practices, and have the opportunity to buy "traditional" artifacts. Twala tourism faced even greater challenges than Il Ngwesi: wildlife numbers were low, except for baboons, and Twala was far from the usual tourist circuit. How then could they attract visitors?

We decided that baboon walks was one option. Twala baboon ecotourism got off to a rocky start but by 2007 was becoming popular with visitors staying in nearby lodges. Unfortunately, post-election violence in 2007/8 caused a temporary decline in tourism. Even at the height of tourism, baboon walks alone might not be enough to attract tourists to Twala. So UNBP developed two additional walks: "Walking with Maasai and their Cattle" and "Touring the Landscape." These proved to be as valuable in maintaining indigenous knowledge as in providing financial benefits via tourism.

Livestock pastoralism has been an important adaptation to the savanna environment. The Maasai way of life revolves around their cattle. "Walking with Cattle" lets visitors learn about Maasai traditional herding techniques and Maasai culture (Nightingale, 2001). This is a hands-on experience where tourists have the opportunity to try herding livestock. Equally important, by valuing indigenous knowledge we were helping to preserve it.

"Touring the Landscape" explores the factors that created today's environment: climate, topography, wildlife, and people. The landscape walk also compares human, livestock, and wildlife use of plants, including their medicinal value for people. Tourists appear to be very interested in this information. Because the local knowledge about the landscape is disappearing, telling the history to visitors keeps it alive for the community.

Each walk is limited to just a few people at a time (maximum of five). Having three walks makes it possible to accommodate larger tourist groups without increasing the number on any one walk. Another advantage is that visitors get both scientific and traditional perspectives on plants, birds, baboons, cattle, and ecology during different walks. There are strict rules about visitor and guide behavior which we carefully monitor and enforce. The guidelines for the walk stipulate a certain dress code for the safety of the tourists and to present a consistent image to the baboons so that they are less wary of the visitors. Visitors on the walk must wear a cap (which is provided as part of the price of the ticket). The cap has a logo and is beaded by women from the Twala Cultural Manyatta. This ensures that the women get some revenue from the baboon walk, as well as advertising our walk, as the visitors continue to wear the cap elsewhere. The guide, the clothing, and the hat are all "passports" for getting close to the baboons.

At about the same time (2006), UNBP introduced baboon walks to an area at the opposite end of Kenya which was developing community-based ecotourism. Ol Kiramatian Group Ranch covers 21 000 km^2 on the floor of the rift valley between Lake Magadi and the Nguruman Escarpment, near the border with Tanzania. Only 60% of the residents are Maasai. Crops are grown in some of the higher and wetter areas. The remaining area is used for grazing or conservation and tourism (Manzolillo Nightingale & Western, 2006). In 1991, the community started a tourism project which failed. In 2004/5 they went into partnership with a private investor to build an eco-lodge and set up a 10 000 acre (approx. 4050 ha) conservation area contiguous to a sanctuary on the neighboring ranch. Local politics and embezzlement dogged the eco-lodge. However, the situation resolved itself and baboon walks began in earnest in 2008. By August 2009 many groups of tourists had taken the baboon walk.

Translating ecotourism into conservation action: what criteria to use?

Ecotourism may have many advantages over regular tourism (Honey, 1999/2008) but whether ecotourism promotes conservation is controversial (Higham, 2007; Walpole & Thouless, 2005) because it is rarely objectively assessed and it is unclear what criteria should be used. Honey (1999/2008) and Koch *et al.* (2002) provide a "scorecard" in their definition of ecotourism: travel to natural destinations, minimal impact, enhances environmental awareness, provides direct financial benefits for conservation and local people (and their empowerment), respects local culture, and supports human rights and democratic movements. The challenge is how to operationalize this variety of success criteria.

At Il Ngwesi and Ol Kiramatian, wildlife returned after the community sanctuary was created (Western et al., 2009). An ecological monitoring system is in place at both Il Polei and Ol Kiramatian to measure the consequences of ecotourism on biodiversity conservation quantitatively. UNBP baboon research keeps track of the demography of study and tourist troops and general population dynamics. For example, we can demonstrate that when communities are not positively engaged, baboons die regardless of the source of the conflict, whether they are raiding crops in agricultural communities, killing small stock belonging to pastoralists, or eating invasive species such as prickly pear cactus and helping the plant to spread (Figure 9.3). The baboon walks also score well compared with reports of other primate tourism (see this volume). There is no direct contact between visitors and baboons and the guides are diligent in controlling the small number of no more than five visitors at a time. We are currently collecting data on activity budgets to compare the baboon groups visited to adjacent troops not visited by tourists.

The impact on conservation attitudes must be assessed both from the perspective of the visitor and that of the community living with primates. Our tourists fill out exit questionnaires and the information helps the guides to fine-tune the experience. So far, all the visitors who have been asked (more than 100) said they would recommend the walk to others. Some even suggested that while they were already in very small groups, they would be willing to pay more to have an exclusive, one-to-one experience with the guide and baboons. Embedding the baboon walks in an ecotourist package that includes walks focused on Maasai knowledge and livestock husbandry, as well as interpreting ecological history, should improve visitor understanding and support conservation efforts by the local community by enhancing their role in these activities. The walks help to link nature and culture in the minds of visitors, illustrating the value of nature in culture and the importance of culture in maintaining nature in pastoral areas of Kenya. Pastoralists guiding tourists in their own areas deploy their skills as first-rate naturalists. They know the local ecology because of their cattle and are able to give visitors a perspective on the environment that they will not get anywhere else. The visitor comes away with a more holistic view of the meaning of Maasai culture in the landscape in which there is also wildlife and some of the world's best scenery. However, visitor satisfaction does not necessarily translate into improved attitudes or behavior among the community toward sustainable tourism or ecosystem conservation, as Hill et al. (2007) demonstrate.

Our style of ecotourism concentrates on delivering benefits to the local community and influencing their attitudes toward conservation more than on influencing visitors' attitudes. Traditional knowledge, including local communities' conservation practices and sensitivities, is being lost at an alarming rate in many parts of the world. The Il Polei and Ol Kiramatian communities still maintain a strong grasp on their culture and are interested in keeping it alive. The walks give local people a market for their special skills in environmental knowledge, medicinal plants, cultural knowledge, and animal behavior. Using this knowledge helps them to preserve it. More importantly, our recognition of these skills gives local people increased pride and ownership in the product and contributes to the value of maintaining that knowledge.

In 2006, UNBP contributed baboon questions to a drought survey that the African Conservation Centre carried out in three different Maasai communities ($n = 156$). All three communities live with baboons. The first, around Amboseli National Park, has had baboon research in place since the late 1960s but without any involvement with the community. The second, at Ol Kirimatian, has a new integrated CBC initiative which recently includes "Baboon Walks." The third is the community around Il Polei, which is UNBP's partner in the ecotourism venture. The results suggest these three communities perceive baboons differently and act differently based on those perceptions. Baboons have conflict with the residents in all three locations but there was a significant difference between locations in the methods that local people used to deal with baboon conflicts ($n = 154$, $\chi^2 = 22.35$, $p < 0.000$). In all areas, nonlethal methods predominated. However, some lethal methods were used in Amboseli and Ol Kirimatian but not at Il Polei. There was also a difference between locations in the answer to the question, "Do baboons have value?" ($n = 154$, $\chi^2 = 46.48$, $p < 0.000$). Respondents at Il Polei felt that baboons had value more than those at the other two sites (Amboseli vs. Il Polei, $n = 98$, $\chi^2 = 20.29$, $p < 0.001$; Ol Kirimatian vs. Il Polei, $n = 94$, $\chi^2 = 37.88$, $p < 0.000$). The same pattern held when respondents were asked "Can you see a way of making baboons useful?" Fifty percent of those at Amboseli, 60% of those at Ol Kirimatian said no but the majority of respondents from Il Polei (66%) felt that baboons could be made useful ($n = 154$, $\chi^2 = 13.00$, $p < 0.011$). Respondents listed uses ranging from research to tourism.

Admittedly, this is a small sample and the baboon walks are relatively new but the data hint at positive influences of baboon walks on the attitudes of local people. There are a few other clues that can be gleaned from Il Polei. Recently the community developed a strategic plan for environmental conservation and management. This plan reveals a high level of community ecological awareness compared with most communities in the rangelands who do not have and do not want such a plan. In addition, a few households are building livestock ex-closures (areas fenced to keep livestock out) to help restore the grassland. Others are reviving traditional techniques for preventing baboons from having access to young livestock and chickens. Finally, the community is considering alternative ways of managing their livestock in the face of the current realities of livestock and human population size and reduced dispersal areas. Their goal is to lower the impact of grazing on their grasslands. This will benefit wildlife and biodiversity as well.

We think these forays into conservation are at least partly the result of our years of community work. Ecotourism is now embedded at Il Polei as one of many different community connections that can contribute to conservation sensitivity and momentum. Missing from most formal rating methods is an understanding of how motivation really works, especially how incentives are translated into motivation and finally into action, in this case conservation action. The focus of many primate ecotourism projects has been on options and opportunities, primarily financial incentives. Some also aim to incorporate education, hoping to increase awareness

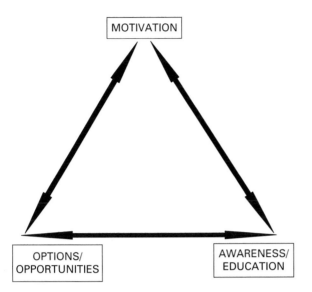

MOTIVATION

OPTIONS/
OPPORTUNITIES

AWARENESS/
EDUCATION

Figure 9.4   The minimum set of factors needed for conservation action.

of conservation issues. However, neither opportunities nor awareness necessarily lead to conservation action. What motivates local people? The history of UNBP community-based activities offers some insights, particularly about the need for a broad-based set of relationships with the community as motivation for possible conservation action (Figure 9.4).

Building relationships

To show how it works, let us examine the other UNBP community projects that began about the time of the baboon ecotourism walks. They are briefly reviewed in Textbox 1.

Textbox 1: Contributions to community motivation for conservation at Il Polei

Building the foundation for conservation motivation

1. Cactus cocktail syrup

An exotic cactus (*Opuntia megacanthus*) was introduced in colonial times and has recently been spreading. In partnership with the community, UNBP experimented with harvesting the fruit, extracting the juice, and using it to make a syrup. The syrup is now in market trials as part of Kenya cocktails. The syrup also makes a healthy "*chai*," possibly an inexpensive addition to traditional black tea for families in the area.

2. Beads for Conservation

Tourists admire and buy traditional Maasai beadwork. There is a potential international market but it is impossible for just a few women to produce commercial quantities. Africa Conservation Centre (ACC), with UNBP's help has formed "Beads for Conservation," connecting women's groups from different areas that have community wildlife sanctuaries and engage in conservation activities. Beads for Conservation may diversify women's income; it also places a new value on the fast-disappearing traditional skills involved in beading.

3. Cricket in the wild

The Maasai community enthusiastically embraced cricket when UNBP recently introduced it. Cricket Warriors give both boys and girls of many ages the opportunity to learn important physical and team skills.

4. The Il Polei "painting club"

The children at Il Polei Primary School are learning to see their world through their own watercolor paintings. The Kenyan school curriculum contains neither art nor environmental education. We use painting as a back door into conservation awareness and as a way to tell stories about local traditions.

5. The future of pastoralism in Kenya

UNBP has joined ACC in a research project examining the future of pastoralism in the dry rangelands of Kenya. The research puts a positive value on pastoralism but accepts its problematic nature.

6. Feedback to the community

Sharing research results with the local community is important. We discuss the results of ecological monitoring and of the surveys on drought strategies and on attitudes toward wildlife and environment. We also help to preserve traditional knowledge through photography, video, and written stories that are done jointly with the community.

7. Facilitating democracy and transparency

UNBP enlisted ACC as a new partner for the Twala Cultural Manyatta. This has already paid dividends for the women's group. The women had lost control of their Cultural Manyatta when the men took it over, hoping to monopolize future income. ACC mobilized the community to hand the Manyatta back to the women. The women have formed a new committee and are now earning money for themselves with the Opuntia syrup, catering, and beadwork.

The Opuntia cocktail syrup activity consolidated the women's group and heightened the women's environmental awareness. Beads for Conservation, although still in its infancy, gave women a glimpse of the lives of Maasai women elsewhere. Curtin and Western (2008) call this "over the horizon learning." It helped the women situate themselves and see beyond their own small world. Cricket team building was a unifying agent across community boundaries. It also brought UNBP into the community's social network and demonstrated the value of activities not based on financial or other direct benefits. The Il Polei painting club, like cricket, broke down old barriers and built up new connections. It also fostered environmental awareness in the children through art. The future of pastoralism research and activities is intricately tied up with natural resource management and the conservation of the area. Studying pastoralism removed us from an adversarial role about lifestyles and is winning additional support for conservation and sustainable development. Feedback lets the community see themselves in context and see the usefulness of both research and their own traditional knowledge. The community really appreciates these activities. Younger community members are particularly enthusiastic about regaining their own traditions. Facilitating democracy could only be done with the help of an experienced partner like the ACC, who helped to restore the cultural manyatta to the women, helped them write a constitution, and helped get them legal control of the surrounding land. Once the women took over, activities around the manyatta began to increase, including efforts to control land degradation and erosion. This points to the importance of democracy and transparency for successful conservation. Individually and together, these activities reconfigured the community conservation landscape, and redefined the community, its perceptions, and UNBP's place and effectiveness in it. Through the variety of engagements, bonds between UNBP and the community increased; this facilitated the development of conservation attitudes.

UNBP had specific conservation goals for the community work undertaken, including the walks: to preserve diversity at the baboon, wildlife, and habitat levels by changing attitudes, and reducing human/baboon conflict. Baboon ecotourism as we developed it in these communities may not be comparable to other primate-focused projects because of factors already discussed. But it has developed clear success criteria (Honey 1999/2008, 2002) and ways of monitoring key indicators, including the impacts of tourism on biodiversity and on people. UNBP's ecological monitoring and baboon research will provide empirical measures of whether UNBP succeeds in linking ecotourism to conservation.

There are concerns about the sustainability of ecotourism as a conservation tool because of its unreliability (Walpole & Thouless, 2005). Perhaps, ecotourism should be viewed as a way to diversify, not replace, basic livelihoods. It should also be seen as part of conflict mitigation (Walpole & Thouless, 2005). At both Twala and Ol Kiramatian, early failures of these ecotourism initiatives led to some discouraging results. Local people need to spread risk by maintaining other livelihoods or other kinds of employment activities.

Conclusions

Savanna baboons are the ultimate generalists. During the 38 years of research reported here, the study groups adapted to a variety of ecological and land use changes as well as translocation. The Baboon Project's changing set of neighbors included private property owners (1971–1979), smallhold agriculturalists (1980–1984), and livestock pastoralists (1984 to present). The challenge throughout these rapid land use changes has been to counteract the negative attitudes produced by human–baboon conflict. The first conflict, crop-raiding, began with a change in land ownership and the conversion of pasture to agriculture in the late 1970s. The second conflict began after translocation in 1984, when baboons began preying on young livestock during dry seasons and droughts. Most recently, local people see baboons as agents in the spread of a prickly pear cactus (*Opuntia stricta*). UNBP's community action is therefore very different today than it was in 1981. This long history has allowed us to build on past experiences; we are now more realistic, more savvy and, as a result more skeptical and perhaps more effective.

The lessons to learn from the UNBP experience go beyond being "primate-focused" because the underlying issues are not only about primates or about tourism. They are about biodiversity conservation in the modern context, about changing human attitudes, behaviors, economies, and societies. Central to the process is cross-cultural translation. In the end, the convergence of many factors simultaneously will determine the future of the primates we care about. Ecotourism is just one possible element.

Our argument, in this chapter, is that the potential of ecotourism for conservation of primates and habitats can be understood only within the framework of community-based conservation (Textbox 2). CBC initiatives have had mixed results, in part because the components and goals were not always clearly defined (Strum, 1994a). For example, what is meant by terms like community, participation, or conservation? The *notion* of "community" as applied to conservation work is frequently based on images from the nineteenth and early twentieth centuries – the community as a "unified, organic whole," territorially fixed, small and homogeneous (Agrawal & Gibson, 2004). The *reality* of "community" could be very different. Communities are often comprised of subgroups with different, conflicting interests or of mobile, transitional groups. The common purpose, or lack thereof, in any "community" will affect the outcome of conservation initiatives.

Textbox 2: Ecotourism in the context of community-based conservation

- **Education** is essential for conservation but not sufficient in itself. Increased awareness and new employment possibilities take time to develop. Most importantly, education does not translate into appropriate conservation action unless the community has options that allow its members to value biodiversity.

- **Local employment** can take many forms and provides much-needed income for members of the community. However, competition over the few valuable jobs can cause conflict. Employment choices need to be tailored to the socio-cultural context. This means knowing enough to make informed decisions. Local para-behaviorists, para-ecologists, resource assessors, or community development officers can also have a positive impact on community attitudes toward primates and biodiversity. Involving local people in research can reshape cultural myths.

- **Respect** often counts as much or more than money in community inter-actions. Mutual respect is the basis of mutual trust which is, in turn, the foundation of any successful community partnership. Beware: this is not as easy as it sounds.

- Communities need **multiple reasons to conserve**. UNBP uses cultural, social, economic, and personal incentives to motivate conservation action. Relying on only one type of value jeopardizes conservation success.

- Promote a **variety of interactions** between the researchers and the community. The goodwill created by our cricket team isn't directly related to conserva-tion but will ultimately provide the strong cross-cutting (local and research) social bonds needed in conservation action.

- The **translation of the community's own cultural experiences** into the ecotour-ism product promotes cross-cultural mutual respect. This community does not see ecotourism as just an economic activity but as a way to keep their culture and traditions alive. Interestingly, this view is being promoted by the youth.

- **Sustainable feedback and information sharing** makes us equal partners. Discussing research results and their potential application for environmental management allows the community to see itself as having some ownership of the projects. This ultimately shifted the community's view of baboons and natural resources in a more positive direction.

- **Encourage "beyond the horizon" learning** (Curtin and Western 2008) as a way for communities to see themselves in context. Horizontal learning effectively promotes new ideas and demonstrates issues

- **Develop success criteria** (Honey 2002) to evaluate the impact of community activities. It is important to decide whether you are interested just in com-munity development or in development for conservation and whether your goals are short term or long term.

- **Monitoring** is the only way to know whether you have met your success cri-teria. Impacts on biodiversity require ecological monitoring; impacts on people require socio-economic monitoring.

In addition to the usual problems with CBC, ecotourism has its own specific dangers. Some ethicists and development specialists view ecotourism critically, as the commercialization of nature (see, for example, Ehrenfeld, 2009). Buscher (2008) argues that neoliberalization, where all values are reduced "to the sphere of the market" (p. 229), has hijacked scientists and conservationists. He claims that few conservation and development projects work and those that do are never "one-dimensional." Both Ehrenfeld and Buscher doubt the ability of conservation approaches like ecotourism to ultimately conserve biodiversity.

At the practical rather than philosophical level, many costs of ecotourism come from inappropriate management of the ecotourism experience (see other chapters, this volume), problems that have solutions. The issue of benefits relative to costs is more complex. They include whether there will be reasonable earnings; where, how, and to whom should these be distributed; and whether ecotourism can become self-sustaining (Honey, 1999/2008; Walpole & Thouless, 2005).

We suggest that even with good revenues, ecotourism is unlikely to be a stand-alone conservation tool. This is because generating conservation motivation in people is complicated, as the Baboon Project's history has demonstrated. Activities seemingly tangential to conservation can play a vital role in providing the motivation to conserve. In our case, each activity contributed to strengthening bonds between the research project and members of the community. Original community divisions became porous and this played a major role in enabling change. Perceptions and actions began to be based on shared concerns for the welfare of the community, wildlife, and the environment. Mutual respect matured and developed into mutual trust. Trust stimulated openness of collaboration and a willingness and concern about conservation that were not present with only financial incentives. There were and will continue to be many small steps forward (and backward).

Unwittingly, UNBP's diverse links to the community circumvented the commercialization dilemma by following McCauley's (2006) suggestion: appealing more to people's hearts than to their wallets. This suggests that the real challenge for ecotourism as a conservation tool is how to build the type of community context in which old (or new) values are supported by enough options and enough human motivation to make a significant conservation impact. Without this social and cultural embedding it should be no surprise that ecotourism does not necessarily lead to conservation of primates or biodiversity.

Acknowledgments

Many individuals contributed to the project data used in this chapter and to the community work that spans more than 30 years. The research was sponsored by the Institute of Primate Research under the National Museums of Kenya and later by the African Conservation Centre and Kenya Wildlife Service. The work was funded most notably by grants from the University of California, San Diego, Academic Senate, 1980–1981; the Fyssen Foundation, 1981; the East African Wildlife Society, 1981–1982; the New York Zoological Society, 1981–1983; the World Wildlife Fund, 1981, 1983; the L. S. B. Leakey Foundation, 1981, 1983, 1988–1992; the H. F. Guggenheim Foundation, 1983–1984; National Geographic Society, 1984–1986; Wildlife Conservation Society 1987–2003; Conservation International 2001–2002;

African Conservation Fund, 2004–2006; EU Tourism Trust Fund, 2005–2007; and the Royal Dutch Embassy (DGIS) 2007–2010. We thank the members of the community who agreed to work with the Baboon Project, including farmers who settled at Kekopey and community partners at Il Polei. We particularly thank David Western for his pioneering approaches to community-based conservation and ecotourism, the UNBP team in the field and in the computer lab, UNBP student interns, and members of African Conservation Centre who have become the main partners in our current community work. Research adhered to the requirements of Kenya under Research Clearance NCST/RRI/12/1/BS001/67 and MOEST/OP/13/001/C1010 (Vol. 1, 2, 3), covering the period from 1972 to 2014. UCSD and funding agencies have not required compliance with institutional animal care committee regulations for my 40 years of research on wild olive baboons.

References

Agrawal, A. and Gibson, C. G. (2004). Enchantment and disenchantment: the role of community in natural resource conservation. In: S. Jones and G. Carswell (eds.) *The Earthscan Reader in Environment, Development and Rural Livelihoods*. London and Sterling, VA: Earthscan, pp. 151–179.

Ayoo, C. (2007). Community-based natural resource management in Kenya, *Management of Environmental Quality: An International Journal*, 18: 531–541.

Ballantine, J. L. and Eagles, P. F. (1994). Defining Canadian ecotourists. *Journal of Sustainable Tourism*, 2: 1–6.

Barrow, E., Gichohi, H., and Infield, M. (2000). *Rhetoric or Reality? A review of Community Conservation Policy and Practice in East Africa*, IIED Biodiversity Group. www.iied.org/bookshop/pubs/7807.html (accessed Sept. 16, 2010).

Barrow, E. and Murphree, M. (2001). Community conservation: from concept to practice. In: D. Hulme and M. Murphree (eds.), *African Wildlife and Livelihoods. The Promise and Performance of Community Conservation*. Oxford: James Currey, pp. 24–37.

Beaumont, N. (2001). Ecotourism and the conservation ethic: recruiting the uninitiated or preaching to the converted? *Journal of Sustainable Tourism*, 9: 317–341.

Blankenship, L. and Qvortrup, S. (1974). Resource management on a Kenya Ranch. *Journal of South African Wildlife Management Association*, 4: 185–190.

Boo, E. (1991). Making ecotourism sustainable: recommendations for planning, development and management. In: T. Whelan (ed.), *Nature Tourism: Managing for the Environment*. Washington, DC: Island Press, pp. 187–199.

Boyd, S. and Butler R. W. (1996). Managing ecotourism: an opportunity spectrum approach. *Tourism Management*, 17: 557–566.

Bramwell, B. and Lane, B. (1993). Interpretation and sustainable tourism: the potential and pitfalls. *Journal of Sustainable Tourism*, 1: 71–80.

Buscher, B. E. (2008). Conservation, neoliberalism and social science: a critical reflection on the SCB 2007 Annual Meeting in South Africa. *Conservation Biology*, 22: 229–231.

Ceballos-Lascurain, H. (2000). Interview by Ron Mader "Ecotourism Champion: a Conversation with Hector Ceballos-Lascurain." www.planeta.com/ecotravel/weaving/hectorceballos.html (accessed Sept. 16, 2010).

Curtin, C. and Western, D. (2008). Grasslands, people and conservation: Over-the-horizon learning: exchanges between African and American pastoralists. *Conservation Biology*, DOI: 10.1111.

Ehrenfeld, D. (2009). *Becoming Good Ancestors: How we can Balance Nature, Community and Technology*. New York: Oxford University Press.

Forthman, D. L. (1986). Controlling primate pests: the feasibility of conditioned taste aversion. In: D. M. Taub and F. A. King (eds.), *Current Perspectives in Primate Social Dynamics*. New York: Van Nostrand Reinhold, pp. 252–273.

Gibson, C. (1999). *Politicians and Poachers. The political economy of wildlife policy in Africa*. Cambridge University Press.

Hall, C. M. and. Boyd, S. W. (eds.) (2005) *Nature Based Tourism in Peripheral Areas: Development or disaster?* Clevedon: Channel View Publications.

Harding, R. S. O. (1976). Ranging patterns of a troop of baboons (*Papio anubis*) in Kenya. *Folia Primatologica*, 25: 143–185.

Higham, J. (ed.) (2007). *Critical Issues in Ecotourism: Understanding a complex tourism phenomenon*. Oxford: Butterworth-Heinemann.

Higham, J. and Carr, A. (2003). Defining ecotourism in New Zealand: differentiating between the defining parameters within a national/regional context. *Journal of Ecotourism*, 2: 17–32.

Hill, J., Woodland, W., and Gough, G. (2007). Can visitor satisfaction and knowledge about tropical rainforests be enhanced through biodiversity interpretation, and does this promote a positive attitude towards ecosystem conservation? *Journal of Ecotourism*, 6: 75–85.

Honey, M. (1999/2008). *Ecotourism and Sustainable Development: Who owns Paradise?* Washington, DC: Island Press.

Honey, M, (ed.). (2002). *Ecotourism and Certification: Setting Standards in Practice*. Washington, DC: Island Press.

Hughes, L. (2006). *Moving the Maasai: A Colonial Misadventure*. New York: Palgrave Macmillan.

Hutton, J., Adams, W. M., and Murumbedzi, J. C. (2005). Back to the barriers? Changing narratives in biodiversity conservation. *Forum for Development Studies*, No. 2.

Koch, E., Massyn, P. J., and Spenceley, A. (2002). Getting started: the experiences of South Africa and Kenya. In: M. Honey (ed.), *Ecotourism and Certification: Settings Standards in Practice*. Washington, DC: Island Press, pp. 232–263.

Manzolillo Nightingale, D. L., and Western, D. (2006). *The Future of the Open Rangelands. An exchange of ideas between East Africa and the American Southwest*. Nairobi: African Conservation Centre and African Centre for Technology Studies.

McCauley, D. J. (2006). Selling out on nature. *Nature*, 443: 1312–1314.

Muchiru, A. N., Western, D., and Reid, R. S. (2008). The role of abandoned pastoral settlements in the dynamics of African large herbivore communities. *Journal of Arid Environments*, 72: 940–952.

Murombedzi, J. C. (1998). *The Evolving Context of Community-based Natural Resource Management in sub-Saharan Africa in Historical Perspective*. Plenary Presentation, World Bank International CBNRM Workshop, Washington DC, May 10–14.

Murphree, M. W. (2000). *Community-based Conservation: Old Ways, New Myths and Enduring Challenges*. Paper presented to the Conference on African Wildlife Management in the New Millennium, College of African Wildlife Management, Mweka, Tanzania, Dec. 13–15.

Nash, R. (1967). *Wilderness and the American Mind*. New Haven, CT: Yale University Press.

Nightingale, D. L. M. (2001). Cattle, culture and environment: a new opportunity for eco-tourism? *Ecotourism Society of Kenya Newsletter*, 3: 2.

Orams, M. B. (1997). Effectiveness of environmental education: can we turn tourists into greenies? *Progress in Tourism and Hospitality Research*, 3: 295–306.

Pennisi, L. A., Holland, S. M., and Stein, T. V. (2004). Achieving bat conservation through tourism. *Journal of Ecotourism*, 3: 195–207.

Seema, P., JoJo, T. D., Freeda, M. S., Santosh, B., Sheetal, P., Gladwin, J., *et al.* (2006). *White Paper on Eco-tourism Policy*, Center for Conservation Governance and Policy, Ashoka Trust for Research in Ecology and the Environment (ATREE), BHC, New Delhi.

Shetler, J. B. (2007). *Imagining Serengeti. A History of Landscape Memory in Tanzania from Earliest Times to the Present*. Athens: Ohio University Press.

Sindiga, I. (1999). *Tourism and African Development: Change and Challenge of Tourism in Kenya* Research Series 14/1999. Leiden: African Studies Centre.

Southgate, C. R. J. (2006). Ecotourism in Kenya: the vulnerability of communities *Journal of Ecotourism*, 5: 80–96.

Strum, S. C. (1975). Primate predation: interim report on the development of a tradition in a troop of olive baboons. *Science*, 187: 755–757.

Strum, S. C. (1987/2001). *Almost Human*. University of Chicago Press.

Strum, S. C. (1994a). Prospects for management of primate pests. *Revue d'Ecologie (La Terre et la Vie)*, 49: 295–306.

Strum, S. C. (1994b). "Lessons learned". In: D. Western, R. M. Wright, and S. C. Strum (eds.), *Natural Connections: Perspectives in Community-based Conservation*. Washington, DC: Island Press, pp. 512–523.

Strum, S. C. (2010). The development of primate raiding: Implications for management and conservation. *International Journal of Primatology*. DOI: 10.1007/s10764–009–9387–5.

Strum, S. C. and Southwick, C. (1986). Translocation of primates. In: K. Benirschke (ed.), *Primates: The Road to Self-sustaining Populations*. New York: Springer-Verlag, pp. 949–958.

Waithaka, J. (2004). The role of community wildlife-based enterprises in reducing human vulnerability: the Il Ngwesi Ecotourism Project, Kenya. In *Africa Environment Outlook Case Studies on Human Vulnerability to Environmental Change*. African Ministerial Conference on the Environment (AMCEN) and United Nations Environment Programme (UNEP), Nairobi, Kenya, pp. 105–121.

Walpole, M. and Thouless, C. (2005). Increasing the value of wildlife through non-consumptive use? Deconstructing the myths of ecotourism and community-based tourism in the tropics. In: R. Woodroffe, S. Thirgood, and R. Rabinowitz (eds.), *People and Wildlife: Conflict or Coexistence?* London: Cambridge University Press, pp. 122–139.

Watkin, J. (2003). *The evolution of Ecotourism in East Africa: From an idea to an industry*. Summary of the Proceedings of the East African Regional Conference on Ecotourism, Nairobi, March 2002. IIED Wildlife and Development Series No. 15. London and Nairobi: IIED and ACC.

Western, D. (1982). The environment and ecology of pastoralists in arid savannahs. *Development and Change*, 13: 183–211.

Western, D. (1994). Ecosystem conservation and rural development: The case of Amboseli. In: D. Western, M. Wright, and S. C. Strum (eds.), *Natural Connections: Perspectives in community-based conservation*. Washington, DC: Island Press, pp. 15–52.

Western, D. and Manzolillo Nightingale, D. L. (2004). Environmental change and the vulnerability of pastoralists to drought. In *Africa Environment Outlook Case Studies on*

Human Vulnerability to Environmental Change. African Ministerial Conference on the Environment (AMCEN) and United Nations Environment Programme (UNEP), Nairobi, Kenya.

Western, D., Russell, S., and Cuthill, I. (2009). The status of wildlife in protected areas compared to non-protected areas of Kenya. *PLOS One*, 4(7): e6140.

Western, D. and Wright, M. (1994). The background to community-based conservation. In: D. Western, M. Wright, and S. C. Strum (eds.), *Natural Connections: Perspectives in community-based conservation*. Washington, DC: Island Press, pp. 1–12.

Wheeler, B. (1993). Sustaining the ego. *Journal of Sustainable Tourism*, 1: 121–129.

Wheeler, B. (2005). Ecotourism/egotourism and development. In C. M. Hall and S. W. Boyd (eds.), *Nature Based Tourism in Peripheral Areas: Development or disaster?* Clevedon: Channel View Publications, pp. 263–271.

10 Mountain gorilla tourism as a conservation tool: have we tipped the balance?

Michele L. Goldsmith

Introduction

The importance of gorilla conservation was recognized early, when the first National Park in Africa was established in 1925. This park, now known as the Virunga National Park, was created primarily to protect the mountain gorilla (*Gorilla beringei beringei*). Population estimates by George Schaller in 1960 averaged between 400 and 500 individuals (Schaller, 1963), but by 1978 only 252–285 remained (Weber & Vedder, 1983). Although the park did offer some protection, instability in the area along with poaching and habitat encroachment continued to threaten the gorilla population. Attempting to stem the decline, The Mountain Gorilla Project (now known as the International Gorilla Conservation Program – IGCP) was launched in 1979 with tourism as one of its main goals.

Gorilla tourism initiatives in areas of high threat started by converting poachers into rangers, to demonstrate that living gorillas were worth more than the price of a hand, foot, or head. For shifting this view of gorillas and getting governments on board, gorilla tourism is credited with saving this endangered species from possible extinction (Harcourt, 1986; McNeilage, 1996; Sholley, 1991; Stewart, 1991; Vedder & Weber, 1990; Weber, 1993; Williamson, 2001). Along with this success, however, came unforeseen costs and recently, some researchers have taken a more critical look at the short- and long-term consequences of gorilla tourism as a conservation tool (e.g. Butynski & Kalina, 1998; Frothmann *et al.*, 1996; Goldsmith, 2005; Homsy, 1999).

Tourism on habituated gorillas now involves three of the four subspecies (excluding the Cross River gorillas), and in each case it tries to help stem hunting and increase revenue. The most profitable of these programs are focused on mountain gorillas in the Volcanoes National Park, Rwanda and Bwindi Impenetrable National Park (BINP), Uganda. Gorilla tourism in Bwindi added $660 million US to Uganda's Gross Domestic Product and accounted for more than 80% of foreign revenues in 2010 (Lanyero, 2011).

Achieving conservation and revenue generation, however, is a delicate balance and some concern has been raised. Macfie and Williamson, in their 2010 IUCN

Primate Tourism: A Tool for Conservation?, ed. Anne E. Russon and Janette Wallis. Published by Cambridge University Press. © Cambridge University Press 2014.

Best Practice Guidelines for Great Ape Tourism and their chapter in this volume, emphasize the need to maintain this balance in favor of conservation. "If great ape tourism is not based on sound conservation principles right from the start, the odds are that economic objectives will take precedence, the consequences of which in all likelihood would be damaging to the well-being and eventual survival of the apes, and detrimental to the continued preservation of their habitat" (Macfie & Williamson, 2010, p. 1).

This chapter assesses whether this balance is being maintained by examining the present state of mountain gorilla tourism. After a brief historical account of gorilla tourism, I review costs and benefits along with efforts underway to reduce associated risks. Based on this review, I offer recommendations to improve the sustainability of gorilla tourism. Although many social, economic, and political issues are associated with tourism, the main focus of this chapter is its impacts on gorilla behavior and wellbeing.

History of gorilla tourism programs

Gorilla tourism has a long history, with the earliest attempts traceable to 1955 in Uganda's Mgahinga Game Reserve. Although tours were unregulated and visitors viewed only non-habituated groups, this first attempt demonstrated strong public interest and suggested gorilla tourism was both possible and could be financially profitable (Butynski & Kalina, 1998). The first attempt to habituate gorillas to human presence (i.e. reduce their fear of humans) for the purpose of tourism was in Kahuzi Biega National Park (KBNP), Democratic Republic of Congo (DRC), in 1966. Adrien Deschryver began habituating two groups of eastern lowland gorillas (*Gorilla beringei graueri*) and started taking tourists to view them in 1972 (Yamagiwa, 1997). These early tourism programs, however, suffered from a lack of organization, structure, and oversight. Large groups of tourists, sometimes as many as 40 people at once, were taken to view the gorillas, suggesting to some that the focus was on revenue rather than conservation (Fawcett *et al.*, 2004; Weber, 1993). The Zaire Gorilla Conservation Project began its formal tourism program on this subspecies in 1984 (Aveling & Aveling, 1989). By 1989, four tourist groups and one research group were being monitored daily (Yamagiwa, 2003).

Most agree that the first truly regulated gorilla tourism program began in 1979 with mountain gorillas in Volcanoes National Park (VNP), Rwanda. This park lies within the larger Virunga National Park, often called the Virunga Conservation Area. This area covers 7 800 km^2 and encompasses all three countries and national parks in which Virunga mountain gorillas are found. In the late 1970s, populations in VNP were highly threatened and declining. In 1978, with problems of poaching and plans to clear a large area of the mountain gorilla range for cattle ranching looming, tourism was proposed as a solution. In 1979, the Mountain Gorilla Project was established with the intention of starting a regulated tourism program (Vedder & Weber, 1990). The project habituated two mountain gorilla groups that were viewed by only a limited number of tourists for a limited amount of time. Tourism progressed rather quickly in this area, since infrastructure and research had already begun at the Karisoke Research Center, established in 1967 by Dian Fossey.

In 1978, prior to any real park protection, thousands of cattle and hundreds of humans roamed freely through the park destroying vegetation and introducing parasites and disease while gorillas continued to be hunted. By 1988, with a well-developed tourism program in place, no cattle were found in the park and gorilla poaching was almost non-existent (Harcourt, 2001). As the tourism program continued to grow, gorilla numbers began to rise and gorilla tourism became a major source of foreign currency for the government, second only to coffee and tea exports. The population continued to increase and in 2003 a census estimated 320 individuals (Gray *et al.*, 2006). By 2010, the Virunga Conservation Area population had increased 26.3% from 2003 estimates and had reached its largest size of 480 individuals (Gray *et al.*, 2011).

The Virunga Conservation Area extends into Uganda, but formal mountain gorilla tourism did not begin in that country until 1994. In 1997, the Ugandan sector became Mgahinga National Park and its one tourist gorilla group was marketed. However, this group moves between the two countries, making consistent tourist visits difficult. A second mountain gorilla population is found outside the Virunga region, approximately 20 km north, in Bwindi Impenetrable National Park, Uganda. This park, gazetted in 1991, started its gorilla tourism program in 1993 with two habituated groups (Mubare and Katendegere) in its Buhoma region. Bwindi's gorilla population was estimated at about 280–300 in the early 1990s and had grown to an estimated 340 individuals by 2006 (McNeilage *et al.*, 2006a). In a 2011 census, which employed different census techniques than earlier counts, the population was found to have grown to 400 individuals (IGCP, 2012a). However, caution should be taken when claims are made that tourism is responsible for this growth as it may simply be an artefact of methodology (i.e. earlier counts may have been underestimated).

The number of mountain gorilla groups habituated for tourism has grown steadily over the years throughout their range in the Virungas and in Bwindi. Figure 10.1 demonstrates this growth in Bwindi, from only two groups totaling 15 individuals in 1993 to nine groups totaling 152 individuals in 2011. Table 10.1 shows the most recent numbers of habituated groups and individuals in the Virunga region, by country, and in Bwindi. As many impacts of tourism can be attributed to the process and state of habituation (e.g. Blom *et al.*, 2004), those groups habituated for research are also included in this table. As of 2010, 73% of the entire Virunga mountain gorilla population has been habituated for either tourism (47%) or research (26%) (Gray *et al.*, 2011). Based on the latest population estimates from 2011, 42% of the entire Bwindi population is habituated for either tourism (38%) or research (4%). Therefore, of the 880 mountain gorillas estimated to survive in the world, 61% are accustomed to the presence of humans, 44% of which are potentially visited on a daily basis by tourists.

Table 10.1 Number of habituated vs. non-habituated mountain gorilla groups and individuals in the Virungas (2010) and Bwindi (2011)

	Habituated for tourism		Habituated for research		Not habituated		Total	
	# Groups	# Ind.	# Groups	# Ind.	# Groups	# Ind.	# Groups	# Ind.
Virunga Region[1]								
DRC	6	84	0	0	11	102	17	186
Rwanda	8	134	9	125	1	9	18	268
Uganda	1	9	0	0	0	1	1	10
Virunga Total	**15**	**227**	**9**	**125**	**12**	**112**	**36**	**464***
Virunga %	**42%**	**47%**	**25%**	**26%**	**33%**	**23%**		
Bwindi Total[2]	**9**	**152**	**1**	**16**	**26**	**232**	**36**	**400**
Bwindi %	**25%**	**38%**	**3%**	**4%**	**72%**	**58%**		
Overall Total	**24**	**379**	**10**	**141**	**38**	**344**	**72**	**864**
Overall %	**33%**	**44%**	**14%**	**16%**	**53%**	**40%**		

[1] Population numbers for the Virunga Conservation Area in DRC, Rwanda, and Uganda based on the 2010 census (Gray *et al.*, 2011) (*480 if 16 undetected infants are added).
[2] Population numbers for Bwindi Impenetrable National Park, Uganda based on the 2011 census (IGCP, 2012a).

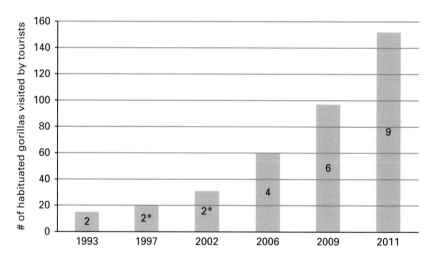

Figure 10.1   The number of gorillas visited by tourists has increased steadily through the years in Bwindi, Uganda.
- Numbers within bars represent the number of groups visited by tourists.
* In 1997 and 2002 three groups were habituated for tourism but only two were open for tourist visits. If added, the total number of gorillas habituated for tourism is 43 in1997 and 39 in 2002.
1997, 2002 & 2006, and 2011 are based on census studies by McNeilage *et al.* (2001), McNeilage *et al.* (2006b), McNeilage *et al.* (2006a), and IGCB (2012a), respectively.

Impacts related to tourism

Behavioral responses to the habituation process

There are numerous possible impacts on gorillas due to tourism, but they all begin with habituation, a process by which wild animals come to accept human observers as a neutral element in their environment (Blom *et al.*, 2004; Tutin & Fernandez, 1991). Like research gorilla groups, tourist gorilla groups need to be habituated to: (1) increase their members' visibility, especially in forest habitats; (2) enable identification of individuals and their relationships with each other; and (3) reduce any effect observers might have on "natural" behavior (Goldsmith, 2005). Some question if observers can ever truly be neutral elements in any animal's environment (Burghardt, 1992; Goldsmith, 2005). Also, it is not always clear when, if ever, the state of habituation is reached. That is, it is impossible to study a state of "neutrality" or to compare behavior without human presence because at least one observer must be present to assess it.

Some primate species, such as baboons, are relatively easy to habituate as they live in open areas and have a long history of interacting with humans (Strum, 1987). Habituating great apes can be more challenging. Early attempts to habituate gorillas involved food provisioning (Osborn, 1957 and Baumgartel, 1960 as cited in Schaller, 1963), but individuals refused the foods provided. Later, Schaller (1963) attempted to use Carpenter's (1965) method of concealment, but he was easily detected when trying to follow the group. Due to the gorillas' "keen eyesight and heightened curiosity they were much less excited when I presented myself out in the open than when I had attempted to hide" (Schaller, 1963, p. 22). With gorillas, signaling your presence is important so as to not surprise them; this is often accomplished with neutral vocalizations such as belching or tongue clicking, or leaf tearing (Williamson & Feistner, 2003).

Although habituation has enabled gorilla tourism to expand and prosper, little attention has been given to the costs these gorillas experience during the process. Great apes' responses to observer presence during the habituation process have been studied for chimpanzees (Johns, 1996), bonobos (van Krunkelsven *et al.*, 1999), and western lowland gorillas (*Gorilla gorilla gorilla*) (Blom *et al.*, 2004; Cipolletta, 2003; Doran-Sheehy *et al.*, 2007; Klailova *et al.*, 2010; Tutin & Fernandez, 1991). Unfortunately, no systematic data have been collected on mountain gorillas' responses to observer presence during the habituation process as it is often considered a means to an end for either research or tourism (Williamson & Feistner, 2003). Due to the lack of data on mountain gorillas, I turn to the few studies conducted on western lowland gorillas.

Western lowland gorillas show a variety of responses when investigators arrive, which appears to be related to their previous exposure to humans, the structure of the habitat, and species-typical behavior when faced with unfamiliar intruders into their environment (Blom *et al.*, 2004; Tutin & Fernandez, 1991). Findings across three different field sites are somewhat similar: gorillas generally flee when

humans first approach and then react with other fear behaviors such as vocalization and aggression (Blom *et al.*, 2004 for Bai Hokou; Doran-Sheehy *et al.*, 2007
for Mondika; Tutin & Fernandez, 1991 for Lope). In Bai Hokou and Lope, reactions differed between sexes, with males responding with aggression more often
than females (Cipolletta, 2003; Tutin & Fernandez, 1991). Doran-Sheehy *et al.*
(2007) found that both males and females in Mondika responded with aggression,
although female responses were delayed. Aggression from females intensified over
time, and adult females routinely attacked and bit humans. The discrepancy in
female behavior between sites appears related to the how long they were followed
(Doran-Sheehy *et al.*, 2007). A later study in Bai Hokou showed adult females
were the most likely age-sex class to direct aggression toward observers (Klailova
et al., 2010). Gorillas are considered nearing habituation when "ignore" becomes
their predominant first response to human presence and daily path lengths shorten
significantly (Cipolletta, 2003).

Concerning process and procedure, forest density can be a major factor (Doran-
Sheehy *et al.*, 2007). For example, gorillas ignored habituators more often in dense
understory than mixed closed forest, and were more tolerant of observers when
in the trees than on the ground (Blom *et al.*, 2004). Human-directed aggression
of a Bai Hokou silverback declined as distance from observers increased: Most
occurred within 2–3 m, reduced significantly by 6–10 m, and virtually disappeared
by 16–20 m (Klailova *et al.*, 2010).

Findings based on western lowland gorillas should be applied with caution to
mountain gorillas. Lowland habitat differs greatly from the more open vegetation
found in montane areas, which allows habituators visual contact while staying
much further from the group than required in denser forest. This might result in a
less stressful situation for mountain gorillas and might explain why they are habituated more quickly (sometimes in less than a year).

Changes in gorilla behavioral ecology and increased human–animal conflict

In addition to the costs induced by the habituation process, there are negative
impacts once gorillas are fully habituated. Literature on the long-term consequences
of habituation on mountain gorilla behavior and well-being is beginning to grow
(e.g. Butynski & Kalina, 1998; Goldsmith *et al.*, 2006; McNeilage, 1996; Wallis &
Lee, 1999) and is sure to increase as this topic becomes more controversial. What is
already obvious is that some changes in behavior that result from habituation can
have negative consequences.

Perhaps one of the most troubling results is that once mountain gorillas lose
their fear of humans, they start spending more time outside their protected areas
and in close contact with local people. Gorilla groups habituated for tourism often
live close to park edges, a decision made to reduce tourist impacts inside the park
by placing most of its infrastructure along the boundary. In Bwindi, this has had
dire consequences because the areas adjacent to the park maintain one of the highest densities of rural human habitation in Uganda. The last national census of the

two main districts surrounding the park, Kabale and Kisoro, found population densities of 281 and 324 persons/km², respectively (Uganda Bureau of Statistics, 2002). With growth rates of 3% in Kabale and 2.8% in Kisoro, population density in 2010 was estimated at 318 and 407 persons/km², respectively. In addition, agricultural lands abut the park boundary so if gorillas venture out of the park it is likely they will destroy crops and interact directly or indirectly with humans. This is especially worrisome because, unlike regulated tourist visits, interactions between local people and gorillas are uncontrolled and create problems regarding safety, health, crop-raiding, and strained relations between local people and park officials.

One reason the Nkuringo gorilla group in Bwindi, first habituated in 1999, was not opened to tourists until 2004 was because its members were spending too much time outside the park (Meder, 2000); they nested (75%) and traveled (55%) more often outside than within the park's boundary (Goldsmith *et al.*, 2006). When outside the park, its members raided crops planted in fields in the surrounding area and in less than 18 months, destroyed more than 842 sweet and plantain banana plants, 90 planted eucalyptus trees, and smaller amounts of sweet potato and sugar cane (Goldsmith *et al.*, 2006). Crop-raiding by mountain gorillas as well as other species in the park comes at a substantial cost to farmers. For example, some say their children need to drop out of school because they can no longer afford the fees or they need to stay at home to guard the fields (Tumusiime & Svarstad, 2011).

Mountain gorillas leaving the park to raid crops are not exclusive to the area around Bwindi; they come into contact with local people around the Virungas as well. Perhaps the best-known "problem" gorilla in the Virungas is the silverback, Mukunda. On many occasions he has had to be anesthetized and brought back to the forest. His wanderings into villagers' fields have increased conflict and created health risks. The problem with gorillas leaving the park has become so serious that in the DRC, officials have installed an electric fence near the park's boundary (Mburanumwe, 2010).

Gorilla behavior is influenced by the presence of tourists

Little direct research has been conducted on behavioral changes in mountain gorillas due to the presence of tourists, primarily due to tight restrictions on visiting and a policy of "no research" on tourist gorilla groups. This has changed recently, as managers and government wildlife officials came to recognize the need for this information. Fawcett *et al.* (2004) compared the behavior of three different groups in the Virungas for one-hour sessions before, during, and after tourist visits and found that all three groups spent significantly less time feeding and more time moving during tourist visits. In addition, in all three groups, certain aggressive behaviors, many directed at humans, increased in frequency during tourist visits and group members increased their proximity to the lead silverback male. Some behaviors observed were correlated with distance from the gorillas and number of tourists present: closer proximity and more tourists had a bigger impact. The authors suggest their behavioral findings indicate increased stress levels in gorillas during tourist visits and recommend that guides be better trained at maintaining the 7 m distance rule between tourists and gorillas.

Muyambi (2005) conducted a similar study in Bwindi and found that gorillas were more vigilant when tourists were present, but that this visual scanning decreased as tourist distance increased. He also found that during tourist visits, the gorillas traveled more, fed less, were more cohesive around the silverback, and displayed more self-directed behaviors such as self-grooming and scratching.

Habituation results in increased risk of disease transmission

Perhaps the greatest risk to habituated gorilla groups is disease transmission, as gorillas and humans are genetically and physiologically similar. Although tourists present disease risks, gorillas are more likely to contract disease from local people who share their habitat (Guerrera et al., 2003), park staff managing tourism (Nizeyi et al., 2002a), researchers studying them (Wallis & Lee, 1999; but see Rothman et al., 2002), military and human traffic (Dudley et al., 2001; Hamilton et al., 2000), cattle (Nizeyi et al., 2002b), and baboons who also spend time outside the park (Hope et al., 2004). However, any activities that encourage close contact with humans raise concern about interspecies transmission of infectious agents (e.g. Butynski & Kalina, 1998; Hastings et al., 1991; Köndgen et al., 2008; Palacios et al., 2011; Woodford et al., 2002); habituation for tourism is one such activity.

Noted changes in gorilla behavior during habituation and while being visited by tourists indicate they experience heightened levels of stress (Blom et al., 2004; Butynski & Kalina, 1998; Doran-Sheehy et al., 2007; Fawcett et al., 2004; Muyambi, 2005). Stress greatly impacts the health of gorillas (Hudson, 1992) and acts to weaken their immune system (Muehlenbein & Wallis this volume), leaving individuals even more vulnerable to disease. Increases in stress levels over time may inhibit reproductive function and cause infertility (Moberg 1985; Wasser et al., 1993) and can damage the hippocampus (Sapolsky et al., 1990).

Specific to mountain gorillas, studies have shown recent increases in endoparasitic loads and infections, many of which are shared with humans, in both the Virungas (Graczyk & Cranfield, 2001; Sleeman et al., 2000) and Bwindi (Ashford et al., 1996; Graczyk et al., 1999; Nizeyi et al., 1999). For example, a human-specific genotype of *Giardia* spp. (Graczyk et al., 2002a) and a single genotype of *Encephalitozoon intestinalis* (Graczyk et al., 2002b) were found to infect both gorillas and people sharing their habitat in Bwindi. Nizeyi et al. (2001) found widespread infections of *Campylobacter* spp. and *Salmonella* spp. in habituated Bwindi gorillas, with the prevalence of these pathogens doubling in just four years. These same researchers isolated a human *Shigella* spp. not previously found in mountain gorillas. Bwindi gorillas also suffer from Cryptosporidiosis, which is found in local cattle (Nizeyi et al., 2002b) and the park staff who are frequently in contact with gorillas (Nizeyi et al., 2002a).

Ectoparasitic conditions, such as scabies, have been a major problem in Bwindi gorillas. There have been multiple outbreaks over the years, one of which resulted in the death of an infant in a Bwindi tourist gorilla group in 1996 (Graczyk *et al.*, 2001; Kalema-Zikusoka *et al.*, 2002; Macfie, 1996; Meder, 2000). Scabies is prevalent in the human population surrounding the park and habituated and non-habituated gorilla groups spending time outside the park have been infected (Mudakikwa, 2001).

Perhaps the greatest threats introduced by tourist visits are the transmission of more serious diseases such as measles and respiratory infections. Both are significant with regard to mountain gorilla mortality (Hastings *et al.*, 1991: Macfie, 1991; Sholley & Hastings, 1989). Infectious diseases, especially respiratory diseases, account for 20% of sudden deaths in mountain gorillas – a rate second only to trauma (Mudakikwa *et al.*, 2001). Measles killed 6 and sickened 27 mountain gorillas in a 1988 outbreak in Rwanda. During the outbreak, the Mountain Gorilla Veterinary Project conducted the only vaccination campaign ever attempted on wild gorillas, eventually dart-vaccinating about 65 gorillas (Hastings *et al.*, 1991). The most recent respiratory outbreak, in 2009, was caused by the human metapneumovirus (HMPV) and resulted in the death of two Rwandan mountain gorillas (Palacios *et al.*, 2011). These studies demonstrate that intensified human–gorilla interactions as a result of habituation and tourism have enhanced the transmission of anthropozoonotic pathogens and threaten the health of these endangered populations.

Increased susceptibility to poachers

Tourist gorilla groups are monitored daily and therefore have greater protection from poachers than non-monitored groups. However, when daily monitoring is stopped, typically due to human conflict and unrest, habituated gorillas become easy targets for poachers. During intense civil conflict in DRC during the mid-1990s, most of the gorillas killed by poachers were habituated, including the silverbacks from each of the four tourist groups (Pole Pole Foundation, 2012). Mushamuka, the longest known habituated silverback in KBNP, was found killed in 1997 (Ilambu, 1998). Researchers from this area suggest it was due to their greater vulnerability (Kasereka *et al.*, 2006; Yamagiwa, 2003). A study conducted on the DRC population, based on interviews with ex-poachers, suggests that when habituated and non-habituated gorillas are exposed to similar poaching pressures, habituated gorillas are the more likely to be killed. The estimated ratio of habituated gorillas killed was as high as 71%, as compared with 42% for non-habituated gorillas. Habituated gorillas were 1.6 times more susceptible to poaching than their non-habituated counterparts (Kasereka *et al.*, 2006).

Poaching events in other areas have been rare. In Bwindi in 1995, four adults from the habituated research group, Kyaguriro, were found killed. It is thought poachers were after an infant because one of the females killed was lactating and no baby was recovered. In 2002, two habituated Virunga female mountain gorillas were found slain and an infant was missing. The infant was later found and confiscated and a few of the poachers were caught during a search by the Uganda Wildlife Authority (UWA) and Uganda People's Defense Force. The most recent killing occurred on June 17, 2011 when the sole blackback gorilla (Mizano) from the Habinyanja tourist group in Bwindi was killed. Reportedly, poachers had been hunting duiker, came upon the gorilla, and speared him to death. IGCP officials asked: "What is a gorilla worth?" Due to lack of evidence presented in court, none of the men were found guilty of illegal poaching. Instead, one was fined $37 US for entering a protected area without permission and possessing illegal devices capable of killing wildlife and the other two were fined $18 US for trying to escape arrest (IGCP, 2012b).

Impacts on group size, social system, and the spread of habituation

Another possible impact of habituation is the artificial inflation of social group sizes. The average group size for the Virunga gorilla population averages 9.2 individuals (Sholley, 1991). However, the most recent census in the Virungas (Gray *et al.*, 2011) found that habituated gorilla groups averaged 14.5 individuals, while non-habituated groups averaged only 8.4 individuals. One of the longest habituated research groups in the Virungas, Pablo, has 47 members. The average growth rate of habituated groups was also notably higher (4.6%) than that of non-habituated groups (0.9%). These differences are most likely due to daily monitoring and medical intervention with habituated groups (McNeilage, 1996; Robbins *et al.*, 2011). Either group members are choosing to stay in their natal group and/or habituated groups are more successful at obtaining females (usually through intergroup conflict). Either way, it seems clear that the added protection afforded to habituated groups has influenced their natural social system.

Finally, a potentially problematic consequence is the spread of habituation into the "wild" non-habituated gorilla population. In gorillas, both males and females emigrate from their natal group. As the number of habituated gorillas continues to increase, so does the likelihood that non-habituated gorillas will find themselves in habituated groups and vice versa. The 2010 Virunga census found 36 social groups, the largest number to date (Gray *et al.*, 2011). This increase is due, in part, to the addition of three new social units created when three different habituated adult males left their groups, acquired females, and formed their own social units. During the same census, genotyping documented the emigration of two adult females from their well-habituated study groups into non-habituated groups. The spread of habituation into non-habituated groups can possibly lower these groups' overall fear of humans, making them more vulnerable to poachers and increasing the chance that they will start crop-raiding.

Efforts to improve sustainability of gorilla tourism

Although tourists are not guaranteed to view gorillas during their trek, being able to find, approach, and watch gorilla groups without them fleeing is a necessary prerequisite for any successful tourism enterprise. The fact that the process of habituation can be stressful and dangerous, however, clearly causes major concerns for conservation. For this reason, it is very important to employ habituation methods that minimize stress and keep fear to a minimum. To achieve this, habituators need to maintain a good distance from the gorillas, keep their numbers small, stop following the gorillas after they flee, and vocally notify them during approach. While researching the habituation process, Blom (2001) and Doran-Sheehy *et al.* (2007) identified a number of factors that influenced the western lowland gorillas' first response after being contacted; therefore, it is important to consider how, when, and where to best approach gorillas to minimize the negative effects of the habituation process. Williamson and Feistner (2003) recommend proper processes and techniques for habituating primates, with special reference to the great apes.

The main challenge after gorillas are habituated for tourism is keeping them inside parks and out of peoples' fields (Goldsmith *et al.*, 2006). In a survey of all the animals venturing out of the park in Bwindi, gorillas were third only to baboons and bushpigs in frequency (Olupot *et al.*, 2009). In 2001–2002, the habituated Nkuringo gorilla group slept inside the park only 23% of the time (Goldsmith *et al.*, 2006) and by 2008–2009 spent only 10% of their time inside the park (Kalpers *et al.*, 2010).

One attempt to keep gorillas from leaving the park is the creation of buffer zones. In 2005, UWA created a 350 m buffer zone along the range of the Nkuringo group. Unfortunately, crops remained in this zone which resulted in gorillas continuing to use the buffer zone in the absence of efforts to shoo them away (i.e. clanging bells and chasing). Once crops in this buffer zone were gone, the gorillas once again began to enter farmers' fields. Management of the buffer zone has been challenging and many proposed solutions to keep the gorillas behind the buffer zone have ended in failure. As of 2010, a small, managed *Mauritius* hedge represents the only successful effort, and IGCP recognizes the urgency to find a proper solution (Kalpers *et al.*, 2010). The use of electric fencing along Park boundaries to keep gorillas from entering fields has been used in DRC and may be piloted soon in Bwindi.

Studies on disease risks to great ape groups used for tourism emphasize the need for more precautions (Homsy, 1999; Wallis & Lee 1999; Woodford *et al.*, 2002). Disease transmission could have devastating consequences for this endangered species (Robbins & Williamson, 2008) and would also aggravate biodiversity loss and human health risks (Daszak *et al.*, 2000). Few, if any, regulations exist regarding health requirements for the local people, rangers, researchers, and tourists that visit mountain gorilla tourist groups. Dr. Michael Cranfield of the Mountain Gorilla Veterinary Program once said, "It is easier for me to go up within 5 m of one of the rarest animals in the world with no health restrictions than it is for me to travel to Cincinnati or UC Davis and go behind the scenes in their primate house" (Ferber, 2000, p. 1278). In response to these concerns, Dr. Janette Wallis and others helped develop regulations to protect the health of wild primates, which were formally approved as a policy statement by the American Society of Primatology on July 11, 2000 (ASP, 2000). A slightly modified version of this statement was also accepted by the International Primatological Society (IPS, 2012).

Table 10.2 Current rules to control mountain gorilla tourism in Rwanda and Bwindi (adapted from Litchfield, 1997)

A minimum distance of 7 m needs to be maintained between tourists and gorillas*

A maximum of 8 tourists are allowed to visit a group*

Tourist gorilla groups are visited only once a day for a maximum of one hour

Tourists who are visibly unwell or declare themselves to be ill cannot visit the gorillas

There is a minimum age of 15 years for tourists (there is no maximum age)

No flash photography allowed

Tourists must remain together in a tight group

During visits no loud noises, pointing, or other sudden or disturbing behaviors are allowed

Eating, drinking, and smoking are not permitted within 200 m of the gorillas

Tourists need to turn away and cover their mouth when coughing or sneezing

Tourists need to bury their feces in a hole at a minimum depth of 30 cm

No trash is to be deposited in the park

Tourists are not allowed to clear vegetation to get a better view

* Originally, distance was 5 m and maximum number of tourists allowed was 6 (Homsy, 1999).

Some risks to gorillas from tourism can be controlled by minimizing contact between gorillas and visiting tourists. For example, current rules for mountain gorilla tourism in Rwanda and Uganda require that only eight tourists visit a group at a time, they can only stay for one hour, and they must remain at least 7 m from the closest gorilla (see Table 10.2 for a complete list of regulations). Even these rules may not be strict enough (Homsy, 1999; Muyambi, 2005). More importantly, and unfortunately often, these rules are not always followed in the Virungas (Fawcett et al., 2004; McNeilage, 1996; Stewart, 1992) or Bwindi (Macfie, 1997; Schmitt, 1997).

Sandbrook and Semple (2006) interviewed tourists from 133 independent tours and found the average closest distance to the gorillas was only 2.76 m, and that for longer periods of time during the visit (greater than 15 minutes) tourists averaged 4.85 m from the gorillas. Although both humans and gorillas are guilty of breaking the 7 m distance rule, closer proximity lasted longer when it was initiated by the tourist. When gorillas do approach, it is most often juveniles who are the most curious age group and may have no history of fearing humans. This, along with their poorly developed immune system, puts them at the greatest risk of contracting diseases.

Close proximity between humans and gorillas may be tough to avoid due to dense foliage and steep topography, which makes retreat difficult when gorillas approach. Retreat is also challenging when the gorilla group being trekked is dispersed over a wide area and the tourists find themselves surrounded. Sandbrook and Semple (2006) suggest that both reasons for close proximity were exacerbated by increasing the visitor limit from the original six to eight people per visit. Graczyk et al. (2001) suggest close proximity and physical contact occur due to "over-habituation" to human presence. This phenomenon is most evident in a recent YouTube video posted December 2011, in which one of the larger tourist gorilla groups, Rushegura, is seen visiting Buhoma Community Rest Camp in Bwindi. A tourist takes video of a man as he is continuously touched and even groomed by several juvenile gorillas, with the silverback sitting only 1 m behind him (Touched by a Wild Gorilla, 2011). Unfortunately, it is almost impossible to regulate or control situations such as this, except by avoiding over-habituation in the first place (Goldsmith, 2005). Another unfortunate consequence is that on December 23, 2011, Good Morning America showed this video with the comment that this was the experience of a lifetime. These portrayals, along with flashy safari brochures and photos highlighting close encounters between people and gorillas, certainly mislead people about acceptable visitor behavior and put the gorillas at even greater risk.

An additional measure that can help reduce disease risk is surveying people to better understand what diseases they may be carrying during their visits. For example, tourists and others potentially in contact with chimpanzees used for tourism in Kibale, Uganda, showed a high prevalence of symptoms of infectious diseases and respiratory ailments and a lack of many current vaccinations. To control the spread of disease and illness, use of respiratory masks is recommended (Macfie & Williamson, 2010) because most pathogens are airborne (Cranfield, 2008). In fact, aerosol pathogen transmission was the original reason for increasing the minimum tourist–gorilla distance from 5 to 7 m (Homsy, 1999). Masks can be an effective barrier to exhaled pathogens but, in addition to other challenges, they are uncomfortable, can be expensive, can fog up glasses, and create waste (Macfie & Williamson, 2010). Since 2009, all tourists and guides visiting mountain gorillas at Mikeno in DRC are required to wear respiratory masks and these masks are beginning to be used more frequently in other regions in the Virungas. However, to date, their use is not required in either Rwanda or Uganda.

Discussion

Tourism was a necessary and successful means of saving the Virunga mountain gorillas from the devastating impacts of encroachment and poaching. The question, now, is whether the success of the earlier tourism programs has been used to justify the continued habituation of more gorillas to generate more and more revenue instead of to protect the gorillas. Have we tipped the balance from salvation to greed? Mountain gorilla tours are often claimed to be a form of ecotourism. Ecotourism requires that travel contribute to the conservation of animals and their habitat and to improving the well-being of local people (TIES, 2005). In addition, it should cause no damage to natural and cultural heritage, include education, and have low visitor impact (Boo, 1990; Ceballos-Lascuráin, 1993). Macfie and Williamson (2010), who address the distinctions between tourism and ecotourism and similarly question, strongly, the use of the term ecotourism as it relates to great apes. Given the evidence presented above, it is clear that both habituation and tourist visits puts the gorillas in harm's way. Macfie and Williamson (2010) suggest that in the best cases, "sustainable tourism" may be a more appropriate term.

Is gorilla tourism sustainable? Dr. Tom Butynski, who left Uganda shortly before tourism began, was highly critical of gorilla tourism as a conservation tool (Butynski & Kalina, 1998). Early tourism in Bwindi, with just three tourism gorilla groups, helped to raise funds, build schools, and the gorillas were protected. At this level it appeared to be a sustainable practice (Goldsmith, 2000). However, as of 2011, nine groups have been habituated for tourism in Bwindi, and 61% of the entire endangered mountain gorilla population is now habituated for either research (17%) or tourism (44%). We know habituated gorillas are more susceptible to disease from a variety of human sources (including livestock) and have become ill or died as a result. The fear remains that one deadly, highly infectious disease will be introduced, travel quickly through the small isolated populations, and leave few survivors. For example, over 5000 western lowland gorillas in the Lossi Sanctuary in Republic of Congo, 90–95% of that area's total population, have died as a result of Ebola (Bermejo *et al.*, 2006). What is most important at this point is not habituating more groups but managing the already habituated groups better. Responsible tourism (Litchfield, 2007) and prioritizing the well-being of the gorillas need to be the main goals.

Presently the cost of a mountain gorilla tracking permit in Uganda is $500 per person, and raising it to $750 US during peak seasons has been discussed. In Bwindi alone, at full capacity, that could mean $36 000 US a day (8 tourists to each of the 9 groups for a total of 72 permits a day). With local communities receiving only a small percentage of park entry fees and no share of the permits, millions of dollars are going to the UWA. Much of this revenue helps keep the rest of Uganda's parks open, which is admirable, but it also implies that the gorillas themselves have become a commodity. While recognizing that mountain gorilla tourism helped save dwindling populations 30 years ago and generates some current benefits, it seems we have reached a point where gorilla tourism has become exploitive and may no longer be the "ape's savior."

If we take an ethical approach, it is clear that preservation, the protection of other species without human interference, should always be preferred to conservation, the active management of other species by humans. In environmental ethics, these two terms are often associated with biocentric and anthropocentric views, respectively. As stewards, we need to decide whether our interests are anthropocentric (instrumental), that is we value the gorillas for what they can do for us, or biocentric (intrinsic), that the gorillas' value is based on their own worth not related to our needs (e.g. Regan, 2004). Tourism is an anthropocentric approach to gorilla conservation that sacrifices habituated gorilla groups in the name of protecting the species but more clearly aims to serve humans. We have now tipped the balance away from protection, and it is time for humans to adopt a biocentric ethic.

Recommendations

Place an immediate moratorium on habituating new Virunga and Bwindi mountain gorilla groups for tourism

The most critical recommendation is to put a stop to habituating more mountain gorilla groups for both research and tourism. This chapter demonstrates the alarming increase in habituated versus non-habituated groups and presents the many associated risks. Gray *et al.*, (2011) state, "given that such a large proportion of the [Virunga] population is habituated, it would be difficult to justify habituation of additional groups because of the associated negative risks involved" (p. 3). In addition, based on findings in mountain gorillas, researchers and conservationists trying to save Nigeria's Cross River gorillas have been quick to recognize the risks and do not support tourism programs based on the gorillas (UNEP, 2009).

Although tourism may have saved the mountain gorilla population from rapid depletion 20–30 years ago, is giving them monetary value still the only way to protect them? With the majority of this small population of mountain gorillas habituated for our use, how is the tourism experience any different from Disney's Wild Animal Kingdom?

Every attempt should be made to regulate the distance between humans and gorillas and surgical masks should be worn by all visitors

Many tourist gorilla groups are over-habituated, which brings them into very close contact with humans (sometimes touching). This greatly increases the risk of disease transmission. The distance between tourist and gorilla should be better regulated and increasing the minimum distance should be considered. Hanes (2012), following Sandbrook and Semple (2006), found that distance is still not enforced; 25 interviewed tourists reported getting even closer (2.2 vs. 2.8 m), and 5 had physical contact with a gorilla. Only the human part of this equation can be controlled, by regulating visitors. The other consideration must be keeping gorillas from approaching humans and that is partly a function of over-habituation. Over-habituation of gorillas is a problem, especially among juveniles and infants. We must devise and test different methods that will communicate to gorillas that approaching humans is not acceptable behavior. In addition to reducing over-habituation, surgical mask use should be made mandatory for all visitors trekking to view the gorillas because of the risks of close contact involved.

The number of tourists visiting groups should be restored to six

The maximum number of tourists visiting a gorilla group should be restored to six (or fewer) for three main reasons: (1) this might help reduce over-habituation and the potential for disease transmission; (2) guides can better control smaller tourist groups and can intervene more promptly when gorillas approach or when rules are broken; and (3) eight tourists actually means up to twenty to twenty-five people at or near the gorilla group (eight tourists plus two guides, two to three rangers/soldiers, and sometimes up to two porters per tourist).

The goal of gorilla conservation should be their long-term well-being and protection without introducing harm

Lifetime commitments need to be made to habituated gorilla groups in the form of infrastructure that will withstand the pressures of political, social, and economic change. If conservation is about saving gorillas then less focus should be put on tourism and more should go into non-invasive measures such as building trust funds to pay park guards and support the development of local people. The Bwindi-Mgahinga Conservation Trust, set up with an endowment in 1994, has the mission of conserving biodiversity in both parks in harmony with the developmental needs of the surrounding human community, and has successfully built schools and hospitals greatly improving the health and well-being of the local community. Given that there are alternatives to tourism, our focus should be on protecting gorillas for their own right, not for ours.

Acknowledgments

I thank both Anne E. Russon and Janette Wallis for their invitation to contribute to this timely book on primate tourism, and for providing a forum to write on a topic I am so passionate about. Their assistance with this manuscript went above and beyond what is expected of any editor and for that I am grateful. I am privileged to have worked in Bwindi Impenetrable National Park for over 17 years and am thankful to the Ugandan Wildlife Authority and the Ugandan National Council of Science and Technology for their continued support and permission. Most importantly, I thank the people of Nkuringo town who continue to welcome me with open arms and always make me feel I'm home.

References

American Society of Primatology (ASP). (2000). *Protecting the Health of Wild Primates.* https://www.asp.org/society/resolutions/primate_health.cfm. (accessed Apr. 10, 2011).

Aveling, C. and Aveling, R. (1989). Gorilla conservation in Zaire. *Oryx*, 23: 64–70.

Ashford, R. W., Lawson, H., Butynski, T. M., and Reid, G. D. F. (1996). Patterns of intestinal parasitism in the mountain gorilla *Gorilla gorilla* in the Bwindi-Impenetrable Forest, Uganda. *Journal of Zoological Society, London*, 239: 507–514.

Bermejo, M., Rodriguez-Teijeiro, J. D., Illera, G., *et al.* (2006). Ebola outbreak killed 5000 gorillas. *Science*, 314 (5805): 1564.

Blom, A. (2001). *Ecological and Economical Impacts of Gorilla-Based Tourism in Dzanga-Sangha, Central African Republic*. Wageningen University, Netherlands: Unpublished doctoral thesis.

Blom, A., Cipolletta, C., Brunsting, A. M., and Prins, H. H. (2004). Behavioral responses of gorillas to habituation in the Dzanga-Ndoki National Park, Central African Republic. *International Journal of Primatology*, 25: 179–196.

Boo, E. (1990). *Ecotourism: The Potentials and Pitfalls*. Washington, DC: World Wildlife Fund.

Burghardt, G. (1992). Human-bear bonding in research on black bear behavior. In H. Davis and D. Balfour (eds.), *The Inevitable Bond: Examining Scientist-Animal Interactions*. Cambridge University Press, pp. 365–382.

Butynski, T. M. and Kalina, J. (1998). Gorilla tourism: A critical look. In E. J. Milner-Gulland and R. Mace (eds.), *Conservation of Biological Resources*. Oxford: Blackwell Press, pp. 294–313.

Ceballos-Lascuráin, H. (1993). *The IUCN Ecotourism Consultancy Programme*. México.

Carpenter, C. R. (1965). The howlers of Barro Colorado Island. In I. DeVore (ed.) *Primate Behavior: Field Studies of Monkeys and Apes*. New York: Holt, Rinehart and Winston, pp. 250–291.

Cipolletta, C. (2003). Ranging patterns of a western gorilla group during habituation to humans in the Dzanga-Ndoki National Park, Central African Republic. *International Journal of Primatology*, 24: 1207–1226.

Cranfield, M. (2008). Mountain gorilla research: the risk of disease transmission relative to the benefit from the perspective of ecosystem health. *American Journal of Primatology*, 70: 751–754.

Daszak, P., Cunningham, A. A., and Hyatt, A. D. (2000). Wildlife ecology – emerging infectious diseases of wildlife – threats to biodiversity and human health. *Science*, 287: 443–449.

Doran-Sheehy, D. M., Derby, A. M., Greer, D., and Mongo, P. (2007). Habituation of western gorillas: The process and factors that influence it. *American Journal of Primatology*, 69: 1–16.

Dudley, J. P., Ginsberg, J. R., Plumptre, A. J., *et al.* (2001). Effects of war and civil strife on wildlife and wildlife habitats. *Conservation Biology*, 16: 319–329.

Fawcett, K., Hodgkinson, C., and Mehlman, P. (2004). *An Assessment of the Impact of Tourism on the Virunga Mountain Gorillas: Phase I – Analyzing the Behavioral Data from Gorilla Groups Designated for Tourism*. Rwanda: Unpublished report, Dian Fossey Gorilla Fund International.

Ferber, D. (2000). Human diseases threaten great apes. *Science*, 289 (5438): 1277–1278.

Frothmann, D. L., Burks, K. D., and Maples, T. L. (1996). Letter to the editor: African great ape ecotourism considered. *African Primates*, 2: 52–54.

Goldsmith, M. L. (2000). Effects of ecotourism on behavioral ecology of Bwindi gorillas, Uganda: Preliminary results. *American Journal of Physical Anthropology*, Supp. 30: 161.

Goldsmith, M. L. (2005). Habituating primates for field study: Ethical considerations for great apes. In T. Turner (ed.), *Biological Anthropology and Ethics: From Repatriation to Genetic Identity*. State University of New York Press, pp. 49–64.

Goldsmith, M. L., Glick, J., and Ngabirano, E. (2006). Gorillas living on the edge: Literally and figuratively. In N. E. Newton-Fisher, H. Notman, J. D. Paterson, and V. Reynolds (eds.), *Primates of Western Uganda*. New York: Springer Publishers, pp. 405–422.

Graczyk, T. K. and Cranfield, M. R. (2001). Coprophagy and intestinal parasites: Implications to human habituated mountain gorillas (*Gorilla gorilla beringei*). *Recent Research Developments in Microbiology*, 5: 285–303.

Graczyk, T. K., Lowenstine, L. J., and Cranfield, M. R. (1999). *Capillaria hepatica* (Nematoda) infection in human-habituated gorillas (*Gorilla gorilla beringei*) of the Parc National de Volcans, Rwanda. *Journal of Parasitology*, 85: 1168–1170.

Graczyk, T. K., Mudakikwa, A. B., Eilenberger, U., and Cranfield, M. R. (2001). Hyperkeratotic mange caused by *Sarcoptes scabiei* (Acariforemes: Sarcoptidae) in juvenile human-habituated mountain gorillas (*Gorilla gorilla beringei*). *Parasitology Research*, 87: 1024–1028.

Graczyk, T. K., Nizeyi, J. B., Ssebide, G., *et al.* (2002a). Anthropozoonotic *Giardia duodenalis* Genotype (Assemblage) A infections in habitats of free-ranging human-habituated gorillas, Uganda. *Journal of Parasitology*, 88: 905–909.

Graczyk, T. K., Nizeyi, J. B., da Sliva, A. J., *et al.* (2002b). A single genotype of *Encephalitozoon intestinalis* infects free-ranging gorillas and people sharing their habitats in Uganda. *Parasitology Research*, 88: 926–931.

Gray, M., Fawcett, K., Basabose, A., *et al.* (2011). *Virunga Massif Mountain Gorilla Census – 2010 Summary Report*. Unpublished report, International Gorilla Conservation Program.

Gray, M., McNeilage, A., Fawcett, K., *et al.* (2006). *Virunga Volcano Range Mountain Gorilla Census, 2003*. Joint organizer's unpublished report, UWA/ORTPN/ICCN.

Guerrera, W., Sleeman, J. M., Ssebide, B. J., *et al.* (2003). Medical survey of the local human population to determine possible health risks to the mountain gorillas of Bwindi Impenetrable Forest National Park, Uganda. *International Journal of Primatology*, 24: 197–207.

Hamilton, A., Cunningham, A., Byarugaba, D., and Kayanja, F. (2000). Conservation in a region of political instability: Bwindi Impenetrable Forest, Uganda, *Conservation Biology*, 14: 1722–1725.

Hanes, A. C. (2012). The 7-metre Gorilla Tracking Regulation. *Gorilla Journal*, 44: 9–11.

Harcourt, A. H. (1986). Gorilla conservation: Anatomy of a campaign. In K. Benirschke (ed.), *Primates: The Road to Self-Sustaining Populations*. New York: Springer-Verlag, pp. 31–46.

Harcourt, A. H. (2001). The benefits of mountain gorilla tourism. *Gorilla Journal*, 22: 36–37.

Hastings, B. E., Kenny, D., Lowenstine, L. J., and Foster, J. W. (1991). Mountain gorillas and measles: Ontogeny of a wildlife vaccination program. *Proceedings of the American Association of Wildlife Veterinarians*, 198–205.

Homsy, J. (1999). *Ape Tourism and Human Diseases: How Close Should We Get? A Critical Review of the Rules and Regulations Governing Park Management and Tourism for Wild Mountain Gorillas (Gorilla gorilla beringei)*. Nairobi, Kenya: Unpublished report, Consultancy for the International Gorilla Conservation Program.

Hope, K., Goldsmith, M. L. and Graczyk, T. (2004). Parasitic health of olive baboons in Bwindi Impenetrable National Park, Uganda. *Veterinary Parasitology*, 122 (2): 165–170.

Hudson, H. R. (1992). The relationship between stress and disease in orphaned gorillas and its significance for gorilla tourism. *Gorilla Conservation News*, 6: 8–10.

Ilambu, O. (1998). *Impact de la guerre d'octobre 1996 sur la distribution spatiale de grands mammifères (gorilles et éléphants) dans le secteur de haute altitude du Parc National de Kahuzi-Biega*. Programme report of Wildlife Conservation Society in Kahuzi-Biega National Park.

International Gorilla Conservation Program (IGCP). (2012a). *Population of Mountain Gorilla in Bwindi Determined by 2011 Census.* www.igcp.org/population-of-mountain-gorillas-in-bwindi-determined-by-census/ (accessed Apr. 4, 2012).

International Gorilla Conservation Program (IGCP). (2012b). *Dismay over light sentencing of mountain gorilla poachers in Bwindi.* www.igcp.org/dismay-over-light-sentencing-of-mountain-gorilla-poachers-in-bwindi/ (accessed Apr. 4, 2012).

International Primatological Society (IPS). (2012). *Protection of Primate Health in the Wild.* www.internationalprimatologicalsociety.org/ProtectionOfPrimateHealthInTheWild.cfm (accessed Jun. 6, 2012).

Johns, B. G. (1996). Responses of chimpanzees to habituation and tourism in the Kibale Forest, Uganda, *Biological Conservation*, 78 (3): 257–262.

Kalema-Zikusoka, G., Kock, R. A., and Macfie, E. J. (2002). Scabies in free-ranging mountain gorillas (*Gorilla beringei beringei*) in Bwindi Impenetrable National Park, Uganda. *Veterinary Record*, 150: 12–15.

Kalpers, J., Gray, M., Asuma, S., *et al.* (2010). *Buffer Zone and Human Wildlife Conflict Management: IGCP Lessons Learned.* CARE/IGCP/EEEGL programs, unpublished report.

Kasereka, B., Muhigwa, J. B. B., Shalukoma, C., and Kahekwa, J. M. (2006). Vulnerability of habituated Grauer's gorilla to poaching in the Kahuzi-Biega National Park, DRC. *African Study Monographs*, 27(1): 15–26.

Klailova, M., Hodgkinson, C., and Lee, P. C. (2010). Behavioral responses of one western lowland gorilla (*Gorilla gorilla gorilla*) group at Bai Hokou, Central African Republic, to tourists, researchers and trackers. *American Journal of Primatology*, 72: 897–906.

Köndgen, S., Kuhl, H., N'Goran, P. K., *et al.* (2008). Pandemic human viruses cause decline of endangered great apes. *Current Biology*, 18: 260–264.

Lanyero, F. (2011). UWA lowers gorilla tracking fees. *Daily Monitor*, www.monitor.co.ug/News/National/-/688334/1164758/-/c1i12wz/-/index.html (accessed Sept. 19, 2011).

Litchfield, K. (1997). *Treading lightly: Responsible Tourism with the African Great Apes.* Pamphlet. Adelaide, Australia: Travellers' Medical and Vaccination Centre Group.

Litchfield, K. (2007). Responsible tourism: A conservation tool or conservation threat? In T. S. Stoinski, H. D. Steklis, and P. T. Mehlman (eds.), *Conservation in the 21st Century – Gorillas as a Case Study*. New York: Springer Verlag, pp. 107–127.

Macfie, E. (1991). The Volcanoes Veterinary Center. *Gorilla Conservation News*, 5: 21.

Macfie, E. (1996). Case report on scabies infection in Bwindi gorillas. *Gorilla Journal*, 13: 19–20.

Macfie, E. (1997). Gorilla tourism in Uganda. *Gorilla Journal*, 15: 16–17.

Macfie E. J. and Williamson, E. A. (2010). *Best Practice Guidelines for Great Ape Tourism. Gland*, Switzerland: International Union for Conservation of Nature and Natural Resources.

Mburanumwe, I. (2010). *Mukunda's Return to Congo from Rwanda and the Construction of an Electric Fence.* http://gorillacd.org/2010/08/03/mukunda%E2%80%99s-return-to-congo-from-rwanda-and-the-construction-of-an-electric-fence/ (accessed June 2, 2011).

McNeilage, A. (1996). Ecotourism and mountain gorillas in the Virunga Volcanoes. In V. J. Taylor and N. Dunstone (eds.), *The Exploitation of Mammal Populations*. London: Chapman and Hall Press, pp. 334–344.

McNeilage, A., Plumptre, A., Brock-Doyle, A., and Vedder, A. (2001). Bwindi Impenetrable National Park, Uganda – gorilla census, 1997. *Oryx*, 5 (1): 39–47.

McNeilage, A., Robbins, M. M., and Gray, M. (2006b). Census of the mountain gorilla *Gorilla beringei beringei* population in Bwindi Impenetrable National Park, Uganda. *Oryx*, 40 (4): 419–427.

McNeilage, A., Robbins, M. M., Guschanski, K., Gray, M., and Kagoda E. (2006a). *Mountain gorilla Census – 2006 Bwindi Impenetrable National Park*. Summary Report. www.igcp.org/library/ (accessed 4 Apr. 2012).

McNeilage, A., Robbins, M. M., and Gray, M. (2006b). Census of the mountain gorilla *Gorilla beringei beringei* population in Bwindi Impenetrable National Park, Uganda. *Oryx*, 40 (4): 419–427.

Meder, A. (2000). Scabies again. *Gorilla Journal*, 21: 8–9.

Moberg, G. P. (1985). Influence of stress on reproduction: measure of well-being. In Moberg, G. P. (ed.) *Animal Stress*. American Physiological Society, Bethesda, MD, pp. 245–267.

Mudakikwa, A. (2001). An outbreak of mange hits the Bwindi gorillas. *Gorilla Journal*, 22: 24.

Mudakikwa, A., Cranfield, M., Sleeman, J., and Eilenberger, U. (2001). Clinical medicine, preventive health care and research on mountain gorillas in the Virunga Volcanoes region. In M. Robbins., P. Sicotte, and K. Stewart (eds.), *Mountain Gorillas: Three Decades of Research at Karisoke*. Cambridge University Press, pp. 341–360.

Muyambi, F. (2005). The impact of tourism on the behaviour of mountain gorillas. *Gorilla Journal*, 30: 14–15.

Nizeyi, J. B., Sebunya, D., Dasilva, A. J., Cranfield, M. R., Pieniazek, N. J., and Graczyk, T. K. (2002a). Cryptosporidiosis in people sharing habitats with free-ranging mountain gorillas (*Gorilla gorilla beringei*). *American Journal of Tropical Medicine and Hygiene*. 66 (4): 442–444.

Nizeyi, J. B., Cranfield, M. R., and Graczyk, T. K. (2002b). Cattle near the Bwindi Impenetrable National Park, Uganda, as a reservoir of *Cryptosporidium parvum* and *Giardia duodenalis* for local community and free-ranging gorillas. *Parasitology Research*, 88: 380–385.

Nizeyi, J. B., Innocent, R. B., Erume, J., *et al.* (2001). Campylobacteriosis, salmonellosis and shigellosis in free-ranging human-habituated mountain gorillas of Uganda. *Journal of Wildlife Diseases*, 37: 239–244.

Nizeyi, J. B, Mwebe, R., Nanteza, A., *et al.* (1999). *Cryptosporidium* sp. and *Giardia* sp. infections in mountain gorillas (*Gorilla gorilla beringei*) of the Bwindi Impenetrable National Park, Uganda. *Journal of Parasitology*, 85: 1084–1088.

Olupot, W., Barigyira, R., and Chapman, C. A. (2009). The status of anthropogenic threat at the people-park interface of Bwindi Impenetrable National Park, Uganda. *Environmental Conservation*, 36 (1): 41–50.

Palacios, G., Lowenstine, L. J., and Cranfield, M. R. *et al.*(2011). Human metapneumovirus infection in wild mountain gorillas, Rwanda. *Emerging Infectious Disease*, 17 (4): 711–713.

Pole Pole Foundation. (2012). *History of Gorilla Tracking*. www.polepolefoundation.org/kbnp.php (accessed Apr. 6, 2011).

Regan, T. (2004). *The Case for Animal Rights: Updated with New Preface*. Berkeley and Los Angeles, CA: University of California Press.

Robbins, M. and Williamson, L. (2008). *Gorilla beringei. IUCN Red List of Threatened Species. Version 2010.4*. (www.iucnredlist.org) (accessed Apr. 6, 2011).

Robbins, M. M., Gray, M., Fawcett, K. A., *et al.* (2011). Extreme conservation leads to recovery of the Virunga mountain gorillas. *PLoS ONE*, 6 (6): 1–10.

Rothman, J. M., Bowman, D. D., Eberhard, M. L., and Pell, A. N. (2002). Intestinal parasites found in the research group of mountain gorillas in Bwindi Impenetrable National Park, Uganda: preliminary results. *Annals of the New York Academy of Science*, 969: 346–349.

Sandbrook, C. and Semple, S. (2006). The rules and the reality of mountain gorilla *Gorilla beringei beringei* tracking: how close do tourists get? *Oryx*, 40 (4): 428–433.

Sapolsky, R. M., Uno, H., Rebert, C. S., and Finch, C. E. (1990). Hippocampal damage associated with prolonged glucocorticoid exposure in primates. *Journal of Neuroscience*, 10 (9): 2897–2902.

Schaller, G. B. (1963). *The Mountain Gorilla: Ecology and Behavior*. University of Chicago Press.

Schmitt, T. M. (1997). Close encounter with gorillas at Bwindi. *J. Berggorilla and Regenwald Direkthilfe*, 14: 12–13.

Sholley, C. R. (1991). Conserving gorillas in the midst of guerrillas. In: *American Association of Zoological Parks and Aquariums, Annual Conference Proceedings*, San Diego, pp. 30–37.

Sholley, C. R. and Hastings, B. (1989). Outbreak of illness among Rwanda's gorillas. *Gorilla Conservation News*, 3: 7.

Sleeman, J. M., Meader, L. L., Mudakikwa, A. M., *et al.* (2000). Gastrointestinal parasites of mountain gorillas (*Gorilla gorilla beringei*) in the Parc National des Volcans, Rwanda. *Journal of Zoo and Wildlife Medicine*, 31 (3): 322–328.

Stewart, K. J. (1991). Editorial. *Gorilla Conservation News*, 5: 1–2.

Stewart, K. J. (1992). Gorilla tourism: Problems of control. *Gorilla Conservation News*, 6: 15–16.

Strum, S. C. (1987). *Almost Human*. New York: Random House Publishers.

TIES (The International Ecotourism Society) (2005). *TIES Global Ecotourism Fact Sheet*, updated in 2006. Washington, DC.

Touched by a Wild Gorilla (2011). www.youtube.com/watch?v=hg2hCuDy2wg, uploaded by aleutiandrem on December 17, 2011 (accessed Jan. 20, 2012).

Tumusiime, D. M. and Svarstad, H. (2011), A local counter-narrative on the conservation of Mountain Gorillas. *Forum for Development Studies*, 38: 239–265.

Tutin, C. E. G. and Fernandez, M. (1991). Responses of wild chimpanzees and gorillas to the arrival of primatologists: behavior observed during habituation. In H. O. Box (ed.), *Primate Responses to Environmental Change*. London: Chapman & Hall Press, pp. 187–197.

Uganda Bureau of Statistics. (2002). *2002 Uganda Population and Housing Census District Reports: Main Report*, p. 7. www.ubos.org/onlinefiles/uploads/ubos/pdf%20documents/2002%20Census%20Final%20Reportdoc.pdf (accessed Apr. 4, 2012).

UNEP (2009). *Cross River Gorilla (Gorilla gorilla diehli): Gorilla Agreement Action Plan*. www.cms.int/species/gorillas/gor_tc1_documents/inf8_2_cr_river_ggd_e.pdf (accessed Jun. 4, 2012).

van Krunkelsven, E., Dupain, J., and van Elsacker, L. (1999). Habituation of bonobos (*Pan paniscus*): First reactions to the presence of observers and the evolution of response over time. *Folia Primatologica*, 70 (6): 365–368.

Vedder, A. and Weber, A. W. (1990). The mountain gorilla project. In A. Kiss (ed.), *Living with Wildlife: Wildlife Resource Management with Local Participation*. Washington, DC: World Bank Technical Publication, 130: 83–90.

Wallis, J. and Lee, D. R. (1999). Primate conservation: the prevention of disease transmission. *International Journal of Primatology*, 20: 803–826.

Wasser, S. K, Sewall, G., and Soule, M. R. (1993). Psychosocial stress as a cause of infertility. *Fertility and Sterility*, 59 (3): 685–689.

Weber, A. W. (1993). Primate conservation and eco-tourism in Africa. In C. S. Potter, J. I. Cohen, and D. Janczewski (eds.), *Perspectives on Biodiversity: Case Studies of Genetic Resource Conservation and Development*. Washington, DC: American Association for the Advancement of Science Press, pp. 129–150.

Weber, A. W. and Vedder, A. (1983). Population dynamics of the Virunga gorillas: 1959–1978. *Biological Conservation*, 26: 341–366.

Williamson, E. A. (2001). Gorillas and eco-tourism. *Gorilla Journal*, 22: 35–36.

Williamson, E. A. and Fawcett, K. A. (2008). Long-term research and conservation in the Virunga mountain gorillas. In R. Wrangham and E. Ross (eds.), *Science and Conservation in African Forests: The Benefits of Long-Term Research*. Cambridge University Press, pp. 213–229.

Williamson, E. A. and Feistner, A. T. C. (2003). Habituating primates: processes, techniques, variables and ethics. In J. M. Setchell (ed.), *Field and Laboratory Methods in Primatology: A Practical Guide*. Cambridge University Press, pp. 25–39.

Woodford, M. H., Butynski, T. M., and Karesh, W. (2002). Habituating the great apes: the disease risks. *Oryx*, 36: 153–160.

Yamagiwa, J. (1997). Mushamuka's story: the largest group and the longest tenure. *Gorilla Journal*, 15: 3–4.

Yamagiwa, J. (2003). Bushmeat poaching and the conservation crisis in Kahuzi-Biega National Park, Democratic Republic of the Congo. *Journal of Sustainable Forestry*, 16 (3/4): 111–130.

11 Evaluating the effectiveness of chimpanzee tourism

James S. Desmond and Jennifer A. Z. Desmond

Introduction

Chimpanzee tourism confronts the same goals and concerns faced by all wildlife tourism initiatives. With the potential to generate funds critical to protecting habitat and help communities see value in the protection of native wildlife, wildlife tourism can greatly benefit endangered populations. However, designing and implementing tourism programs to benefit chimpanzee conservation specifically presents unique challenges. Unlike other species, chimpanzees' nearly identical genetic similarity to humans puts them at an especially high risk of contracting human infectious diseases (Muehlenbein *et al.*, 2010; Woodford *et al.*, 2002), making careful evaluation of the benefits and limitations of chimpanzee tourism programs critical.

In this chapter we will discuss chimpanzee tourism by examining its history and development, its benefits and limitations for chimpanzee conservation, various procedures and protocols, and the unique challenges faced at three sites: Gombe and Mahale Mountains National Parks in Tanzania and Kibale National Park in Uganda. We will evaluate the overall costs and benefits of chimpanzee tourism as they relate to chimpanzee conservation and weigh the risk factors such as disease, physiological stress, and increased human conflict against the benefits such as habitat protection and increased protection from hunting and poaching.

Chimpanzee tourism sites

Chimpanzee tourism exists across the species' range with programs in Tanzania, Uganda, Democratic Republic of Congo, and Côte d'Ivoire. This chapter concentrates on three well-established tourism sites with wild chimpanzee tourism programs. The focus on Gombe National Park and Mahale Mountains National Park in Tanzania and Kibale National Park in Uganda is based on their long-standing tourism programs and the extensive research conducted at these sites.

Primate Tourism: A Tool for Conservation?, ed. Anne E. Russon and Janette Wallis. Published by Cambridge University Press. © Cambridge University Press 2014.

Tanzania: Gombe National Park

Gombe National Park has the longest running chimpanzee research program in the world and was one of the first to institute chimpanzee tourism as a conservation measure (Collins & Goodall, 2008). Studies on chimpanzees began in 1960 in the area known then as Gombe Stream Game Reserve, established in 1943 by the colonial government to protect chimpanzee habitat (Pusey *et al.*, 2007). Within the reserve's first year of research, primatologist Jane Goodall discovered that chimpanzees make and use tools, as well as hunt for and eat meat. These groundbreaking discoveries and their coverage in *National Geographic* (Goodall, 1963) led to enormous interest in chimpanzees from both the scientific world and the general public, leading to an expansion in research projects and ultimately the development of chimpanzee-related tourism activities. Ongoing research, combined with Jane Goodall's worldwide fame and advocacy for chimpanzee protection and welfare, has continued to fuel interest in their study and increased demand for chimpanzee tourism. For chimpanzees, a direct conservation benefit of early research was the designation of Gombe Stream Game Reserve as a national park in 1968 (Collins & Goodall, 2008; Pusey *et al.*, 2008). National park status is the single most important factor in saving the Gombe chimpanzee population from extirpation: areas outside the park have been almost completely deforested and converted to farmland while natural chimpanzee habitat such as forest cover and woody vegetation inside the park has increased (Pintea, 2007).

Three chimpanzee communities are currently known to range in Gombe: Kasakela, the most studied and well-known group, and the only group visited by tourists; Mitumba, residing in the northern part of the park, initially habituated for tourism purposes in the late 1980s but ultimately only visited by researchers; and Kalende, an unhabituated group living in the southern section of the park. From the late 1980s up to 2008, Mitumba numbers dropped from approximately 32 to 25 individuals and Kalende numbers from 30–40 to 11. The Kasakela community, considered the most highly habituated, has increased from approximately 40 up to 62 individuals over the same time period (Pusey *et al.*, 2008).

Many of the threats to chimpanzees documented throughout their range have affected Gombe's chimpanzees including habitat loss, wildlife and bushmeat trades, disease, poaching, and human–wildlife conflict (Mittermeier & Cheney, 1987; Wallis & Lee, 1999). Decreased populations in the Mitumba and Kalende communities are primarily attributed to habitat loss, spread of disease, and killing by humans, as these communities historically ranged outside the park where deforestation has limited their access to suitable habitat for foraging (Pusey *et al.*, 2008). In addition these communities have had increased conflict with humans through poaching, crop-raiding, and hunting, especially in the southern portion of the park where the Kalende community has been adversely impacted (Greengrass, 2000a). In comparison, the Kasakela community's entire range is within the protected boundaries of the park and, as a result, their population has not suffered as much from the effects of habitat destruction and conflict with humans. In addition, this community has been followed for research and tourism for over 50 years, providing them with a higher level of protection against habitat loss, poaching, and hunting.

Disease epidemics, some of which may have had human origins, have negatively impacted chimpanzee populations at Gombe (Lonsdorf *et al.*, 2006; Wallis & Lee, 1999; Williams *et al.*, 2008). From 1960 through 2006, five epidemics accounted for 29% of all chimpanzee deaths in the Kasakela community. While none of these can be attributed to humans with certainty, there is a high likelihood that four of the five epidemics were of human origin (Wallis & Lee, 1999; Williams *et al.*, 2008). A respiratory outbreak in 2000 resulted in several changes in practices within the park. Food provisioning immediately ceased and stricter health guidelines were implemented for both tourists and researchers, including required quarantine periods and vaccinations for researchers, time-limited tourist visits with individual chimpanzee groups (one hour), and the enforcement of a minimum distance rule (10 m) between all human observers and chimpanzees (Collins, 2003). In addition, the Gombe Stream Research Center instituted a health monitoring program which included observation of chimpanzees for potential health issues and collection of fecal and urine samples for analysis, allowing researchers to identify incipient disease outbreaks to assist management decision making and enable intervention when and if necessary or appropriate (Lonsdorf *et al.*, 2006).

Another risk factor for Gombe's chimpanzees habituated for research and tourism is conflict with humans living adjacent to the park. Despite the high level of habituation of the Mitumba community and the proximity of agricultural land, there are no published reports of crop-raiding in the agricultural areas adjacent to Mitumba's home range. Alternatively, crop-raiding by some members of the unhabituated Kalende community was observed outside the southern boundary of the park (Greengrass, 2000b). Since Kalende chimpanzees had not been habituated and had more reason to fear humans due to increased hunting and poaching of the park's wildlife at the time (Greengrass, 2000a), their crop-raiding was most likely due to habitat loss. Therefore, while habituation of chimpanzees decreases fear of humans, in Gombe it does not appear to cause an increase in crop-raiding activity. Around Gombe, and elsewhere where chimpanzee habitat is adjacent to humans, crop-raiding is more likely a function of habitat loss and proximity to favorable crops than a decreased fear of humans resulting from habituation (Hockings, 2009; McLennan, 2008).

Because of the low populations of the Mitumba and Kalende communities, a disease outbreak could have devastating effects. Unfortunately, this puts the primary focus on the park's Kasakela group for all tourism, and most research, activities. Since this community has suffered repeated disease epidemics that likely originated with humans conducting research inside the park, stricter health and behavioral protocols were developed and implemented in order for the park to continue tourism and research activities with this community while also minimizing risks to the chimpanzees' health.

Tanzania: Mahale Mountains National Park

Mahale Mountains National Park, 160 km south of Kigoma in Tanzania, on the shores of Lake Tanganyika, has a similar history to Gombe. Toshisada Nishida began habituating chimpanzees in 1965 for research and quickly realized the land would need protected status to safeguard the ecosystem. Twenty years later, Mahale was granted national park status and tourism commenced in 1989 (Nishida & Nakamura, 2008). Since then, tourism has increased steadily, from fewer than 100 visitors in the first year to over 1000 by 2003 (Nakamura & Nishida, 2009; Nishida & Nakamura, 2008).

Regulations for chimpanzee tourism at Mahale are similar to those at Gombe and stipulate a maximum group size of six tourists and one guide, with visitors allowed to remain in the presence of one habituated chimpanzee group for a maximum of one hour and no less than 10 m away. Additional regulations include no eating in the presence of chimpanzees, no defecation in the forest, no use of flash photography, and no belongings left unattended. The park allows up to three tourist groups staggered throughout the day. In addition, Mahale is the only chimpanzee tourism program that requires visitors to wear surgical masks in the presence of chimpanzees (Hanamura et al., 2008; Williamson & Macfie, this volume). This regulation was established following a flu-like epidemic in the park's habituated group that killed 12 chimpanzees and was confirmed to be of human origin (Kaur et al., 2008).

A recent study at Mahale examined adherence to park tourist regulations in order to better understand risks to the chimpanzees (Nakamura & Nishida, 2009). The study found that on 23% of 121 observation days, the number of visiting tourists exceeded the three-group limit, including one day with nearly 40 individuals visiting one chimpanzee group. Other regulations were also violated, such as the 10 m distance rule and one hour time limit, potentially increasing disease risk and stress levels in the chimpanzees. In addition, anecdotal observations prior to the study revealed that over 20 people sometimes visited the chimpanzees at once, including multiple tourist groups and the site's researchers. This not only violated regulations designed to protect the chimpanzees but also led to problems between tourists and researchers and disrupted studies focused on the chimpanzees' natural behaviors (Nishida & Nakamura, 2008).

These studies highlight the problems inherent in great ape tourism and in combining tourism and research activities. Resulting recommendations included the establishment of a centralized booking system to eliminate overbooking, a chimpanzee tracking fee to increase revenue from fewer tourists, and higher park fees during peak tourism seasons to better distribute visitation throughout the year (Nakamura & Nishida, 2009; Nishida & Mwinuka, 2005). The habituation of additional groups of chimpanzees to be used strictly for tourism, a practice employed with mountain gorilla and chimpanzee tourism sites in Uganda, was also suggested, to reduce the burden of visitation on just one group (Nishida & Nakamura, 2008).

Uganda: Kibale National Park

Kibale National Park was granted national park status in 1993. As with Gombe National Park, long-term research played an important role in achieving this status. What began as the Kibale Forest Project, started by Thomas Struhsaker in 1970 to study red colobus monkeys, became the Makerere University Biological Field Station in 1987, designed to continue and expand multiple field studies begun in the early 1970s (Struhsaker, 2008).

As in Gombe and Mahale in Tanzania, chimpanzee tourism in Kibale began as a direct result of research. Kibale was found to contain one of the highest levels of primate biodiversity in the world, generating great interest in tourism in the area. At the time the park had only one habituated group of chimpanzees, the Kanyawara community, used only for research purposes. To avoid additional stress on this community, the Kibale Chimpanzee Project proposed chimpanzee tourism in Kanyanchu, on the opposite side of the park. Not only would this limit disturbance to the Kanyawara chimpanzees, it would also potentially provide revenue for the park and surrounding villages as well as opportunities for further research on issues such as the impact of tourism on the forest and the chimpanzees (Mugisha, 2008). In 1992, the Kibale Chimpanzee Project started the Kanyanchu ecotourism program.

When gazetted in 1993, Kibale was met with resistance. Designation of national park status restricted forest access, which negatively impacted local communities who had previously used the forest for medicinal purposes, hunting, firewood, and timber extraction (Ferraro & Kramer, 1997; Naughton-Treves, 1997). In response to community concerns at Kibale and other national parks, the Uganda Wildlife Authority (UWA) developed its Integrated Conservation and Development Program (ICDP) with the goal of linking wildlife protection to economic benefits, hoping to encourage a change in attitude toward wildlife conservation and protected areas. Under UWA's ICDP program, a portion of the revenues collected by parks must be shared with local communities (Archabald & Naughton-Treves, 2001). In Kibale, 20% of park entrance fees are distributed to the communities surrounding the park. Visitation to Kibale has risen steadily since its start, with 1297 visitors in 1992 increasing to 9482 in 2010 (UBOS, 2011). A survey in 1996 found that wildlife viewing was the primary interest of 85% of visitors to Kibale (Obua & Harding, 1996). For the period 2006–2010, based on an average of the annual number of visitors and current entrance fees, an estimated $55 000 US was distributed annually to local communities totaling $275 000 US over five years. Considering the fact that chimpanzee tracking is the primary focus of tourism in the park, increased visitation and resulting revenues are likely directly linked with this activity.

Historically, funds generated from park entrance fees were used to build schools or health clinics. However, a recent study found that investing in infrastructure to deter crop-raiding was a more effective way to ensure community protection of the park. Villages in which park funds were used to deter crop-raiding perceived a direct benefit from the park and had a lower incidence of illegal resource extraction, poaching, and human disturbance in protected areas; this reduced the risk of disease transmission and snare injuries, a clear conservation benefit to chimpanzees (MacKenzie, 2012).

Weighing the benefits and limitations of chimpanzee tourism in conserving chimpanzees

Chimpanzee tourism comes with clear benefits and inherent risks. For tourists to view wildlife closely in its natural setting, animals must feel comfortable while being tracked and observed. The first step of any chimpanzee tourism program is habituation, the process of desensitizing animals to the presence of humans. Habituation requires humans to spend a great deal of time in close proximity to their subjects. Tourism programs mean that chimpanzees will be exposed to the consistent presence of humans. While increased presence of humans can greatly benefit chimpanzee conservation in deterring illegal activity and encouraging the protection of habitat, it also presents a serious risk of disease transmission and other risks like physiological stress, loss of fear of humans, and altered foraging and behavior patterns (Williamson & Feistner, 2003).

Controlled human presence

The protection provided by the controlled presence of humans, whether through tourism or research, can greatly contribute to the conservation of chimpanzees when properly managed. Currently, chimpanzee tourism in East Africa exists primarily in protected areas. In all three chimpanzee tourism sites discussed in this chapter, designation of national park status came as a direct result of the efforts of researchers concerned about the future of their research subjects. With the widespread deforestation across sub-Saharan Africa over the past two decades and projected human population growth rates, tropical forests are under threat (Chapman & Peres, 2001; Junker et al., 2012). Designation of national park status to tropical forests is by far the most effective conservation strategy for protection of biodiversity, which includes chimpanzees (Bruner et al., 2001). Without legal protection, forests are more readily available for conversion to agricultural land with demand increasing as human populations continue to grow.

Even within protected areas, chimpanzees remain vulnerable without proper enforcement of their boundaries, as humans inevitably encroach on the land via illegal logging, hunting, poaching, and land reclamation (Junker et al., 2012; Struhsaker et al., 2005). A distinct benefit of chimpanzee tourism is the human presence it provides when it is properly controlled. Studies in Taï Forest and Serengeti National Parks evaluating the impact of human presence, including researchers, guides, rangers, and tourists, found that research and tourist activities deterred illegal activity within parks (Campbell et al., 2011; Hilborn et al., 2006; Köndgen et al., 2008). Gombe National Park's most highly habituated chimpanzee community has fluctuated but, ultimately, increased in population size while the park's unhabituated community has steadily declined. Gombe's unhabituated Kalende community suffered a precipitous drop in population in the mid to late 1990s, going from approximately 30–40 individuals to fewer than 20 (Greengrass, 2000b; Pusey *et al.*, 2008). This dramatic drop in population may have been due to multiple factors including disease, movement of females from one community to another, habitat loss around the park, and hunting. However, the most likely cause of this rapid loss is the influx of Congolese refugees into communities at the southern end of the park leading to increased hunting and poaching, clearly evident during chimpanzee surveys of the area in 1999 (Greengrass, 2000b). The Mitumba community, originally targeted for tourism but ultimately visited only by researchers, suffered a decline due to disease outbreaks early in the habituation process. However, with the implementation of stricter health protocols, this group has also benefited from human presence and has since remained stable (Pusey *et al.*, 2008).

In addition to providing an important presence, chimpanzee tourism can provide revenue to sustain law-enforcement activities within protected areas (Davenport *et al.*, 2002) making it an even more valuable conservation tool. The most important factor determining the conservation success of great apes in Africa is strong law enforcement in protected areas (Tranquilli *et al.*, 2012). Without proper law enforcement, protected areas suffer from widespread poaching, hunting, and loss of biodiversity (Hilborn *et al.*, 2006; Kuehl *et al.*, 2008).

Infectious disease

The greatest risk to habituated chimpanzees is contracting infectious diseases from humans (Köndgen *et al.*, 2008). Chimpanzees targeted for habituation have most likely had little prior contact with humans or the wide range of pathogens that visitors may bring into their environment, especially those from international locations (Adams *et al.*, 2001; Muehlenbein *et al.*, 2010). Examples of chimpanzees contracting diseases of human origin are well documented in Mahale National Park and Taï Forest National Park in Côte d'Ivoire and humans are a likely source of disease outbreaks in Gombe National Park (Hanamura *et al.*, 2008; Pedersen & Davies, 2009; Wallis & Lee, 1999; Williams *et al.*, 2008; Woodford *et al.*, 2002). Chimpanzees in these parks have experienced epidemics that have negatively impacted their local populations. In Taï Forest National Park, five respiratory outbreaks from 1999 through 2006 across three habituated groups of chimpanzees resulted in 15 deaths due to two common human paramyxoviruses (Köndgen *et al.*, 2008). Such disease outbreaks could have long-lasting effects on populations. However, these outbreaks cannot necessarily be attributed to tourists and certainly, in the case of Gombe, many occurred prior to the existence of tourism so were most likely transmitted by researchers. Close proximity between humans and wild chimpanzees, whether for tourism or research, increases the risk of disease transmission.

While infectious disease will always be a serious threat to chimpanzee populations, when proper protocols are followed, disease risks due to tourism and its infrastructure can be greatly minimized. To mitigate the risk of disease transmission, it is important to institute and enforce rules on proper visitor quarantine, minimum visitor to chimpanzee distance, and number of tourists visiting per day (Homsy, 1999; Wallis & Lee, 1999). Most chimpanzee tourism sites have such regulations in place, but their strict enforcement can be difficult (Nakamura & Nishida, 2009; Sandbrook & Semple, 2006).

The greatest health risk to chimpanzees that experience close proximity with humans is posed by respiratory diseases, as aerosolized droplets can travel approximately 7 m and up to three times further in certain wind conditions (Macfie & Williamson, 2010). To better protect against respiratory disease transmission, IUCN best practices guidelines on great ape tourism recommend the use of N95 surgical respirator masks as a requirement for tourists (Macfie & Williamson, 2010; Williamson & Macfie, this volume). Adopting the use of these masks does not eliminate the need to enforce a minimum distance rule but provides an added level of protection for chimpanzees being observed. Arguments against the use of masks include the difficulty in educating tourists on proper usage, the need to habituate chimpanzees to the masks, complications in supplying the proper type of masks, the costs associated with providing masks (they cannot be reused), and ensuring proper waste disposal (Macfie & Williamson, 2010). Only Mahale National Park currently requires tourists to wear N95 surgical respirator masks when in close proximity with chimpanzees, but this practice could be adopted at all sites.

Typically, regulations ask that all visitors inform park staff and withdraw voluntarily from the visit if they are feeling unwell, and require park officials to deny entrance to the forest to visitors who are visibly ill. Again, enforcement of this regulation can be difficult, with tourists often hiding illness and staff uncomfortable imposing the rule (Adams et al., 2001; Muehlenbein et al., 2010). Some additional recommendations to help ensure the health of chimpanzees involved in tourism activities include requiring tourists to provide proof of vaccination for diseases such as measles, polio, and influenza and enforcing quarantine periods prior to visiting great ape tourism sites (Macfie & Williamson, 2010). Park staff, guides, rangers, and researchers are also regularly in close proximity with habituated chimpanzees, making the institution of employee health programs, such as the program implemented by the Mountain Gorilla Veterinary Project in Rwanda, an important step in protecting the chimpanzees visited (Ali et al., 2004). It is critical that chimpanzee tourism and research sites are uncompromising in their commitment to the establishment, implementation, and enforcement of recommended protocols and regulations, to protect chimpanzees from associated health risks and spread of disease.

Physiological stress

The presence of humans has the potential to cause physiological stress. Stress is a normal response to a perceived threat, the "fight or flight" response, but repeated exposure to stress-inducing stimuli can produce chronic stress. Immunosuppression, reduced reproduction, impaired cognition, and stunted growth are all potential impacts of chronic stress (Woodford *et al.*, 2002).

Historically, the stress of primates during habituation has been measured by behavioral responses (Blom *et al.*, 2004) but recently researchers have been able to use non-invasive sampling techniques to examine stress hormones (cortisol, testosterone) in feces. A study of chimpanzees in Kibale National Park evaluated stress hormone levels relative to parasite richness (Muehlenbein, 2006). Parasite richness may be used as a rough indicator of immune response, where higher levels of parasite richness correspond to immunosuppression. Increased stress hormone levels correlated with increased parasite richness indicate that stress can contribute to immune suppression (Muehlenbein, 2006). This finding is important to chimpanzee tourism given tourism's potential to introduce new pathogens brought in by visitors whose microbial fauna is unfamiliar to the chimpanzees' and, if paired with chronic stress, increases susceptibility to infectious disease. Habituated groups of gorillas have been found to carry higher parasite burdens than those that are unhabituated (Kalema-Zikusoka *et al.*, 2005) and certain bacterial species have increased in habituated groups since tourism was initiated (Nizeyi *et al.*, 2001). The combination of immunosuppression due to increased stress and exposure to new pathogens could exacerbate disease risk.

A recent study investigated the impact of tourism on two habituated wild orangutans by measuring stress hormone (glucocorticoid) levels in their feces before and after tourist visits. Glucocorticoid levels rose and then fell back to normal levels following tourist visits and did not induce chronic stress (Muehlenbein *et al.*, 2012). A similar study measuring stress hormones in relation to tourism does not exist for chimpanzees. However, while chimpanzee populations at long-term field study and tourism sites across their range show signs of stress during the initial stages of habituation, there is no evidence of long-term or chronic stress. Chimpanzees at these sites show a significant decrease in stress over time, with behavioral indicators such as increased vocalizations and displaying by adult males decreasing over time with increased exposure to humans (Johns, 1996; McLennan & Hill, 2010). Further study of the relationship between stress and tourism with chimpanzees could offer better information for the management of chimpanzee tourism sites and serve as a tool in determining whether a chimpanzee community may or may not be suited for tourism activities.

Behavior

Chimpanzees have been shown to alter their behavior during the habituation process and in the presence of large groups of humans, leading to changes in foraging and ranging habits with a subsequent increase in energy expenditure (Blom *et al.*, 2004; Johns, 1996). In addition, habituation can alter inter-group dynamics and behavior, leading to unnatural behaviors such as including humans as a "tool" in social interactions, which can disrupt the social structure of the group (Williamson & Feistner, 2003).

A decreased fear of humans could potentially lead to an increased tendency to raid crops, higher susceptibility to poaching and hunting threats, and attacks on humans (Adlemanl & Goldsmith, 2007; in this volume see chapters by Berman *et al.*, Dellatore *et al.*, Goldsmith, Kauffman, Kurita, Strum & Manzolillo Nightingale). It is important to note, however, that, while this may be the case with gorillas (Goldsmith, this volume) and some other primates, no credible evidence exists that shows habituation of chimpanzees has any correlation with increased crop-raiding. Instead, crop-raiding by chimpanzees appears to be related to forest fragmentation, degraded habitat, and the crop selection of farmers living adjacent to the forest (Hill, 1997; McLennan, 2008).

Conclusion

Chimpanzee tourism has existed in Africa for over 30 years with its roots in Gombe National Park in Tanzania. The research community has played an instrumental role in establishing protected areas and raising awareness about the importance of endangered chimpanzees and other wildlife, and in opening the door for the development of chimpanzee tourism programs (Wrangham, 2008). The presence of research and tourism programs in protected areas has helped decrease poaching, illegal timber extraction, and unlawful use of the forest. However the risk of disease transmission to chimpanzee populations from visitors, if not managed properly, has the potential to outweigh these benefits.

It is unclear whether chimpanzee tourism has realized its conservation promise. Chimpanzee tourism programs carry risks, the greatest of which is infectious disease. However, disease risks can be mitigated through the implementation and enforcement of strict regulations for researchers and tourists recommended by IUCN (Macfie & Williamson, 2010). These include minimum distance rules, wearing surgical masks, disinfecting shoes and clothing before entering the forest, time limits, and restricted numbers of tourists visiting chimpanzee groups. Separating research and tourism sites when possible can also help reduce these risks by reducing the number of visitors to the same groups, resulting in decreased stress and disease risk. In addition, sites can implement health monitoring programs to allow management and staff to react appropriately and rapidly to potential health risks or signs of an epidemic, as was done in Gombe National Park (Lonsdorf *et al.*, 2006). The funds to support these efforts could be provided in part by tourism revenues (Davenport *et al.*, 2002).

A clear benefit of chimpanzee tourism is the protection it provides to chimpanzees habituated for tourism programs. The presence of rangers, guides, tourists, and other humans following these chimpanzees on a daily basis helps protect them from falling victim to illegal hunting and poaching. In addition, chimpanzee tourism often provides the revenue used to sustain law-enforcement activities, the most important factor in the success of great ape conservation in protected areas (Tranquilli *et al.*, 2012).

While there is an expectation that local communities will benefit financially from the development of chimpanzee tourism activities, several studies have shown these benefits to be modest (Archabald & Naughton-Treves, 2001; MacKenzie, 2012). However, when communities realize a direct benefit from revenues generated from tourism, chimpanzee conservation is positively impacted through decreased poaching and encroachment as well as a greater willingness by community members to protect chimpanzee habitat (MacKenzie, 2012). In addition, local communities often see tourism as responsible for increased crop-raiding by chimpanzees. Presenting research which shows that crop-raiding is more likely caused by habitat loss and a proximity to favored crops may promote greater support of tourism programs by nearby communities. Attention to community plays an important role in chimpanzee conservation and in guaranteeing the future of chimpanzee habitats, as attitudes toward the park by local communities has been cited as a critical component for conservation success (Struhsaker et al., 2005).

In conclusion, it has yet to be determined whether chimpanzee tourism's benefits to chimpanzee conservation outweigh its risks. Gombe's Kasakela community, habituated for over 50 years, has lived with the constant presence of researchers over several generations and the presence of tourists for over 30 years. There have been several epidemics over that time, most likely caused by human disease, but despite these devastating losses, the population of this group has increased (Pusey et al., 2008). In Gombe, the protection provided by the presence of research and tourism programs in reducing habitat loss and poaching threats now may outweigh the disease risks. With the implementation of stricter protocols for researchers and tourists (Collins, 2003), disease risk has been minimized and the park's chimpanzee population has ultimately benefited from the presence of tourists and researchers.

A great deal of work remains to be done to evaluate the effectiveness of chimpanzee tourism as a conservation tool. While it has been shown to be successful in some cases, to truly evaluate its effectiveness in conserving chimpanzees, tourism programs must take an evidence-based approach when determining the best strategies for enhancing wildlife protection and chimpanzee conservation. Establishment of research-based practices, protocols, and regulations is the key to ensuring that chimpanzee tourism is beneficial to chimpanzee conservation going forward.

Chimpanzee tourism programs have the potential to deliver benefits to all stakeholders. However, for a program to be successful it must first safeguard the health and welfare of chimpanzees. Chimpanzee tourism can be a valuable and effective conservation tool where research-based practices are instituted, disease risks are minimized, and government and institutional support exists.

References

Adams, H. R., Sleeman, J. M., Rwego, I. et al. (2001). Self-reported medical history survey of humans as a measure of health risk to the chimpanzees (*Pan troglodytes schweinfurthii*) of Kibale National Park, Uganda. *Oryx*, 35 (4): 308–312.

Adlemanl, S. J. and Goldsmith, M. L. (2007). Habitat use of habituated versus unhabitu-ated gorilla groups in Bwindi Impenetrable National Park. *American Journal of Physical Anthropology*, 132 (S44): 60.

Ali, R., Cranfield, M., Gaffikin, L. *et al.* (2004). Occupational health and gorilla conserva-tion in Rwanda. *International Journal of Occupational and Environmental Health*, 10 (3): 319–325.

Archabald, K. and Naughton-Treves, L. (2001). Tourism revenue-sharing around national parks in Western Uganda: early efforts to identify and reward local communities. *Environmental Conservation*, 28 (2): 135–149.

Blom, A., Cipolletta, C., Brunsting, A. M. H. *et al.* (2004). Behavioral responses of gorillas to habituation in the Dzanga-Ndoki National Park, Central African Republic. *International Journal of Primatology*, 25 (1): 179–196.

Bruner, A. G., Gullison, R. E., Rice, R. E. *et al.* (2001). Effectiveness of parks in protecting tropical biodiversity. *Science*, 291 (5501): 125–128.

Campbell, G., Kuehl, H., Diarrassouba, A. *et al.* (2011). Long-term research sites as refugia for threatened and over-harvested species. *Biology Letters*, 7 (5): 723–726.

Chapman, C. A. and Peres, C. A. (2001). Primate conservation in the new millennium: the role of scientists. *Evolutionary Anthropology*, 10: 16–33.

Collins, A. D. (2003). Health guidelines for visiting researchers in Gombe National Park to minimize risk of disease transmission among primates. *Pan Africa News*, 10: 1–3.

Collins, A. D. and Goodall, J. (2008). Long-term research and conservation in Gombe National Park, Tanzania. In: Wrangham, R. W. and Ross, E. (eds.), *Science and Conservation in African Forests: The Benefits of Long-term Research*. Cambridge University Press, pp. 158–171.

Davenport, L., Brockelman, W. Y., Wright, P. C. *et al.* (2002). Ecotourism tools for parks. In: Terborgh, J., van Schaik, C., Davenport, L., and Rao, M. (eds.), *Making Parks Work: Strategies for Preserving Tropical Nature*. Washington DC: Island Press, pp. 279–306.

Ferraro, P. J. and Kramer, R. A. (1997). Compensation and economic incentives: Reducing pressure on protected areas. In: Kramer, R., van Shaik, C., and Johnson, J. (eds.), *Last Stand: Protected Areas and the Defense of Tropical Biodiversity*. New York: Oxford University Press, pp. 187–211.

Goodall, J. M. (1963). My life among wild chimpanzees. *National Geographic Magazine*, 124 (2): 272–308.

Greengrass, E. (2000a). The sudden decline of a community of chimpanzees at Gombe National Park. *Pan Africa News*, 7: 5–7.

Greengrass, E. (2000b). The sudden decline of a community of chimpanzees at Gombe National Park: A supplement. *Pan Africa News*, 7 (2): 25–26.

Hanamura, S., Kiyono, M., Lukasik-Braum, M. *et al.* (2008). Chimpanzee deaths at Mahale caused by a flu-like disease. *Primates*, 49 (1): 77–80.

Hilborn, R., Arcese, P., Borner, M. *et al.* (2006). Effective enforcement in a conservation area. *Science*, 314 (5803): 1266–1266.

Hill, C. M. (1997). Crop-raiding by wild vertebrates: the farmer's perspective in an agricul-tural community in western Uganda. *International Journal of Pest Management*, 43 (1): 77–84.

Hockings, K. J. (2009). Living at the interface human-chimpanzee competition, coexistence and conflict in Africa. *Interaction Studies*, 10 (2): 183–205.

Homsy, J. (1999). *Ape Tourism and Human Diseases: How close should we get? A critical review of the rules and regulations governing park management and tourism for the wild mountain gorilla (Gorilla gorilla beringei)*. Nairobi, Kenya: International Gorilla Conservation Program.

Johns, B. G. (1996). Responses of chimpanzees to habituation and tourism in the Kibale Forest, Uganda. *Biological Conservation*, 78 (3): 257–262.

Junker, J., Blake, S., Boesch, C. *et al.* (2012). Recent decline in suitable environmental conditions for African great apes. *Diversity and Distributions*, 18 (11): 1077–1091.

Kalema-Zikusoka, G., Rothman, J. M., Fox, M. T. *et al.* (2005). Intestinal parasites and bacteria of mountain gorillas (*Gorilla beringei beringei*) in Bwindi Impenetrable National Park, Uganda. *Primates*, 46 (1): 59–63.

Kaur, T., Singh, J., Tong, S. *et al.* (2008). Descriptive epidemiology of fatal respiratory outbreaks and detection of a human-related metapneumovirus in wild chimpanzees (*Pan troglodytes*) at Mahale Mountains National Park, western Tanzania. *American Journal of Primatology*, 70 (8): 755–765.

Köndgen, S., Kühl, H., N'Goran, P. K. *et al.* (2008). Pandemic human viruses cause decline of endangered great apes. *Current Biology*, 18 (4): 260–264.

Kuehl, H., Kouame, P. N. G., and Boesch, C. (2008). Alarming decline of West African chimpanzees in Côte d'Ivoire. *Current Biology*, 18 (19): R903–R904.

Lonsdorf, E. V., Travis, D., Pusey, A. E. *et al.* (2006). Using retrospective health data from the Gombe chimpanzee study to inform future monitoring efforts. *American Journal of Primatology*, 68: 897–908.

Macfie, E. J. and Williamson, E. A. (2010). *Best Practice Guidelines for Great Ape Tourism. Occasional Paper of the IUCN Species Survival Commission*. Gland, Switzerland: IUCN.

MacKenzie, C. A. (2012). Trenches like fences make good neighbours: revenue sharing around Kibale National Park, Uganda. *Journal for Nature Conservation*, 20 (2): 92–100.

McLennan, M. R. (2008). Beleaguered chimpanzees in the agricultural district of Hoima, western Uganda. *Primate Conservation*, 23 (1): 45–54.

McLennan, M. R. and Hill, C. M. (2010). Chimpanzee responses to researchers in a disturbed forest–farm mosaic at Bulindi, western Uganda. *American Journal of Primatology*, 72 (10): 907–918.

Mittermeier, R. A. and Cheney, D. L. (1987). Conservation of primates and their habitats. In: Smuts, B. B., Cheney, D. L., Seyfarth, R. M., Wrangham, R. W., and Struhsaker, T. T., (eds.), *Primate Societies*. University of Chicago Press, pp. 477–490.

Muehlenbein, M. P. (2006). Intestinal parasite infections and fecal steroid levels in wild chimpanzees. *American Journal of Physical Anthropology*, 130 (4): 546–550.

Muehlenbein, M. P., Ancrenaz, M., Sakong, R. *et al.* (2012). Ape conservation physiology: fecal glucocorticoid responses in wild *Pongo pygmaeus morio* following human visitation. *Plos One*, 7 (3): e33357.

Muehlenbein, M. P., Martinez, L. A., Lemke, A. A. *et al.* (2010). Unhealthy travelers present challenges to sustainable primate ecotourism. *Travel Medicine and Infectious Disease*, 8 (3): 169–175.

Mugisha, A. (2008). Potential interactions of research with the development and management of ecotourism. In: Wrangham, R. W. and Ross, E. (eds.), *Science and Conservation in African Forests: The Benefits of Long-term Research*. Cambridge University Press, pp. 115–127.

Nakamura, M. and Nishida, T. (2009). Chimpanzee tourism in relation to the viewing regulations at the Mahale Mountains National Park, Tanzania. *Primate Conservation*, 24 (1): 85–90.

Naughton-Treves, L. (1997). Farming the forest edge: vulnerable places and people around Kibale National Park, Uganda. *Geographical Review*, 87 (1): 27–46.

Nishida, T. and Mwinuka, C. (2005). Introduction of seasonal park fee system to Mahale Mountains National Park: a proposal. *Pan Africa News*, 12 (2): 17–19.

Nishida, T. and Nakamura, M. (2008). Long-term research and conservation in the Mahale Mountains, Tanzania. In: Wrangham, R. W. and Ross, E. (eds.), *Science and Conservation in African Forests: The Benefits of Long-term Research*. Cambridge University Press, pp. 173–183.

Nizeyi, J. B., Innocent, R. B., Erume, J. *et al.* (2001). Campylobacteriosis, salmonellosis, and shigellosis in free-ranging human-habituated mountain gorillas of Uganda. *Journal of Wildlife Diseases*, 37 (2): 239–244.

Obua, J. and Harding, D. M. (1996). Visitor characteristics and attitudes towards Kibale National Park, Uganda. *Tourism Management*, 17 (7): 495–505.

Pedersen, A. B. and Davies, T. J. (2009). Cross-species pathogen transmission and disease emergence in primates. *EcoHealth*, 6 (4): 496–508.

Pintea, L. (2007). *Applying Satellite Imagery and GIS for Chimpanzee Habitat Analysis and Conservation*. PhD Thesis. St. Paul, MN: University of Minnesota.

Pusey, A. E., Pintea, L., Wilson, M. L. *et al.* (2007). The contribution of long-term research at Gombe National Park to chimpanzee conservation. *Conservation Biology*, 21 (3): 623–634.

Pusey, A. E., Wilson, M. L., and Collins, D. A. (2008). Human impacts, disease risk, and population dynamics in the chimpanzees of Gombe National Park, Tanzania. *American Journal of Primatology*, 70 (8): 738–744.

Sandbrook, C. and Semple, S. (2006). The rules and the reality of mountain gorilla (*Gorilla beringei beringei*) tracking: how close do tourists get? *Oryx*, 40 (4): 428–433.

Struhsaker, T. T. (2008). Long-term research and conservation in Kibale National Park. In: Wrangham, R. W. and Ross, E. (eds.), *Science and Conservation in African Forests: The Benefits of Long-term Research*. Cambridge University Press, pp. 27–37.

Struhsaker, T. T., Struhsaker, P. J., and Siex, K. S. (2005). Conserving Africa's rain forests: problems in protected areas and possible solutions. *Biological Conservation*, 123 (1): 45–54.

Tranquilli, S., Abedi-Lartey, M., Amsini, F. *et al.* (2012). Lack of conservation effort rapidly increases African great ape extinction risk. *Conservation Letters*, 5 (1): 48–55.

UBOS. (2011). *Visitors to National Parks (Citizens and Foreigners), 2006 – 2010*. Uganda Bureau of Statistics.

Wallis, J. and Lee, D. R. (1999). Primate conservation: The prevention of disease transmission. *International Journal of Primatology*, 20 (6): 803–826.

Williams, J. M., Lonsdorf, E. V., Wilson, M. L. *et al.* (2008). Causes of death in the Kasekela chimpanzees of Gombe National Park, Tanzania. *American Journal of Primatology*, 70 (8): 766–777.

Williamson, E. A. and Feistner, A. T. C. (2003). Habituating primates: processes, techniques, variables and ethics. In: Setchell, J. M. and Curtis, D. J. (eds.), *Field and Laboratory Methods in Primatology*. Cambridge University Press, pp. 25–39.

Woodford, M. H., Butynski, T. M., and Karesh, W. B. (2002). Habituating the great apes: the disease risks. *Oryx*, 36 (02): 153–160.

Wrangham, R. W. (2008). Why the link between long-term research and conservation is a case worth making. In: Wrangham, R. W. and Ross, E. (eds.), *Science and Conservation in African Forests: The Benefits of Long-term Research*. Cambridge University Press, pp. 1–8.

Part IV

Neotropical primates

12 The impact of tourist group size and frequency on Neotropical primate behavior in Tambopata, Peru

Chloe Hodgkinson, Christopher Kirkby, and Eleanor J. Milner-Gulland

Introduction

Nature-based tourism, defined as "all tourism directly dependent on the use of natural resources in a relatively undeveloped state, including scenery, topography, water features, vegetation and wildlife" (Ceballos-Lascurain, 1996), is frequently described as one of the fastest growing sectors of the global tourism sector (Balmford et al., 2009). Its non-extractive nature may lead to the assumption that it is inherently sustainable, yet nature tourism typically occurs in fragile environments and may open up previously undiscovered destinations to the mass market. The mass market then repeatedly and actively seeks out resident wildlife (Jacobson & Figueroa-Lopez, 1994). Often little is known of nature tourism's true impacts on the biological and physical environment and only rarely are these quantified (Blanc et al., 2006; Roe et al., 1997). This study explored the effects that the presence of tourists can have on the behavior of Neotropical primates of the Tambopata-Candamo Reserved Zone, southeast Peru, and then discusses the implications of these results for nature tourism management in the area.

Numerous studies have shown the strong effect that human presence, in any capacity, may have on wild animal behavior. Cheetahs (*Acinonyx jubatus*) may alter their feeding time to coincide with periods of low tourist density (Gakahu, 1992); bottlenose dolphins (*Tursiops truncatus*) have been found to spend significantly less time resting in the presence of tourist boats (Constantine et al., 2004); polar bears (*Ursus maritimes*) increased their vigilance in the presence of vehicles (Dyck & Baydack, 2004); and Asian rhinoceros (*Rhinoceros unicornis*) responded to the presence of elephant-borne tourists by spending more time on alert, at the cost of feeding (Lott & McCoy, 1995). More clearly associated with longer-term impacts, yellow-eyed penguin chicks (*Megadyptes antipodes*) and juvenile hoatzins (*Opisthocomus hoazin*) that were visited regularly by tourists displayed lowered body weight, which is directly correlated with an increased risk of mortality (McClung et al., 2004; Mullner et al., 2004). Similarly, a long-term study of wood turtle (*Clemmys insculpta*) numbers showed a clear population decline coinciding with the growth of human activity levels in their area (Garber & Burger, 1995).

Primate Tourism: A Tool for Conservation?, ed. Anne E. Russon and Janette Wallis. Published by Cambridge University Press. © Cambridge University Press 2014.

A range of factors have been suggested to be influential in determining the responses of wildlife to human presence. Underlying behavioral characteristics of the species, such as timing of activity periods and form of escape, as well as the age and sex of the individuals involved may be significant (Dyck & Baydack, 2004; Tutin & Fernandez, 1991; Yalden, 1991). The number of humans present and their proximity to the wildlife have been found to be influential (Albert & Bowyer, 1991; Duchesne *et al.*, 2000; Gakahu, 1992; Rodgers & Smith, 1997; Treves & Brandon, 2005) as have the frequency and predictability of the human disturbance (Cassirer *et al.*, 1992; Neuhaus & Mainini, 1998; Scott *et al.*, 2001; Speight, 1973). All have important implications for management.

Disturbance of wildlife by humans may be reduced by habituating the animals to human presence, defined as the acceptance of humans as a neutral element in their environment (Tutin & Fernandez, 1991). Habituation may be attempted deliberately or occur incidentally, but it is usually achieved by repeatedly exposing the animals in question to human presence in a non-threatening manner. Habituation efforts, which are costly in terms of both time and money, are rarely conducted deliberately for the purposes of nature tourism, although they may occur incidentally or for other purposes such as research.

Primates are often assumed to be relatively easy to habituate to human presence (Griffiths & van Schaik, 1993). Significant behavioral alterations have nonetheless been found to be shown in the presence of tourists, including increased primate group spread (Treves & Brandon, 2005), altered ranging patterns (Goldsmith *et al.*, 2006), disrupted social behavior (de la Torre *et al.*, 2000; O'Leary & Fa 1993), and increased vocalization rates (Johns, 1996) (in this volume, see Berman *et al.*, de la Torre, Dellatore *et al.*, Goldsmith, Kauffman, Wright *et al.*). Time budgets can also be seriously disrupted, with mountain gorilla groups visited by tourists spending significantly more time moving and less time feeding (Steklis *et al.*, 2004; in this volume de la Torre, Dellatore *et al.*, Wright *et al.*).

If the effects of tourism on wildlife are to be managed, they must be identified and assessed. Primates' tendency to occur at relatively high densities in small areas, their high popularity with tourists, and their sensitivity to human presence make them potentially good bio-indicators of the impact of tourism on an area (de la Torre *et al.*, 2000; Kinnaird & O'Brien, 1996).

For these reasons, this study focused on the primates most commonly found in the Tambopata-Candamo Reserved Zone, southeast Peru, an area that has recently undergone rapid expansion in its tourism industry. Our investigation concentrated on the effects of tourist group size and tourist group frequency on primate behavior, two variables that are relatively easy to regulate and should form part of any management plan. We used study walks to observe and record the reaction of the four most commonly seen primate species in the area, namely the red howler monkey (*Alouatta seniculus*), the Bolivian squirrel monkey (*Saimiri boliviensis*), the brown capuchin monkey (*Cebus apella*), and the saddleback tamarin (*Saguinus fuscicollis*). All are widespread and common in the Amazonian area.

Alouatta seniculus, among the largest of the New World primates, is known for its characteristic long calls, usually produced at dawn and late afternoons. They are found most frequently in the mid-to-upper forest strata where they spend much of the day resting (Gaulin & Gaulin, 1982; Mittermeier & van Roosmalen, 1981). *Saimiri boliviensis* is a small monkey weighing less than 1 kg, found most frequently in large groups in the lower levels of the forest strata (Terborgh, 1983). *Cebus apella*, a medium-sized monkey weighing 2.5–4 kg, is found mainly in mid-forest strata. Both *S. boliviensis* and *C. apella* spend a large proportion of their time either foraging or moving between foraging sites (Robinson & Janson, 1987). *Saguinus fuscicollis*, a small, inquisitive primate weighing less than 500 g, tends to move and feed in the lower level in groups of 8–12 individuals (Terborgh, 1983).

We recorded the behavior of all four species during encounters with tourist groups with respect to initial reaction, overall behavior throughout each encounter, duration of contact, and termination of contact. Primate behavior was then analyzed for its relation to primate group size and height in the canopy, tourist group size, distance between tourists and monkeys when first encountered, and trail use frequency. A self-administered questionnaire was also distributed to visitors to the area, to explore their attitude to tourist group size.

Methods

Study area

This study was located in and around the Tambopata-Candamo Reserved Zone situated in the departments of Madre de Dios and Puno, southeast Peru. The zone was created in 1990 by the Peruvian government. Covering approximately 274 690 ha, the reserve features high levels of biodiversity including over 200 species of mammals (Dunstone & O'Sullivan, 1996). The area experiences a mean temperature of 27°C with an average of 2500 mm rainfall each year and a wet season lasting from October to April. The reserve is zoned into different land use types and includes a buffer zone in which the majority of tourism activities take place. Although it hosted little tourism in the past, due to such factors as difficulties of access, weak marketing, and political instability (Dunstone & O'Sullivan, 1996), the area underwent a rapid expansion of its tourist industry in the early 1990s. This has resulted in a dramatic increase in the number of large static lodges, which form the focus of most of the tourism activities in the region. This rapid expansion has led to a number of tourism-related problems, yet there remains an overall lack of coordinated and rigorous monitoring of the tourism operations. As a result, there is significant uncertainty regarding the magnitude of tourism-related impacts and few regulations aimed at controlling them.

We conducted fieldwork from May to July 2002 at two lodges, to allow a comparison between two different sites in the region. Explorers' Inn Amazon Lodge

(EI) (12° 50' 15"S, 69° 17' 30"W) had been in operation for 26 years and featured a trail system of 35 km covering a wide range of habitats classified as permanently and seasonally flooded swamp forest, sandy clay, and ultra-sand forest. There was no evidence of hunting in the area (Kirkby *et al.*, 2000). Ecoamazonia Lodge (ECO) (12° 31' 45"S, 68° 56' 10'"W) had been in operation for 10 years, and featured 15 km of trails that wound primarily through permanently and seasonally waterlogged swamp forest. Levels of hunting in the area were estimated to be low around the time of the study (Kirkby *et al.*, 2000). Because visitor numbers were gradually decreasing at EI and increasing at ECO, at the time of the study both lodges received approximately equal number of visitors, around 1500 people per year (DRITINCI-MD, 2001). The main tourist attraction for both lodges was guided walks along the forest trails around the lodges. Tourists were at no point allowed to leave the trail. The four target primate species were present in sufficient numbers around both lodges to provide frequent encounters with tourists on trail walks.

Data collection

Data were collected by CH, who spent a total of five weeks at each lodge, accompanying tourist groups from both lodges on guided walks. All walks took place in the morning, usually between the hours of 8:00 and 14:00, along established trails. Occasionally more than one organized walk took place at the same lodge, in which case the largest tourist group was accompanied. On days when there were no organized tourist walks, solo walks were conducted by CH. Upon an encounter with a primate group at any point during walks, the following variables were noted: time of day, location of trail, tourist group size, tourist group distance from primate group upon first encounter, the height of the primates in the canopy on first encounter, and an estimation of the primate group size. Tourist group size included the guide and the researcher. Distances were measured by tape measure after the primate group had left the area, to reduce inaccuracies of estimation. Height was estimated by eye following a period of training to improve accuracy.

Behavioral data were collected using instantaneous focal sampling techniques at 30-second intervals. Upon first observation of the primate group, a focal member was chosen. Its behavior was then noted both initially and then at 30-second intervals until termination of contact, which was judged to be when the observer could no longer see any member of the primate group. Behavioral categories recorded are shown in Table 12.1. In addition, the total duration of the contact was recorded (first observation to termination of contact), as well as who terminated the contact by moving away first (human or monkey group).

The focal animal was selected as the first sighted individual, because of practical difficulties associated with using other criteria (e.g. quickly aging and sexing members of some monkey species). If this individual was lost from view for more than 30 seconds, observations were switched to the individual nearest to where the previous focal animal was last seen. We accept this may introduce a certain amount

Table 12.1 Focal individual response categories (adapted from Tutin & Fernandez, 1991)

Category of behavior	Definition
Flight (fast/slow)	Moving away from the observer
Avoidance	Moving higher in the canopy
Hiding	Moving behind vegetation
Monitoring	Staring or constant glances at a human observer although position remains relatively constant
Ignoring	No discernible response. Continuation of previous activities
Curiosity	Moving toward a human observer

of bias, for example individual monkeys monitoring humans may have been more likely to be selected than those engaged in other, less obvious activities. However, in this study we were interested in relative rather than absolute behavior and this potential bias is assumed to affect all primate group–human encounters during this study. Instantaneous rather than continuous sampling was employed because it is simpler and less demanding than continuous sampling, so it tends to be more reliable (Martin & Bateson 1993).

Trail use frequencies were calculated at both lodges by dividing the total number of walks on each trail during the study by the total number of possible days. All tourist walks incorporated a minimum of two trails, either partially or full length. Both partial and full-length use was counted.

It is of course impossible to compare the behavior of primate groups in the presence of humans with their behavior in the absence of humans. Therefore we concentrated on how these primate groups' behavior varied as a function of selected variables outlined above.

A structured, self-administered questionnaire was designed to investigate the attitudes of tourists to tourist group size. Questions were asked regarding the number of walks completed, the number of monkey groups viewed, tourist group sizes, and satisfaction with these. Respondents were also asked their opinion about optimum tourist group size. A combination of closed and open response questions was employed. The questionnaire was translated into three languages, Spanish, English, and French, and handed to respondents on their last afternoon at the lodge.

Data analysis

Encounter data were analyzed in terms of number of overall sightings, initial reactions, overall behavior, duration of the contact, and termination of the contact. All encounters were assumed to be independent because it was unusual for more than one primate group to be encountered per day, and there was always a minimum of one hour between sightings when this did occur. For statistical analysis, the monkey groups' initial reactions to encountering tourist groups were coded on an ordinal "fear ranking" scale. Fleeing was ranked as 1, avoiding 2, hiding 3, monitoring 4, ignore 5, and curiosity 6. Overall behavior of the focal individual was analyzed in terms of the proportion of instantaneous samples spent in each behavior category per focal observation session. For duration of contact, initial behavior, and overall behavior, Spearman's Rank Correlation was used to assess relationships between these behaviors and with the following variables: tourist group size, tourist group distance from primate group upon first encounter, the height of the primates in the canopy on first encounter, and an estimation of primate group size. For termination of contact, primate sightings were divided into those terminated by the tourist group and those terminated by the monkey group. Chi-squared tests were then used to determine whether either group (monkey/tourist) was more likely to terminate the contact. Data collected during solo walks were used only in the analysis of initial and overall behavior.

For questionnaire data, a Mann–Whitney U test was used to investigate relationships between visitor group size and satisfaction.

Results

Tourist walk summary statistics

Tourist group sizes ranged from 4 to 13 (mean = 5.97, SD = 3.69) individuals at ECO and 4 to 11 (mean = 6.04, SD = 3.38) at EI. At ECO, tourist walks used one of three trails, with trail use frequencies for these ranging from 0.1 to 0.8 walks per day. At EI 18 trails were used over the study period, with trail use frequencies displaying a much wider range of use, varying from an average of 0.02 to 1.1 walks per day.

There was no significant difference between the two lodges in the average number of primate sightings per tourist walk ($t = 0.016$, $df = 60$, $p = 0.987$, see Table 12.2). No relationship was found between tourist group size and whether primates were viewed at all on a walk (ECO: Mann–Whitney U = 717.5, $p = 0.979$, EI: Mann–Whitney U = 305.5, $p = 0.711$). Findings were similar for each individual species.

Initial reaction to encountering a tourist group

We present results for each species as the percentage of encounters in which each behavioral category occurred (see Table 12.3). The results for each species were not significantly different between lodges, so data were pooled. There was a strong species division in the initial reaction to tourists. *Saguinus fuscicollis* tended to avoid tourists (48%), typically by gaining height in the canopy, or monitor tourists (48%). *Alouatta seniculus* had a strong tendency to monitor (66%) or ignore tourists (25%). *Cebus apella* displayed a range of initial reactions ranging from monitoring (36%) to fleeing (45%). *Saimiri boliviensis* tended to flee (66%). We found positive correlations between fear ranking and trail use frequency for both *C. apella* (Spearman's $\rho = 0.635$, $p = 0.027$) and *A. seniculus* (Spearman's $\rho = 0.781$, $p = 0.005$), that is the more frequently the trail was used, the less "fearful" their behavior.

Table 12.2 Tourist walk summary statistics

		ECO	EI
No. walks completed		36	26
No. primate viewings	*Alouatta seniculus*	6	6
	Saimiri boliviensis	9	0
	Cebus apella	10	2
	Saguinus fuscicollis	10	12
No. primate viewings (total)		35	20
Average no. sightings per walk		0.97	0.77

Table 12.3 Initial reactions of each primate species given as the percentage of the total number of tourist group encounters in which each type of behavior occurred

	Fear rank	*Saguinus fuscicollis*	*Alouatta seniculus*	*Cebus apella*	*Saimiri boliviensis*
Curiosity	6	0	0	0	0
Ignore	5	0	25	0	0
Monitor	4	48	66	36	22
Hide	3	0	0	0	0
Avoid	2	48	0	18	11
Flee	1	4	8	45	66
N		22	12	12	9

Note: *N* – total number of tourist group encounters sampled.

Overall behavior during tourist group encounters

We conducted analyses separately for each species, pooling results from both lodges (Table 12.4). First, in these four species, no primate group encountered at either lodge showed full habituation to tourist groups, judged as individuals consistently displaying no discernible response to the presence of tourists. As trail use frequency increased, the time *C. apella* spent fleeing significantly decreased (Spearman's ρ = −0.596, p = 0.041) and monitoring significantly increased (Spearman's ρ = 0.587, p = 0.045). Time spent avoiding the tourists, that is gaining height in the canopy, was negatively related to the height *C. apella* were in the canopy upon detection (Spearman's ρ = −0.761, p = 0.006). *Saguinus fuscicollis* showed a significant positive correlation between the amount of time spent monitoring a tourist group and the size of that tourist group, that is the larger the tourist group, the longer the monkeys spent monitoring it (Spearman's ρ = 0.470, p < 0.05, see Figure 12.1). We found no significant relationships between the size of tourist groups or trail use frequency and overall behavior for *A. seniculus* or *S. boliviensis*, between tourist group

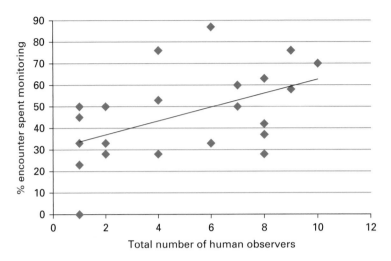

Figure 12.1   Correlation between tourist group size and the percentage of the contact time *Saguinus fuscicollis* spend monitoring (Spearman's ρ = 0.470, *p* < 0.05).

Table 12.4 Percentages of overall behavior upon contact with tourists, given as the percentage of the total number of encounters sampled

	Fear rank	*Saguinus fuscicollis*	*Alouatta seniculus*	*Cebus apella*	*Saimiri boliviensis*
Curiosity	6	2	0	1	0
Ignore	5	8	46	5	6
Monitor	4	45	47	26	8
Hide	3	0	0	5	0
Avoid	2	16	1	4	1
Flee	1	28	5	59	85
N		22	12	12	9

Note: *N* – total number of tourist group encounters sampled.

size and overall behavior for *C. apella* or between trail use frequency and the overall behavior of *S. fuscicollis*.

Duration of contact

Only observations that were terminated by the monkeys were used for this section of the analysis, as the duration of tourist-terminated contacts was likely to be strongly influenced by external factors and interest focuses on the primates' responses. As a consequence, sufficient data were available only for *C. apella* and *S. fuscicollis*. *Cebus apella* showed a strong positive relationship between trail frequency use and the duration of the contact (Spearman's ρ = 0.7, *p* = 0.011). No other trends were found.

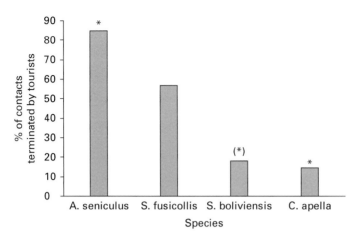

Figure 12.2 Percentage of observations of each primate species that were terminated by the tourists rather than by the primate group. Asterisks indicate a significant deviation from the null hypothesis of random encounter termination, using chi-squared tests, where (*) = $p < 0.1$, * = $p < 0.05$.

Termination of contact

There were no significant differences in contact termination between lodges for any species, therefore data were pooled. Both *S. boliviensis* and *C. apella* were significantly more likely to terminate contact before the tourists chose to move away ($\chi^2 = 4.455$, $df = 1$, $p = 0.035$ and $\chi^2 = 4.000$, $df = 1$, $p = 0.046$, respectively; see Figure 12.2). However, with *A. seniculus*, the tourists were significantly more likely to terminate the contact ($\chi^2 = 6.231$, $df = 1$, $p = 0.013$). No termination difference was detected for *S. fuscicollis* ($\chi^2 = 0.391$, $df = 1$, $p = 0.532$).

Associations between who terminated the contact and a range of explanatory variables were tested for *S. fuscicollis* only, because of insufficient variation in who terminated contact in the other species. Tourist group distance, initial height of primate in the canopy, and trail use frequency were non-significant, but the monkeys were more likely to terminate the contact when the tourist group was smaller ($\chi^2 = 5.490$, $df = 1$, $p = 0.036$).

Tourist response to group size

In total, 55 tourist questionnaires were completed: 42 at EI and 13 at ECO. Visitors to EI reported completing between 1 to 7 trail walks (mean = 3.42) during their stay, viewing an average of 1.32 monkey groups in total. Visitors to ECO reported going on fewer forest walks and completing 1 to 3 (mean = 2.00), but they still viewed an average of 1.25 monkey groups.

For tourist satisfaction and tourist group size, there were no significant differences between lodges so data were pooled. When asked to rate their satisfaction with tourist group size, no respondents considered their group "much too small" or "too small"; 53% considered their group as "satisfactory" and 47% as "too large" or "much too large." Those who considered their group size as satisfactory walked with an average of 7.42 people in their group, while those who considered their group "too big" or "much too big" had completed walks with an average tourist group size of 9.35. The average optimum group sizes suggested by tourists for trail walks were 7.1 people (ECO) and 5.8 (EI). Satisfaction was also significantly negatively related to actual group size (Mann–Whitney U = 237, $p < 0.05$).

Discussion

In the rush to cater to the rising demands for ecotourism, the impact on wildlife often seems to have been neglected, with the emphasis all too frequently placed on current revenue rather than long-term sustainability (e.g. Berman et al., 2007; Isaacs, 2000; Liu et al., 2001; Sandbrook & Semple, 2006). Nowhere is this better illustrated than by the frequent failure by park managers to take into account the impact of tourist group size and trail use frequency on fauna, variables which have been shown to be important in determining primates' reactions to observers in this and other studies (Johns, 1996; Kinnaird & O'Brien, 1996; Treves & Brandon, 2005). However, it is also clear from this study that primate species can react differently to tourist groups in the Tambopata-Candamo Reserved Zone. Results for individual species should be considered in light of their known underlying behavioral characteristics.

Saguinus fuscicollis appeared to show the greatest degree of habituation, echoing the results of Kirkby et al. (2000), who concluded that this species, found in significantly higher numbers in areas with higher tourist pressure, benefits in some way from the presence of tourists. Saguinus fuscicollis are known to flee when faced with perceived threats (Romero, 1999); nevertheless, the majority of their encounters with tourists were terminated by the tourists. It could therefore be argued that their staying in the vicinity of tourists indicates the monkeys' acceptance that the tourists do not pose an immediate threat. In addition, the monkeys' initial reaction to human observers was to increase their height in the canopy or to monitor the tourists, again suggesting low levels of fear. Interestingly S. fuscicollis were found to be less likely to terminate the encounter when tourist groups were larger. While the reason for this curious result must remain a matter of speculation, one possible explanation may be that larger groups of people are more easily recognizable as tourists and perhaps therefore seen as less threatening, whereas smaller groups of tourists could more easily be mistaken for hunters.

Alouatta seniculus also displayed some degree of habituation to tourist presence. This is deemed typical of this species when the group in question has been in frequent and non-harmful contact with humans (Mittermeier et al., 1981). On only two occasions were they observed to move away from human observers and terminate the contact. The majority of encounter time was divided between monitoring the tourists, a typical threat response for this species, or ignoring them. The ignore response, judged to be an indication of low levels of fear of humans, was more common on more heavily used trails.

This study found *S. boliviensis* extremely likely to terminate tourist contacts and liable to move away immediately upon contact. Interpretation of this result is somewhat complicated by the highly active nature of *S. boliviensis*. Group movement is usually continuous, and stationary feeding at a single location for more than 10 second intervals is rare (Klein & Klein, 1976). Without further and more detailed monitoring, it is therefore unclear how much of their movement is due to foraging and how much to the presence of the observers. However, it should be noted that no focal individual moved toward a human observer ("curiosity"), which would have been expected if their movements were directed by foraging only.

Cebus apella showed the most sensitivity to tourist presence, terminating 86% of encounters and spending the majority of encounter time moving away from human observers. Again, it proved very difficult to separate how much of this was natural foraging behavior (Terborgh, 1983) and how much was a direct result of the observers' presence. *Cebus* groups were, however, less likely to terminate contact and spent less time fleeing the human observers when encounters took place on more heavily used trails.

Tourist group size was not found to affect these primates' responses except for *S. fuscicollis*, for whom larger tourist groups were associated with more positive behavior. However, this study did not take place at peak tourist season and none of the tourist groups sampled reached the size limit of 20 people (for ECO) (maximum observed group size = 13). Therefore group size may affect primates' responses when numbers are higher, and this should clearly be assessed.

The most important factor affecting both *C. apella* and *A. seniculus* responses appeared to be the frequency of trail use. This suggests either that the primate groups whose ranges overlap the most frequently used trails have become more habituated to human presence, or that the same group behaves differently on different trails, depending on how heavily used they are, due to differences in the predictability of the encounter. More research would have to be done to demarcate the ranges of the primate groups found in the vicinity of the trails at both lodges to evaluate these two hypotheses.

The tourist group's distance from the primates was not found to be a major factor in determining primate response and it would have proved difficult to control. No tourist groups left the trails, and therefore movement toward the group was limited by how close to the path the primates were. However, the stress of a tourist encounter appeared to increase if tourists frequently altered their position (pers. obs.). This could be minimized by staying in one place as soon as monkeys have been located and resisting the temptation to pursue them.

In terms of assessing the impact of tourism on the primates in the Tambopata-Candamo Reserved Zone, this study focused on specified short-term behavioral responses to a few very specific variables. The behavioral changes observed here may seem relatively minor in terms of negative impacts, particularly in comparison with the serious tourism-related impacts seen elsewhere, such as increased levels of intra-group aggression and competition and increased rates of infanticide (Berman *et al.*, 2007, this volume; Matheson *et al.*, 2006; Orams, 2002). Yet the increased activity involved in avoiding the attention of tourists may well increase the daily energy expenditure of individuals, potentially affecting both growth and reproductive rates (Treves & Brandon, 2005). It would be extremely interesting to extend this study to explore the population dynamics of primate groups in the area, looking at factors such as birth rates, birth weights, and levels of infant mortality. While such a study would require a much greater investment in terms of resources, it would provide valuable insight as to the longer-term impacts of nature tourism on primates in the area.

Recommendations for tourism management

This study strongly suggests that tourist pressure be focused on a small number of designated trails, because the most frequently used trails were shown to be positively related to the degree of habituation of both *C. apella* and *A. seniculus*. Some studies propose that the impact on wildlife can be minimized by dispersing tourists over a greater area (Dunstone & O'Sullivan, 1996; Roe *et al.*, 1997). However in this case, the impact of tourists on the primates in the vicinity of the two lodges might best be minimized by the regular use of a few main trails for the tourists, with walks scheduled at the same time every day, thereby allowing the primate groups greater predictability in their tourist encounters. This "zoning" of tourist pressure should in theory allow a small area of the reserve to bear the impact of tourism while producing the revenue to protect a much wider, undisturbed area. It would also allow sensitive animals to choose to stay further away from the tourists, particularly as tourist groups are not permitted to leave the trails. However, this result should not be used to justify exceeding any trail's tourist "carrying capacity," such as suggested by Kirkby *et al.* (2000). Management plans for trails should include careful control of tourist pressure. Given that nature tourism must be sustainable both environmentally and economically and that tourist questionnaire responses in this study suggested a decline in visitor satisfaction for tourist groups exceeding eight people, calculation of tourist group size should also consider visitor preferences for smaller tourist group sizes.

Acknowledgments

We would like to extend our grateful thanks to Sr. Juan Toledo and Max Gunter, for allowing us to work at Ecoamazonia and Explorers' Inn, respectively. Thanks are also due to the staff from both these lodges who gave their time both generously and cheerfully. This research was aided immeasurably in the field by the logistical support of TReeS, Peru. In the preparation of this chapter we would like to thank Dr. Katie Fawcett for supporting its initiation. This research complies with the standards of the Imperial College Ethics Committee and adhered to the legal requirements of Peru, the country in which the research was conducted.

References

Albert, D. M. and Bowyer, R. T. (1991). Factors related to grizzly bear-human interactions in Denali National Park. *Wildlife Society Bulletin*, 19: 339–349.

Balmford, A., Beresford, J., Green, J., Naidoo, R., Walpole, M., *et al.* (2009) A global perspective on trends in nature-based tourism. *PLoS Biology*, 7 (6): art. no. e1000144. doi:10.1371/journal.pbio.1000144.

Berman, C. M., Li, J. H, Ogawa. H., *et al.* (2007). Primate tourism, range restriction, and infant risk among *Macaca thibetana* at Mt. Huangshan, China. *International Journal of Primatology*, 28: 1123–1141.

Blanc, R., Guillemain. M., Mouronval, J., *et al.* (2006). Effects of non-consumptive leisure disturbance to wildlife. *Revue d'Ecologie (Terre et Vie)*, 61: 117–133.

Cassirer, E. F., Freddy, D. J., and Ables, E. D. (1992). Elk responses to disturbance by cross-country skiers in Yellowstone National Park. *Wildlife Society Bulletin*, 20: 375–381.

Ceballos-Lascurain, H. (1996). *Tourism, Ecotourism and Protected Areas*. Gland, Switzerland: IUCN.

Constantine, R., Brunton, D. H., and Dennis, T. (2004). Dolphin-watching tour boats change bottlenose dolphin (*Tursiops truncatus*) behavior. *Biological Conservation*, 117: 299–307.

de la Torre, S., Snowdon, C. T., and Bejarano, M. (2000). Effects of human activities on wild pygmy marmosets in Ecuadorian Amazonia. *Biological Conservation*, 94: 153–163.

DRITINCI-MD. (2001). The Tambopata-Candamo Reserved Zone of Southeastern Peru: A biological assessment. *Conservation International*, RAP working paper, November 1994.

Duchesne, M., Côté, S., and Barrette, C. (2000). Responses of woodland caribou to winter ecotourism in the Charlevoix Biosphere Reserve, Canada. *Biological Conservation*, 6: 311–317.

Dunstone, N. and O'Sullivan, J. N. (1996). The impact of ecotourism development on rainforest mammals. In: V. J. Taylor and N. Dunstone (eds.), *The Exploitation of Mammal Populations*. London: Chapman and Hall, pp. 313–332.

Dyck, M. G. and Baydack, R. K. (2004). Vigilance behavior of polar bears (*Ursus maritimus*) in the context of wildlife-viewing activities at Churchill, Manitoba, Canada. *Biological Conservation*, 116: 343–350.

Gakahu, C. G. (1992). Tourist attitudes and use impacts in Maasai Mara National Reserve (Kenya). *Proceedings of a Workshop Organized by Wildlife Conservation International at Maasai Mara National Reserve, March 1991*. Nairobi, Kenya: Wildlife Conservation International.

Garber, S. D. and Burger, J. (1995). A 20-yr study documenting the relationship between turtle decline and human recreation. *Ecological Application*, 5: 1151–1162.

Gaulin, S. J. C. and Gaulin, C. K. (1982). Behavioural ecology of *Alouatta seniculus* in Andean cloud forest. *International Journal of Primatology*, 3: 1–32.

Goldsmith, M., Glick, J., and Ngabirano, E. (2006). Gorillas living on the edge: Literally and figuratively. In: N. E. Newton-Fisher, H. Notman, J. D. Paterson, and V. Reynolds (eds.), *Primates of Western Uganda*. New York: Springer, pp. 405–420.

Griffiths, M. and van Schaik, C. P. (1993). The impact of human traffic on the abundance and activity periods of Sumatran rainforest wildlife. *Conservation Biology*, 7: 623–626.

Isaacs, J. C. (2000). The limited potential of ecotourism to contribute to wildlife conservation. *Wildlife Society Bulletin*, 28: 61–69.

Jacobson, S. K. and Figueroa Lopez, A. (1994). Biological impacts of ecotourism; Tourists and nesting turtles in Tortugero National Park, Costa Rica. *Wildlife Society Bulletin*, 22: 414–419.

Johns, B. G. (1996). Responses of chimpanzees to habituation and tourism in the Kibale forest, Uganda. *Biological Conservation*, 8: 257–262.

Kinnaird, M. F. and O'Brien, T. G. (1996). Ecotourism in the Tangkoko DuaSudara Nature Reserve: Opening Pandora's box? *Oryx*, 30: 65–73.

Kirkby, C., Farfan, A. C., Doan, T. M., *et al.* (2000) *Project Tambopata; Tourism development and the status of Neotropical lowland wildlife in Tambopata, South-eastern Peru: Recommendations for tourism and conservation*. Peru: TReeS.

Klein, L. L. and Klein, D. J. (1976). Neotropical primates: Aspects of habitat usage, population density and regional variation in La Macarena, Columbia. In R. W. Thorington and P. G. Heltne (eds.), *Neotropical Primates: Field Studies and Conservation*. Washington, DC: National Academy of Science, pp. 70–78.

Liu. J., Linderman, M., Ouyang, Z., *et al.* (2001). Ecological degradation in protected areas: the case of Woolong Nature Reserve for giant pandas. *Science*, 292: 98–101.

Lott, D. F. and McCoy, M. (1995). Asian rhinos *Rhinoceros unicornis* on the run? Impact of tourist visits on one population. *Biological Conservation*,73: 23–26.

Martin, P. and Bateson, P. (1993). *Measuring Behavior: An Introductory Guide*. 2nd edn, Cambridge University Press.

Matheson, M. D., Sheeran, L. K., Li, J. H., and Wagner, S. J. (2006). Tourist impact on Tibetan macaques. *Anthrozoos: A Multidisciplinary Journal of the Interactions of People and Animals*, 19:158–168.

McClung, M. R., Seddon, P. J., Massaro, M., and Setiawan, A. N. (2004). Nature-based tourism impacts on yellow-eyed penguins *Megadyptes antipodes*: Does unregulated visitor access affect fledging weight and juvenile survival? *Biological Conservation*, 119: 279–285.

Mittermeier, R. A., Rylands, A., Coimbra-Filho, A., and da Fonseca, G. (1981). *Ecology and Behaviour of Neotropical Primates*. Rio de Janeiro: Academia Brasileira de Ciencias.

Mittermeier, R. A. and van Roosmalen X. (1981). Preliminary observations of habitat utilization and diet in 8 Surinam monkeys. *Folia Primatologica*, 36: 1–39.

Mullner, A., Linsenmair, K. E., and Wikelski, M. (2004). Exposure to ecotourism reduces survival and affects stress response in hoatzin chicks (*Opisthocomus hoazin*). *Biological Conservation*, 118: 549–558.

Neuhaus, P. and Mainini, B. (1998). Reactions and adjustment of adult and young alpine marmots *Marmota marmota* to intense hiking activities. *Wildlife Biology*, 4: 119–123.

O'Leary, H. and Fa, J. A. (1993). Effects of tourists on Barbary macaques at Gibralter. *Folia Primatologica*, 61: 77–91.

Orams, M.B. (2002). Feeding wildlife as a tourism attraction: a review of issues and impacts. *Tourism Management*, 23: 281–293.

Robinson, J. G. and Janson, C. H. (1987). Capuchins, squirrel monkeys, and Atelines: Socioecological convergence with Old World Primates. In B. B. Smuts, D. L. Cheney, R. M. Seyfarth, *et al.* (eds.), *Primate Societies*. University of Chicago Press, pp. 69–82.

Rodgers, J. A. and Smith, H. T. (1997). Buffer zone distances to protect foraging and loafing waterbirds from human disturbance in Florida. *Wildlife Society Bulletin*, 25: 139–145.

Roe, D., Leader-Williams, N., and Dalal-Clayton, B. (1997). *Take only Photographs, Leave Only Footprints: The Environmental Impacts of Wildlife Tourism*. IIED Wildlife and Development Report 10.

Romero, J. I. A. (1999). *Datos sobre la conducta de Saguinus fuscicollis (Callithricidae: Primates), en la Reserva Cusco Amazónico.* [MSc thesis]. National University of San Antonio Abad, Cusco.

Sandbrook, C. and Semple, S. (2006). The rules and the reality of mountain gorilla *Gorilla beringei beringei* tracking: how close do tourists get? *Oryx*, 40: 428–433.

Scott, G, Knight, R. L., and Miller, C. K. (2001). Wildlife responses to pedestrians and dogs. *Wildlife Society Bulletin*, 29: 124–132.

Speight, M. C. D. (1973). *Outdoor Recreation and its Ecological Effects; A Bibliography and Review.* University College, London, 4.

Steklis, H. D., Hodgkinson, C., Fawcett, K., *et al.* (2004). The impact of tourism on mountain gorillas. *Folia Primatologica*, 75: 40.

Terborgh, J. (1983). *5 New World Primates: A Study in Comparative Ecology.* New Haven, CT: Princeton University Press.

TIES. (2009). *What is Ecotourism? Definition and Ecotourism Principles.* www.ecotourism.org/ (accessed March 15, 2009).

Treves, A. and Brandon, K. (2005). Tourist impacts on the behavior of black howling monkeys (*Alouatta pigra*) at Lamanai, Belize. In J. Paterson, and J. Wallis (eds.), *Commensalism and Conflict: The Human-primate Interface.* Norman, OK: American Society of Primatologist, pp. 146–167.

Tutin, C. E. G. and Fernandez, M. (1991). Responses of wild chimpanzees and gorillas to the arrival of primatologists. In: H. Box (ed.), *Primate Responses to Environmental Change.* London: Chapman & Hall, pp. 187–197.

Yalden, D. W. (1991). Recreational disturbance of large mammals in the Peak District. *Journal of Zoology*, 221: 293–326.

13 Interactions between tourists and white-faced monkeys (*Cebus capucinus*) at Manuel Antonio National Park, Quepos, Costa Rica

Laurie Kauffman

Introduction

Tourism, particularly wildlife-based tourism, is one of the fastest growing industries in the world. Tourism currently accounts for 10.8% of the global gross domestic product and it is expected to grow by 4.6% annually over the next ten years (World Travel and Tourism Council, 2005). Twenty to forty percent of international tourists are interested in wildlife-based experiences and this high market share results in fierce competition among tour operators to provide more wildlife tourism opportunities (Reynolds & Braithwaite, 2001). Many countries and organizations are counting on this demand to grow local economies sustainably while protecting the environment. However, more and more research is finding that wildlife-based tourism is not a panacea for habitat loss. In promoting nature as a tourist attraction, natural surroundings can be negatively affected (Knight & Cole, 1995; Ziffer, 1989). Introducing people into an environment will have some effect, be it production of litter, construction of trails and buildings, changes in the local culture, or changes in the behavior of the wildlife. Effects of tourism on wildlife can vary from feeding and activity budget changes (Ikiara & Okech, 2002) to reproductive changes (Jacobson & Lopez, 1994), ranging changes (Fairbanks & Tullous, 2002), and deaths of individual animals (Mathieson & Wall, 1982; Roe *et al.*, 1997). Indeed, one review found that 81% of studies investigating the effects of wildlife tourism on animals classified tourism impacts as harmful (Knight & Cole, 1995).

Successful sustainable tourism must strike a balance between preserving the environment, enriching local communities, and providing memorable experiences to tourists. These disparate goals may not always be compatible. Research results demonstrate that tourist satisfaction increases as tourists are allowed more control over their encounters with wildlife (Reynolds & Braithwaite, 2001; Sparks, 1994) and closer contact with wildlife (Farber & Hall, 2007; Fredrickson & Anderson, 1999; Valentine *et al.*, 2004). Tourists also prefer experiences during biologically

Primate Tourism: A Tool for Conservation?, ed. Anne E. Russon and Janette Wallis. Published by Cambridge University Press. © Cambridge University Press 2014.

active times, such as birth or breeding seasons (Reynolds & Braithwaite, 2001). These close encounters can result in danger for the tourists, such as being bitten by the animals (Albert & Bowyer, 1991; Brennan *et al.*, 1985; Fairbanks & Tullous, 2002), and danger for the animals in the form of increased chances of disease transmission from human to animal (Kaur & Singh, 2008; Köndgen *et al.*, 2008; Wallis & Lee, 1999).

There are currently few published studies of the effects of tourism on primates. Most studies of primate-based tourism center on great apes in Africa, where tourism was established with the purpose of protecting these apes and appears to be having some success (Blom *et al.*, 2004; McNeilage, 1996). As a consequence of this research, tourists visiting these areas are restricted in number and behavior. With careful control, planning, and habituation, negative impacts on the great apes can be minimized (Blom *et al.*, 2004; Johns, 1996). One researcher states that tourism has contributed to the economic viability of gorilla conservation in Rwanda and the Democratic Republic of Congo by, in effect, having the gorillas pay for their own protection (McNeilage, 1996). While African great ape-centered tourism is tightly monitored and controlled and may be contributing to conservation, tourism involving other primates has proven detrimental to the populations visited. In Belize, the presence of tourists at a Mayan archaeological site was associated with increases in howling monkeys' scattering and fleeing behavior, time on the ground, vigilance toward tourists, vocalization rates, and infant mortality (Grossberg *et al.*, 2003). Other studies on primates and tourism have demonstrated reductions in the primates' home range size, daily travel routes, and amount of time spent feeding in response to tourism (Kinnaird & O'Brien, 1996; Koganezawa & Imaki, 1999; Saj *et al.*, 1999).

This chapter examines the effect of increased tourism on one troop of white-faced capuchin monkeys (*Cebus capucinus*) in Manuel Antonio National Park, Costa Rica's smallest but most visited park. Costa Rica is one of the world's great nature tourism hot spots and success stories. Nature tourism in Costa Rica has grown rapidly (ICT, 2002) because of its vast natural resources (it contains 5% of the world's biodiversity), its conservation of 25% of its land in protected areas (Honey, 2003), and its friendliness and accessibility to the United States. After a description of tourism in Costa Rica, I will present data on behavioral changes seen in these capuchin monkeys as a result of increasing tourism from 1987 to 1998, and relate those changes to the viewpoints of tourists visiting the park.

Tourism in Costa Rica

Costa Rica created its first national park in 1970 (Boza, 1993) and became a major nature tourism destination shortly thereafter. Currently there are 25 national parks in Costa Rica (World Conservation Union and UNEP-World Conservation Monitoring Centre, 2009). From 1975 to 1995 the number of foreign visitors doubled (Honey, 2003) and by the end of the twentieth century tourism had surpassed bananas and coffee as the country's largest industry (ICT, pers. comm.; Lumsdon & Swift, 1998). The importance of natural resources to this tourism can be seen in that, in 1992, 75% of the one million tourists to Costa Rica visited its national parks (Kaimowitz & Segura, 1996) and 60% of these stated that they came for eco-tourism (Honey, 2003).

Manuel Antonio National Park illustrates the popularity of national parks with tourists. The park was created in 1972 and, at only twice the size of New York's Central Park, is the smallest yet most visited park in Costa Rica (Rachowiecki, 1997). Its attractions include white sandy beaches, hiking trails, and abundant wildlife including a population of at least 60 white-faced monkeys (*C. capucinus*). The park had 50 000 visitors in 1987, 116 000 visitors in 1997, and more recently reported receiving about 150 000 visitors a year (Baker, 2007; Manuel Antonio National Park staff, pers. com.). People who have been involved with the park for a number of years report that the park and the surrounding area changed greatly between 1987 and 1998 in terms of development and population.

One of the park's main attractions is Manuel Antonio beach – a white sandy coastline on a calm cove. Many tour groups visit the area with picnic lunches, which quickly attract the attention of white-faced capuchin monkeys. The monkeys attempt to steal the tourists' food despite tour guides standing guard. They have also learned to open backpacks, which they unzip, and steal food while the owners are enjoying a swim. There are many garbage cans around the park that become very full, and monkeys take advantage of these. In addition to obtaining human food in these ways, the monkeys receive food directly from tourists, who actively feed them despite a sign at the park entrance stating that feeding monkeys is forbidden.

Current research

The study reported here took advantage of a unique opportunity to investigate the impact of tourists on white-faced capuchin monkeys in Manuel Antonio National Park. In 1987, a student on the Associated Colleges of the Midwest Field Research program completed a study of the home range, habitat use, activity patterns, and food habits of the white-faced monkeys living in the park (Litvak, 1987). In 1998, I investigated the same parameters to see how the monkeys had changed over the intervening 11 years. I also used tourist surveys to investigate opinions and knowledge about the monkeys.

This paper investigates if and how white-faced capuchin monkeys in Manuel Antonio have been affected by tourism and how the knowledge and opinions of tourists might affect their behavior toward the monkeys. It was expected that when presented with easy-to-obtain, high-calorie food sources by tourists, monkeys would preferentially choose this food over their regular diets. Furthermore, because monkeys that obtain food from humans in a proscribed area will have to travel less to search for food, the size of their home range was expected to decrease. Based on observations of monkey behavior and interviews with tourists, this chapter investigated the effects of increased tourism on the monkeys and the knowledge and opinions of the tourists by testing the following hypotheses:

1) Monkeys increased the amount of tourist food eaten and decreased other food eaten from 1987 to 1998.
2) Monkeys decreased the size of their home range from 1987 to 1998.
3) Tourists think it is acceptable to feed monkeys.
4) Tourists visit the park to see monkeys.
5) Tourists have positive view of monkeys.

Methods

This study had two components. First, I collected behavioral data on a monkey troop that had high levels of contact with tourists. Second, I surveyed tourists visiting the park about their attitudes toward the monkeys.

Study site

The study was conducted at Manuel Antonio National Park, Puntarenas Province, Costa Rica. Manuel Antonio is located 7 km southeast of the town of Quepos. The park covers 683 ha of land (Rachowiecki, 1997) described as an ecotone between a tropical wet forest and a tropical moist forest (Tosi, 1969). Its average annual rainfall is 3820 mm; the rainiest months are from June to November and the driest from December to May (Institute Meteorológico National, pers. comm.). Currently, the park mostly consists of secondary forest in different stages from very young to semi-mature because a storm in 1993 destroyed much of the existing forest. Other vertebrates observed at the study site include coatis (*Nasua narica*), howling monkeys (*Alouatta palliata*), squirrel monkeys (*Saimiri oerstedi citrinellus*), lesser anteaters (*Tamandua mexicana*), agoutis (*Dasyprocta punctata*), armadillos (*Dasypus novemcintus*), two-toed sloths (*Choloepus hoffmani*), three-toed sloths (*Bradypus variegatus*), boa constrictors (*Boa constrictor*), black iguanas (*Ctenosaura sirnilis*), and raccoons (*Procyon lotor*).

Study animals

White-faced capuchin monkeys, of the Cebidae family, are diurnal and arboreal. They live in a variety of habitats throughout Central America, including deciduous dry forest on the Pacific coast of Costa Rica and the wet lowlands on the Atlantic Coast (Freese, 1983). Capuchins weigh from 2.5 to 3.5 kg, with males being slightly larger than females. Their diet is omnivorous and flexible, with food choices changing according to season (Chapman, 1988; Fragaszy *et al.*, 2004). White-faced capuchins spend 70–80% of the day foraging and traveling, with the remainder mostly spent resting (Freese, 1983). It is believed that white-faced capuchins are important in seed dispersal and their main predators are humans, raptors, boa constrictors, and felids (Fragaszy *et al.*, 2004; Freese, 1983; Moscow & Vaughan-Dickhaut, 1987; Oppenheimer, 1982). In captivity, these monkeys often live to 40 years of age and the oldest recorded age was nearly 54.75 years (Fragaszy *et al.*, 2004; Hakeem *et al.*, 1996). The species has an interbirth interval of about 19 months, leading to the estimate that a female could produce at most 22 infants in her lifetime (Fragaszy & Bard, 1997).

At least three troops of capuchins were known to range in the park. I concentrated on the troop that had the most contact with tourists and whose range encompassed Manuel Antonio beach. Capuchins possessing the same home range were previously studied in March and April of 1987 and I compared my results to this earlier research (Litvak, 1987). While we cannot be certain that the troop followed in this research was made up of any of the same individuals as the 1987 troop, the two troops shared a core home range and sleeping sites. Other researchers working in the park believe the troops are the same (G. Wong, pers. comm.). The current study troop consisted of fourteen to seventeen individuals – five to six females and nine to eleven males. Three of the males were infants and two were juveniles. All the females were either adults or juveniles. I identified the troop each day by the characteristics of some members (one with a scarred face, one old female, etc.), group composition, and location.

Survey informants

I chose informants from tourists who were on Manuel Antonio beach and near the monkeys. As I had no systematic way of knowing who was in the park, I opportunistically (Bernard, 1995; Weiss, 1995) chose tourists who were on and around the beach where the study monkeys spent most of their time. These tourists were likely to have interacted with the monkeys and formed opinions about them.

Data collection

Data for this study were collected from March 2 to April 25, 1998. This is the same timeframe as sampled in Litvak's study, so the effects of seasonality should be minimized.

Behavioral data

I collected behavioral data using focal animal sampling but I also used ad lib data collection (Altmann, 1974) to note when I saw monkeys other than the focal monkey being fed by tourists or going through the garbage. I observed the monkeys during the daylight hours each day, from about 05:00 to 17:00. When I found the troop each morning, I chose the monkey closest to me as my focal and recorded all its behaviors and their durations for ten minutes. I recorded the behaviors using the ethogram in Table 13.1. Parts of the ethogram (travel, rest, and foraging) are after

Table 13.1 Ethogram of white-faced monkey behaviors recorded in this study

Behavior	Description
Travel	Any active movement from tree to tree, on an individual tree, or on the ground without any apparent foraging activity.
Rest	Any low-energy activity, including reclining over a limb, sitting passively, sleeping, or scratching.
Grooming	One animal grooms another.
Eating tourist food	An animal appears to eat food that does not occur naturally in the park. This includes food stolen from tourists, food fed to the monkeys, and food taken from the garbage.
Feeding fruit	An animal appears to eat fruit that occurs naturally in the park.
Feeding insects	An animal appears to eat insects.
Foraging	An animal either visually or actively searches for food without apparent swallowing. This includes opening up dead leaves, looking at fresh leaves, and picking or pulling apart fruit, branches, or trunk with hands or teeth. It also includes looking into garbage or recycling cans and approaching or looking through tourists' bags.
Aggressive interaction	An animal displays, growls, threatens, bites, bares teeth, or acts in any other aggressive manner toward another animal, monkey, or person.
Non-aggressive interaction	Any interaction between a monkey and another monkey, a monkey and another animal, or a monkey and a person that involves contact and is not aggressive or grooming. This includes playing, nursing, copulating, touching, etc.
Fed by human	A tourist offers food to a monkey and the food is taken by the monkey.

Litvak (1987) with modifications to allow for foraging to occur in the garbage and in tourists' bags. I then took a five-minute rest and repeated the process, choosing the monkey next closest to me as the next focal. In some cases the focal monkey would move out of sight before the ten-minute sampling period was completed. When this occurred, I would note that the monkey had moved out of sight, wait a fixed amount of time, and then start a new sample with a new focal. If the focal sample period that had been completed prior to the subject's disappearance was less than four minutes, I waited one minute; if it was four to seven minutes, I waited two minutes; and if it was seven or more minutes, I waited five minutes. I also recorded the distance the focal capuchin was from any visible tourists in meters. For my purposes, a tourist was any person who was not the observer or a person who worked or lived in the park, such as a ranger. Guides were counted as tourists because they were generally in the company of tourists and did not live in the park.

I estimated the capuchins' home range by using paths and familiar landmarks to register the location of the troop every time I observed them. I transferred these data to a scale map of the park, joined the points to form the minimum convex polygon, and measured the area inside the polygon to estimate the monkeys' home range in hectares.

Survey data

Using a questionnaire, I interviewed each participating tourist to determine their opinions of the monkeys, the importance of the monkeys in their decision to visit the park, and whether they thought it was good to feed the monkeys. I conducted interviews in Spanish or in English, depending on the informant's preference.

Results

Monkeys

I collected 79.4 hours of behavioral observation on the monkeys in the study troop, representing 25 days' observation and 724 focal samples (average 42.5 focal samples per individual in the troop, assuming 17 individuals). The size of the troop's home range was estimated at 38 ha. Based on these observations, monkeys in this troop spent 13.4% of total time observed traveling, 28.6% resting, 43.9% foraging, and 14.1% feeding. They spent an average of 80.9 minutes a day on Manuel Antonio beach (58% of daily observation time) and were within 20 m of tourists for 67% of the total observation time. They mostly consumed tourist food (46%), followed by insects (28%), then fruit (26%). Resting was most frequent from 10:00 to 14:00. Foraging was constant throughout the day, with peaks at 08:00–09:00 and at 15:00–16:00. Travel was relatively constant, with a low from 12:00–13:00. Feeding on fruit peaked at 07:00–08:00 and 16:00–17:00 and was lowest at 13:00–14:00.

Ad lib samples showed daily averages of 11.6 instances of eating tourist food, consisting of an average of 5.6 instances per day of foraging in the garbage, 3.1 of being fed by tourists, 1.7 of foraging in tourists' bags, and 1.2 of stealing. I observed monkeys in the garbage on 24 of the 25 observation days and saw tourists feed them on 20 of the 25 days.

I compared these results to those obtained by Litvak (1987). Comparisons are shown in Table 13.2. Chi-squared tests were used to test for differences between the time budgets found in 1987 and 1998 and the distribution of foods eaten by capuchins in those years. Both tests were significant ($p < 0.002$, $df = 6$ for time budget, $p < 0.01$, $df = 4$ for types of food eaten), indicating that the data from the current study differ from those expected from the 1987 study. These comparisons support both hypotheses about the monkeys' behavior: Over about ten years since Litvak's (1987) study, the monkeys increased the amount of tourist food eaten, decreased the amount of other foods eaten, and decreased the size of their home range.

Tourists

I interviewed 50 tourists – 31 females and 19 males. The majority (52%) were from the United States, with Costa Rica the second highest represented at 14%. For 66% of the informants, it was their first visit to the park.

Table 13.2 Comparing the current study's findings to Litvak (1987)

Measure	Current study	Litvak (1987)
Troop		
Size	14–17	20–22
Composition	females (including juvenile and adults): 5–9	
	adult males: 4–6	
	juvenile males: 2	
	infants: 3	Unknown
Activity budget		
Travel %	12.8	37.1
Forage %	41.9	20.5
Rest %	27.3	18.1
Feed %	13.4	13.7
Ranging		
Home range size (ha)	38	60
Daily time on beach	> 1 hr	5–10 min
Dietary composition		
# fruit species	11	23
% Fruit	26	79.3
% Insects	28	13.1
% Tourist foods	46	<0.3

To test the hypothesis that tourists think it is acceptable to feed the monkeys, I asked informants, "Do you think feeding the monkeys is a good thing?" and "Why or why not?" Forty-two (84%) of the respondents said it was not good to feed the monkeys, four (8%) thought it was good, and four (8%) did not know. When asked why, 64% of tourists who thought it was not good said it was unnatural and made the monkeys "too tame," 20% said it made the monkeys aggressive, 13% said the monkeys had enough food in nature, and 3% said it was not allowed. Concerning tourists' reasons for visiting the park, only 17/50 (34%) mentioned the monkeys as the main reason for their visit.

Concerning tourists' views of the monkeys, 48% had a favorable view (i.e. said they were good in some way) and 16% had a negative view. Those with a favorable view described the monkeys as "cute," "friendly," "entertaining," and "interesting". Tourists with a negative view described the monkeys as "aggressive," "too tame," "badly behaved," and "sad." The remaining 36% were ambivalent and said that the monkeys were "tame" or that seeing them was an unusual/unique experience.

These results do not support the hypotheses that tourists think it is acceptable to feed the monkeys or that the monkeys are the main draw for tourists. The third hypothesis predicted that tourists would have a positive view of the monkeys; a minority (16%) had a negative opinion but not quite half (48%) had a clearly positive view, so this hypothesis was confirmed.

Discussion

Monkeys

The current study troop differs from Litvak's (1987) study in the percentage of time they spent on each behavior, the size of their home range, and the amount of time they spent on the beach. All these changes took the form predicted if they were caused by tourist influences. Could any of these changes be explained by the 1993 storm that destroyed many trees in the park? While this disturbance could have contributed to changes in the capuchins' ranging patterns, the fact that the monkeys now focus their travel so much on the beach and eat so much food gained from people indicates that the increase in tourism is still greatly influencing the monkeys' behavior. The capuchins may have more quickly increased their dependence on tourist-provided food due to the destruction of important fruit trees, but without the presence of tourists the monkeys may have just moved into different areas of the park or concentrated more on fallback foods.

The current troop has also changed the makeup of its diet since Litvak's study. It is generally reported that the bulk of this species' diet is either fruit or insects (Chapman & Fedigan, 1990; Fragaszy et al., 2004), although great variability exists (Fragaszy et al., 2004). It is therefore important that this study found this troop of white-faced capuchins to eat mostly tourist food, followed by insects and finally fruit.

This high percentage of tourist food in the diet can be linked to the other changes found. Since these white-faced monkeys got most of the tourist food from Manuel Antonio beach, they spent greater amounts of time there and probably therefore traveled less than they did during Litvak's study. This extra time was available for foraging and resting. The amount of tourist food eaten is also consistent with the smaller size of the current troop's home range. Research has found that home range size is related to food availability (Campos et al., 2014; Oates, 1987), thus the abundance of food in the area of Manuel Antonio beach can explain why the study troop centered its home range in this area. Moreover, as food was so abundant in a small area this troop did not need to travel as much, and therefore decreased its home range size. The earlier troop probably had a larger home range because they used more forest foods overall. During Litvak's study, the monkey troop that frequented the beach ate fruit from 23 species while that in the current study ate fruit from only 11 species. As both studies were conducted in March and April, the dry season, it is unlikely that seasonal differences in the availability of fruit caused this result. Also, the current troop ate mostly insects and tourist food, both of which do not necessitate a lot of travel.

These changes to feeding and ranging behavior could have important effects on the monkeys such as increases in intergroup and intragroup behavior, infant mortality, and the health of individual animals. The establishment of a small but valuable home range with higher food availability could result in increased inter-group and intragroup aggression and resultant injury or death. Previous studies have found that provisioned primate groups do show increased within-group aggression when compared with non-provisioned groups (Altmann & Muruthi, 1988; Berman et al., 2007, this volume; Hill, 1999). Additionally, Berman et al (2007) found increased infant mortality due to aggression in Tibetan macaques managed for tourism. Furthermore, disease transmission from humans to monkeys could also have serious effects. Recent research demonstrates the ease with which diseases can be transmitted from humans to nonhuman primates and how such diseases could have serious effects on a population (Köndgen et al., 2008; Wallis & Lee, 1999). Eating richer foods such as the high-sugar fruits and junk food provided by tourists could also lead to pre-diabetic symptoms in individual monkeys (see Sapolsky, this volume).

Finally, we should ask what might happen if humans leave the park. As the monkeys are so habituated, they would be very susceptible to the introduction of hunting to the area. Furthermore, there is the question as to whether the monkeys could adjust to living without tourist food if it was removed. Primates such as white-faced capuchins may depend on social learning to recognize edible foods (Custance et al., 2002; Fragaszy et al., 2004; Perry, 2011). To recognize a food as edible, a monkey may have to see a conspecific eat it. If most of the capuchins at Manuel Antonio cease to use naturally occurring fruits and insects as edible in favor of tourist-provided food, future generations may not be able to return to a more natural diet if tourist food was completely and suddenly removed.

These changes observed in the monkeys' feeding and ranging patterns could also have significant and wide-reaching ecological impacts. White-faced capuchins are important seed dispersers and predators of insects (Wehncke et al., 2003). Others have found that poaching pressure can affect seed dispersal by mammals (Wang et al., 2007; Wright et al., 2000). By ceasing to eat fruit from their habitat as well as reducing their travel distances, white-faced capuchins are essentially "removed" from the ecosystem as far as seed dispersal goes. This could have cascading effects for multiple plant species in the forest, along with the animals who consume these species.

Humans

Tourists were split on their opinion of the monkeys. Nearly half (48%) had a positive opinion of the monkeys, 36% were ambivalent, and 16% had a negative opinion. Those with negative views said the monkeys were "aggressive" or "too tame," but these are characteristics that are most probably products of increased tourism in the park. The monkeys were aggressive when they were not able to access the tourists' food; the tameness was a result of their exposure to humans, possibly the association of humans with food. The fact that 52% of the informants had ambivalent or negative views of the monkeys indicates that the monkeys could become detrimental to tourism in the park. Tourists are unlikely to want to visit an attraction where they meet an animal they view as a pest. Indeed, although I did not witness any attacks by monkeys on tourists, I was told that before my study began at least one child had been bitten by a capuchin and people in the area were beginning to be concerned about the capuchins' aggressiveness. Finally, combined with the fact that the monkeys are not the biggest draw to the park, their aggressiveness adds to the danger that they will become a drawback to the park. Tourists also stated that the capuchins were not the main reason for visits to the park, so negative views of the monkeys could affect their desire to return or recommend the site to others.

Most tourists did not think it was good to feed the monkeys and none admitted to feeding them. While most tourists are then getting the message that feeding the monkeys is wrong, even a few instances of tourists feeding the monkeys can have a large impact. Further, at least some of these tourists must have lied, so the number of instances of tourists feeding monkeys is larger than reported. Tourists' reasons for thinking it was not good to feed the monkeys can offer insights into what may be useful in an educational campaign. Typically, they stated that the monkeys had enough food in the wild and that feeding them would make them aggressive. More education focusing on this point may help stop tourists' feeding of monkeys.

Conclusions

This research indicates that increased tourism in Manuel Antonio National Park has affected one population of white-faced capuchin monkeys. These monkeys have changed their diet, activity patterns, and ranging behaviors, all of which could have consequences for their own population and the ecology of the park. The fact that the monkeys' diet has come to consist of 46% tourist food effectively removes them as seed dispersers, possibly leading to change in the whole ecological community. Additionally, changes in the monkeys' behaviors may undermine the sustainability of tourism in the park, if more tourists develop negative views of the monkeys because of their aggression and tameness. If the park is protected partly to attract tourists and tourist income, and if tourists stop visiting because of disagreeable interactions with monkeys, the income generated by tourism might not be enough to keep the park operational. Such changes could affect the white-faced monkeys themselves and could well affect other species of wildlife in the park with similarly detrimental consequences.

Further education of tourists about the fact that feeding the monkeys will lead to unwanted behaviors like aggressiveness and stealing may help reduce the amount of tourist food eaten by the monkeys, although most tourists already realized that feeding the monkeys was undesirable. The monkeys actually gained most of their access to tourist food through the garbage and stealing from picnics and backpacks. Restricting access to these sources of food would probably have a greater impact on reducing the changes in the monkeys than would ceasing active feeding.

While tourism has changed many aspects of this troop of white-faced capuchins in Manuel Antonio, tourists still have positive experiences with them. Most tourists understood that feeding the monkeys is not an acceptable behavior, and most claimed not to do so. Tourists do not want to see monkeys that are overly tame. This means there is an excellent chance that the monkeys' access to tourist food can be changed, and they will remain a pleasant attraction within the park. Access to tourist food could be decreased by better fitting garbage can lids, picnics limited to specific monkey-proof areas, and increased signage indicating the dangers of feeding the monkeys. The continued existence of the park as a protected area will also result in benefit for the capuchins, who will be free of the threats of hunting and habitat destruction.

Acknowledgments

Research for this project was carried out under the auspices of the Associated Colleges of the Midwest Tropical Field Research Program (ACM) and with approval of all protocols by ACM and by the management of Manuel Antonio National Park. All research adhered to the legal requirements of Costa Rica. I would like to thank all the support staff of ACM, including my adviser, Grace Wong. I would also like to thank everyone at Manuel Antonio National Park for permission to complete this study and logistical assistance while I was in the park. Thanks to my fellow field workers, Lesley Rae and Rachel Gronau for their help and support. Thanks to Rick Moore and the editors for comments on earlier drafts of this chapter. A final thanks to Vicki Bentley-Condit who supported me in earlier explorations of these data.

References

Albert, D. and Bowyer, R. (1991). Factors related to grizzly bear-human interactions in Denali National Park. *Wildlife Society Bulletin*, 19: 339–349.

Altmann, J. (1974). Observational study of behavior. *Behaviour*, 48: 1–41.

Altmann, J. and Muruthi, P. (1988). Differences in daily life between semiprovisioned and wild-feeding baboons. *American Journal of Primatology*, 15: 213–221.

Baker, C. P. (2007). *Moon Costa Rica*. Emeryville, CA: Avalon Travel Publishing.

Berman, C. M., Li, J. H., Ogawa, H., *et al.* (2007). Primate tourism, range restriction, and infant risk among *Macaca thibetana* at Mt. Huangshan, China. *International Journal of Primatology*, 28: 1123–1141.

Bernard, H. R. (1995). *Research Methods in Anthropology: Qualitative and Quantitative Approaches*. New York: AltaMira Press.

Blom, A., Cipolletta, C., Brunsting, A. M. H., and Prins, H. H. T. (2004). Behavioral responses of gorillas to habituation in the Dzanga-Ndoki National Park, Central African Republic. *International Journal of Primatology*, 25: 179–196.

Boza, M. A. (1993). Conservation in action: past, present, and future of the National Park system of Costa Rica. *Conservation Biology*, 7: 239–247.

Brennan, E. J., Else, J. G., and Altmann, J. (1985). Ecology and behaviour of a pest primate: vervet monkeys in a tourist-lodge habitat. *African Journal of Ecology*, 23: 35–44.

Campos, F. A., Bergstrom, M. L., Childers, A., *et al.* (2014). Drivers of home range characteristics across spatiotemporal scales in a Neotropical primate, *Cebus capucinus. Animal Behaviour*, 91: 93–109.

Chapman, C. (1988). Patterns of foraging and range use by three species of Neotropical primates. *Primates*, 29: 177–194.

Chapman, C. A. and Fedigan, L. M. (1990). Dietary differences between neighboring *Cebus capucinus* groups: local traditions, food availability or responses to food profitability. *Folia Primatologica*, 54: 177–186.

Custance, D. M., Whiten, A., and Fredman, T. (2002). Social learning and primate reintroduction. *International Journal of Primatology*, 23: 479–499.

Fairbanks, W. S. and Tullous, R. (2002). Distribution of pronghorn (*Antilocapra americana* Ord) on Antelope Island State Park, Utah, USA, before and after establishment of recreational trails. *Natural Areas Journal*, 22: 277–282.

Farber, M. E. and Hall, T. E. (2007). Emotion and environment: visitors' extraordinary experiences along the Dalton Highway in Alaska. *Journal of Leisure Research*, 39: 248.

Fragaszy, D. M. and Bard, K. (1997). Comparison of development and life history in *Pan* and *Cebus. International Journal of Primatology*, 18: 683–701.

Fragaszy, D. M., Visalberghi, E., and Fedigan, L. M. (2004). *The Complete Capuchin: The Biology of the Genus Cebus*. New York: Cambridge University Press.

Fredrickson, L. M. and Anderson, D. H. (1999). A qualitative exploration of the wilderness experience as a source of spiritual inspiration. *Journal of Environmental Psychology*, 19: 21–39.

Freese, C. H. (1983). *Cebus capucinus* (Mono cara blanca, white-faced capuchin). In: D. H. Janzen (ed.), *Costa Rican Natural History*. University of Chicago Press, pp. 458–460.

Grossberg, R., Treves, A., and Naughton-Treves, L. (2003). The incidental ecotourist: measuring visitor impacts on endangered howler monkeys at a Belizean archaeological site. *Environmental Conservation*, 30: 40–51.

Hakeem, A., Sandoval, G. R., Jones, M., and Allman, J. (1996). Brain and life span in primates. In: J. Birren and K. Schaie (eds.), *Handbook of the Psychology of Aging* 4th edn. San Diego: Academic Press, pp. 78–104.

Hill, D. A. (1999). Effects of provisioning on the social behaviour of Japanese and rhesus macaques: implications for socioecology. *Primates*, 40: 187–198.

Honey, M. (2003). Giving a grade to Costa Rica's green tourism. Calificando el turismo verde de Costa Rica. *NACLA Report on the Americas*, 36: 39–46.

ICT. (2002). *Costa Rican Institute of Tourism: Annual Report of Statistics, 2001*. San Jose, Costa Rica: Costa Rican Institute of Tourism.

Ikiara, M. and Okech, C. (2002). *Impact of tourism on environment in Kenya: status and policy*. Kenya Institute for Public Policy Research and Analysis.

Jacobson, S. K. and Lopez, A. F. (1994). Biological impacts of ecotourism: tourists and nesting turtles in Tortuguero National Park, Costa Rica. *Wildlife Society Bulletin*, 22: 414–419.

Johns, B. G. (1996). Responses of chimpanzees to habituation and tourism in the Kibale Forest, Uganda. *Biological Conservation*, 78: 257–262.

Kaimowitz, D. and Segura, O. (1996). The political dimension of implementing environmental reform: lessons from Costa Rica. In: R. Costanza, O. Segura, and J. Matinez-Alier (eds.), *Getting Down to Earth: Practical Applications of Ecological Economics*. Washington, DC: Island Press, pp. 439–453.

Kaur, T. and Singh, J. (2008). Up close and personal with Mahale chimpanzees – a path forward. *American Journal of Primatology*, 70: 729–733.

Kinnaird, M. F. and O'Brien, T. G. (1996). Ecotourism in the Tangkoko Dua Saudara Nature Reserve: opening Pandora's box? *Oryx*, 30: 65–73.

Knight, R. L. and Cole, D. N. (1995). Wildlife responses to recreationists. In: R. L. Knight and K. J. Gutzwiller (eds.), *Wildlife and Recreationists*, Washington, DC: Island Press, pp. 51–70.

Koganezawa, M. and Imaki, H. (1999). The effects of food sources on Japanese monkey home range size and location, and population dynamics. *Primates*, 40: 177–185.

Köndgen, S., Kühl, H., N'Goran, *et al.* (2008). Pandemic human viruses cause decline of endangered great apes. *Current Biology*, 18: 260–264.

Litvak, M. (1987). *Home Range, Habitat Usage, Troop Movement, Activity Patterns and Food Habits of White-Faced Monkeys (Cebus capucinus) in Manuel Antonio National Park*. San Jose: Associated Colleges of the Midwest Field Research Program.

Lumsdon, L. M. and Swift, J. S. (1998). Ecotourism at a crossroads: the case of Costa Rica. *Journal of Sustainable Tourism*, 6: 155–172.

Mathieson, A. and Wall, G. (1982). *Tourism: Economic, Physical and Social Impacts*. Upper Saddle River, NJ: Prentice Hall.

McNeilage, A. (1996). Ecotourism and mountain gorillas in the Virunga Volcanoes. In: V. J. Taylor and N. Dunstone (eds.), *The Exploitation of Mammal Populations*. London: Chapman and Hall, pp. 334–345.

Moscow, D. and Vaughan-Dickhaut, C. (1987). Troop movement and food habits of white-faced monkeys in a tropical-dry forest. Movimiento de la tropa y hábitos alimentarios de los monos carablanca en un bosque tropical seco. *Revista de Biología Tropical*, 35: 287–297.

Oates, J. F. (1987). Food distribution and foraging behavior. In: B. Smuts, D. L. Cheney, R. M. Seyfarth, *et al.* (eds.), *Primate Societies*. University of Chicago Press, pp. 197–209.

Oppenheimer, J. R. (1982). *Cebus capucinus*: Home range, population dynamics, and interspecific relationships. In E. Leigh, A. Rand, and D. Windsor (eds.), *The Ecology of a Tropical Forest: Seasonal Rhythms and Long Term Changes*. Washington, DC: Smithsonian, pp. 253–272.

Perry, S. (2011). Social traditions and social learning in capuchin monkeys (*Cebus*). *Philosophical Transactions of the Royal Society B*, 366: 988–996.

Rachowiecki, R. (1997). *Costa Rica*. London: Lonely Planet.

Reynolds, P. C and Braithwaite, D. (2001). Towards a conceptual framework for wildlife tourism. *Tourism Management*, 22: 31–42.

Roe, D., Leader-Williams, N., and Dalal-Clayton, D. B. (1997). *Take only Photographs, Leave only Footprints: The Environmental Impacts of Wildlife Tourism*. London: International Institute for Environment and Development.

Saj, T., Sicotte, P., and Paterson, J. D. (1999). Influence of human food consumption on the time budget of vervets. *International Journal of Primatology*, 20: 977–994.

Sparks, B. (1994). Communicative aspects of the service encounter. *Journal of Hospitality & Tourism Research*, 17: 39.

Tosi, J. A. (1969). *Mapa ecologico de Costa Rica*. San José.

Valentine, P. S., Birtles, A., Curnock, M., *et al.* (2004). Getting closer to whales – passenger expectations and experiences, and the management of swim with dwarf minke whale interactions in the Great Barrier Reef. *Tourism Management*, 25: 647–655.

Wallis, J. and Lee, D. (1999). Primate conservation: the prevention of disease transmission. *International Journal of Primatology*, 20: 803–826.

Wang, B. C., Sork, V. L., Leong, M. T., and Smith, T. B. (2007). Hunting of mammals reduces seed removal and dispersal of the Afrotropical tree. *Antrocaryon klaineanum* (Anacardiaceae). *Biotropica*, 39: 340–347.

Wehncke, E. V., Hubbell, S. P., Foster, R. B., and Dalling, J. W. (2003). Seed dispersal patterns produced by white-faced monkeys: implications for the dispersal limitation of Neotropical tree species. *Journal of Ecology*, 91: 677–685.

Weiss, R. S. (1995). *Learning from Strangers: The Art and Method of Qualitative Interview Studies*. New York: Free Press.

World Conservation Union, UNEP-World Conservation Monitoring Centre (2009). *World Database on Protected Areas*. Cambridge, UK: WCMC.

World Travel and Tourism Council. (2005). *World Travel and Tourism: Sowing the Seeds of Growth*. London: World Travel and Tourism Council.

Wright, S. J., Zeballos, H., Dominguez, I., *et al.* (2000). Poachers alter mammal abundance, seed dispersal, and seed predation in a Neotropical forest. *Conservation Biology*, 14: 227–239.

Ziffer, K. (1989). *Ecotourism: The Uneasy Alliance*. Washington, DC: Conservation International.

14 Effects of tourism on Ecuadorian primates: is there a need for responsible primate tourism?

Stella de la Torre

Introduction

Ecuador is one of the smallest and more biodiverse countries in the Neotropics (Ministerio del Ambiente *et al.*, 2001; Mittermeier *et al.*, 1997). The 20 primate species that inhabit forest remnants east and west of the Andes are important components of this diversity and are strongly affected by habitat loss, fragmentation, hunting, and illegal trade (de la Torre, 2010; Tirira, 2011). In the recent evaluation for the Red Data Book of Ecuadorian Mammals, 53% of the country's 21 primate taxa (20 species, 21 subspecies) are considered threatened (19% vulnerable, 24% endangered, and 10% critically endangered), 42% are near threatened, and 5% are data deficient (Tirira, 2011) (see Table 14.1).

The deforestation rate in Ecuador, estimated at about 190 000 ha/year, is one of the highest in South America (Ministerio del Ambiente *et al.*, 2001). It reflects the magnitude of the impact of habitat loss and fragmentation that Ecuadorian primate populations are facing. However, considering that the real geographic ranges of these species are not yet fully known, our understanding of the impacts of habitat loss and fragmentation on each of them is not as profound as it should be. The effects of hunting and illegal trade are even more difficult to quantify, but add to the impacts of the two main habitat threats, explaining the alarming conservation status of primates in the country.

Several strategies have been proposed to improve primate conservation status in Ecuador, including new studies of the geographic distributions of the taxa, detailed evaluations of the impacts of human activities on primate populations, creation of reserves in areas with viable populations of threatened species, and programs of environmental education. Their application has, in most cases, been limited by socio-economic constraints and lack of political support (de la Torre, 2010). Perhaps the most popular and applicable strategy in present times is ecotourism, which some consider a form of passive and sustainable use of natural resources (Mullner & Pfrommer, 2001; Whiteman, 1996; Yu-Douglas *et al.*, 1997). Since there is no national action plan for the conservation of Ecuadorian primates, ecotourism

Primate Tourism: A Tool for Conservation?, ed. Anne E. Russon and Janette Wallis. Published by Cambridge University Press. © Cambridge University Press 2014.

Table 14.1 Ecuadorian primates, distribution (East or West of the Andes) and conservation status in Ecuador (Ecuador Red Data Book, Tirira, 2011)

Species	Common name	West	East	Conservation status
Cebidae – Hapalinae				
Callithrix pygmaea	Pygmy marmoset		X	VU A4acd
Saguinus fuscicollis	Saddle-back tamarin		X	NT A4acd
Saguinus graellsi	Graell's tamarin		X	VU A4acd
Saguinus tripartitus	Golden-mantle tamarin		X	VU A4acd
Cebidae – Cebinae				
Cebus albifrons aequatorialis	Western white-fronted capuchin	X		CR A2cd
Cebus albifrons cuscinus	Eastern white-fronted capuchin		X	NT A4acd
Cebus macrocephalus	Large-headed capuchin		X	NT A4acd
Cebus capucinus	White-faced capuchin	X		EN A4acd
Saimiri sciurus	Squirrel monkey		X	NT A4acd
Pitheciidae				
Callicebus discolor	Dusky titi monkey		X	NT A4acd
Callicebus lucifer	Yellow-handed titi monkey		X	VU A4acd
Pithecia aequatorialis	Equatorial saki		X	NT A4acd
Pithecia monachus	Monk saki		X	NT A4acd
Aotidae				
Aotus lemurinus	Owl monkey		X	DD
Aotus vociferans	Owl monkey		X	NT A4acd
Familia Atelidae				
Alouatta palliata	Mantled howling monkey	X		EN A4acd
Alouatta seniculus	Red howling monkey		X	NT A4acd
Ateles belzebuth	White-bellied spider monkey		X	EN A2cd
Ateles fusciceps	Brown-headed spider monkey	X		CR A2cd
Lagothrix lagotricha	Common woolly monkey		X	EN A4acd
Lagothrix poeppigii	Poeppig's woolly monkey		X	EN A4acd

Notes:
Conservation status criteria from www.iucnredlist.org/documents/redlist_cats_crit_en.pdf

Critically Endangered (CR)
A2: Reduction in population size based on an observed, estimated, inferred, or suspected population size reduction of 80% over the last ten years or three generations, whichever is the longer, where the reduction or its causes may not have ceased OR may not be understood OR may not be reversible, based on:
(c) a decline in area of occupancy, extent of occurrence, and/or quality of habitat;
(d) actual or potential levels of exploitation.

Endangered (EN)
A2: Reduction in population size based on an observed, estimated, inferred, or suspected population size reduction of 50% over the last ten years or three generations, whichever is the longer, where the reduction

Table 14.1 (*cont.*)

or its causes may not have ceased OR may not be understood OR may not be reversible, based on:
(c) a decline in area of occupancy, extent of occurrence, and/or quality of habitat;
(d) actual or potential levels of exploitation.

A4: Reduction in population size based on an observed, estimated, inferred, projected, or suspected
     population size reduction of 50% over any ten-year or three-generation period, whichever is longer
     (up to a maximum of 100 years in the future), where the time period must include both the past and
     the future, and where the reduction or its causes may not have ceased OR may not be understood
     OR may not be reversible, based on:
(a) direct observation;
(c) a decline in area of occupancy, extent of occurrence, and/or quality of habitat;
(d) actual or potential levels of exploitation.

Vulnerable (VU)
A4: Reduction in population size based on an observed, estimated, inferred, projected, or suspected
     population size reduction of 30% over any ten-year or three-generation period, whichever is longer
     (up to a maximum of 100 years in the future), where the time period must include both the past and
     the future, and where the reduction or its causes may not have ceased OR may not be understood OR
     may not be reversible, based on:
(a) direct observation;
(c) a decline in area of occupancy, extent of occurrence, and/or quality of habitat;
(d) actual or potential levels of exploitation.

Near Threatened (NT)

Data deficient (DD).

initiatives focusing directly on the conservation of this group are virtually non-existent.

The promotion of tourism is one of the national goals of the government in Ecuador, and strategies are being developed to increase the number of tourists that arrive to the country each year beyond the current one million (Ministerio de Turismo, 2010; SENPLADES, 2007). Among the different attractions for tourists, natural environments and biodiversity have been increasingly important in terms of the number of tourists, economic investment, and profits they attract (Mindo Cloudforest Foundation, 2006; Mullner & Pfrommer, 2001). Primates are among the natural attractions for tourists in several areas of Amazonian Ecuador, like the Cuyabeno Reserve and the Yasuni National Park, both with primate communities of ten sympatric species, although to date no specific model for primate tourism has been developed (de la Torre *et al.*, 1995; Marsh, 2004).

In this chapter, I review the impact of tourism on primate populations in Amazonian Ecuador, presenting the main results of studies that were carried out in the 1990s with three species: pygmy marmosets (*Callithrix pygmaea*), Graell's tamarin (*Saguinus graellsi*), and red howling monkeys (*Alouatta seniculus*). I then analyze the recommendations of these studies to reduce the negative effects of tourism on these primates, as individuals and as populations, and evaluate the possibility of implementing these recommendations in local and national programs of responsible primate tourism that may result in positive contributions to primate conservation.

Impacts of tourism on pygmy marmosets

The effects of tourism and human presence on the behavior and reproduction of pygmy marmosets were evaluated in the Cuyabeno Reserve, a protected area of 600 000 ha of tropical rainforest in northeastern Ecuador. Pygmy marmosets live in gallery forests on the margins of rivers and lakes of the reserve (Figure 14.1). Field work was carried out from September 1996 through May 1998. Six groups of marmosets were observed in two sites that differed in the number of tourists and use of motor boats. La Hormiga Island was consistently visited by 17 tourist agencies that used motor boats in almost every trip in the surrounding Laguna Grande Lake. The forest around the Zancudococha Lake, approximately 100 km southeast of La Hormiga Island, was visited by only one tourist agency and, thus, had a low rate of tourism compared with Laguna Grande; in addition, motor boats were not allowed (de la Torre *et al.*, 2000).

Scan samples, using instantaneous and one-zero sampling every 20 minutes, provided data on group size, composition, observability, activities, home range, use of strata, and calling behavior. Grooming, play, and physical aggression were recorded using all-occurrences sampling. A reproductive coefficient of each group was obtained by dividing the number of observed newborn infants in a group by the expected maximum number (two litters, of two infants each, were expected per group per year). Infant survival rate was estimated by the number of infants that survived to their fourth month divided by the total number of observed newborn infants. Tourism pressure was estimated by a combination of the number of tourists in an area, the distance of the groups to the tourists, and the presence or absence of motor boats. Combined estimates were then ranked from one (farthest from tourists, low-tourism area, with no motor boats) to six (closest to tourists, high-tourism area, with motor boats) (de la Torre *et al.*, 2000).

Reductions in social play and in the use of the lower stratum of the forests were significantly correlated with higher levels of tourism pressure (social play Spearman's rho = -0.99, $p = 0.03$, $n = 6$; use of lower stratum rho = -0.9, $p = 0.04$, $n = 6$). The strong negative correlation between tourism pressure and play behavior in pygmy marmosets points to an inhibitory impact of this human activity on the social behavior of the groups. The change in their use of forest strata may be related to changes in their patterns of exudate feeding and insect foraging. Tourism workers also captured marmosets from two of the groups studied at La Hormiga for the illegal pet market. These capture events were associated with significant reductions in observability, vocalization rates, and use of lower strata in the affected marmoset groups. There were no significant correlations between tourism pressure and reproductive coefficients or infant survival rate. However, significant differences were found in reproductive coefficients, with groups affected by capture showing lower reproductive coefficients than remaining groups (mean reproductive coefficients 0.31 ± 0.6 and 0.88 ± 0.12, respectively). Additionally, of five newborn infants observed in the groups affected by capture during the whole study period, 60% survived to four months of age. In contrast, of 12 newborn infants observed in

Figure 14.1 Pygmy marmoset, *Callithrix pygmaea* – infant (credits: Pablo Yépez, Stella de la Torre).

the remaining groups, 92% survived to four months of age. The behavioral changes found may have allowed the marmosets to avoid contact with humans and were possibly related to differences in the reproductive performance of the groups (de la Torre *et al.*, 2000).

Impacts of tourism on Graell's tamarins

A study was carried out in the Cuyabeno Reserve from 1995 through 1997 to evaluate the effects of tourism on the flight behavior of groups of Graell's tamarins on tourist trails and control groups on trails with no tourism (Figure 14.2) (Mullner & Pfrommer, 2001). One observer searched for groups of tamarins at times of the day when tourist presence was low (07:00 to 10:00 and after 14:00). The observer simulated the behavior of tourists once a group was encountered, approaching and following the group for short distances, as a basis for estimating tourist effects on flight behavior.

The results of this study suggested that tamarins in areas with tourism were not habituated to tourists even after years of exposure. The flight response of these groups was not significantly different from that of the control groups. In both areas, tamarins showed a flight frequency of almost 100% when disturbed by a human observer. The authors recorded that, after moving away from their initial area, a group did not return to it for at least two hours. They also noted that in months of high tourism traffic on forest trails (mean = 57 tourist groups/month), the frequency of encounters with tamarins was significantly lower than in months of low tourism traffic (mean = 27 tourist groups/month) (Wilcoxon test $p < 0.05$). The authors did not make explicit inferences about the impact of the flights and area displacements on the physiology and behavior of the individuals, but did relate these changes to a reduction in sightings of tamarins in tourism areas that had already been reported by tourist agencies.

Figure 14.2   Graell's tamarin, *Saguinus graellsi* – juvenile (credits: Pablo Yépez, Stella de la Torre).

Effects of tourism on red howling monkeys

De la Torre *et al.* (1999) carried out a pilot study to evaluate the effects of tourism on the howling behavior of groups of red howling monkeys in forests around two similar-size lakes of the Cuyabeno Reserve that differ in the number of tourists and the use of motor boats: Laguna Grande Lake (high tourism rate and use of motor boats) and Zancudococha Lake (low tourism rate and no motor boats) (Figure 14.3). Field work was carried out in 1997. Morning censuses of howling groups in the forests around the lakes were carried out from a fixed point considered to be the center of the lake, in the rainy season (when the number of tourists in both areas was higher). The direction of the howling groups to the center of the lakes was recorded with a compass. The distance of the howling groups to the center of the lakes was estimated in three categories: far (> 2 km), middle (1 km approx.), and close (400–600 m approx.). These estimations were validated with direct observations of some of the groups.

The estimated mean distance to the howling groups was significantly different between the two lakes (Mann–Whitney Z = −2.08, p = 0.037). Groups at Laguna Grande howled further from the shores (mean = 839 m ± 103) than groups at Zancudococha (mean = 478 m ± 129). Considering the similarities between the two lakes in their size and forest composition, this result pointed to a possible effect of motor boats on the calling behavior of the species, possibly through a masking effect due to the similar frequencies of howls and motors. Masking occurs when noise interferes with the animal's ability to hear a sound of interest and is more pronounced when the sounds have similar frequencies (Brenowitz, 1986). It was suggested that groups in the Laguna Grande area moved away from the lake shores to

Figure 14.3   Red howling monkey, *Alouatta seniculus* – infant (credits: Pablo Yépez, Stella de la Torre).

howl or that they reduced their howling frequency to avoid this masking effect. The consequences of these behavioral changes are not known. Previous studies have suggested that the primary function of howling is to deter outside males from entering a troop and gain access to breeding females (Sekulic, 1982). Thus, changes in the howling rate may affect inter-male competition between troops and ultimately the population dynamics (de la Torre *et al.*, 1999).

Discussion

At the time these studies on primate tourism were carried out (in the 1990s), the Cuyabeno Reserve was a protected area with the second highest level of tourism in continental Ecuador (Mullner & Pfrommer, 2001). Tourism in the Cuyabeno Reserve is likely to be representative of nature tourism in Ecuadorian Amazonia, where large rivers and lakes are attractions and means of transportation and recreation for tourists. Tours are complemented by trail walks in different forest types. Several tourist operators tag their programs as ecotourism, but only a few are actively working to fulfill the standards of responsible travel to natural areas – that is, travel that contributes to the conservation of the environment and the improvement of the well-being of local people (Kruger, 2005). Most tourist programs, like the ones in the Cuyabeno Reserve, are better classified as "nature-based tourism," focused on travel to natural places but with limited environmental and social responsibility. It is possible, then, that the impact of tourism on primate populations that these studies reported may now also occur in other areas, since tourism has increased and more areas are being visited. However, no other studies have been carried out on this topic. Behavioral observations of pygmy marmoset groups in two tourist camps in the Napo and Arajuno Rivers, south of the Cuyabeno Reserve, are showing similar behavioral changes in groups that were more exposed to tourism (de la Torre, unpub. data).

Tourism has been less intense in forests west of the Andes since very few remnants of natural ecosystems exist (Ministerio del Ambiente *et al.*, 2001). In some of these remnants (e.g. Cerro Blanco, near Guayaquil), one to three primate species have been recorded (Albuja & Arcos, 2007) but no studies on the effects of tourism on those species have been carried out.

In conclusion, there are very few studies of the impacts of tourism on Ecuadorian primate populations and this pattern holds for most areas in the Neotropics. Findings are nonetheless similar. Studies carried out in Belize and Costa Rica with howling monkeys (*Alouatta* spp.) found that when tourists were present versus absent, the distances between *A. pigra* group members were significantly larger and monkeys climbed higher in trees. *A pigra* groups subjected to intense tourism also showed higher infant mortality that appeared to be related to these behavioral changes (Treves & Brandon, 2005). In *A. palliata*, the potential for exchange of parasites with local human populations through contaminated water or food was suggested (Stuart *et al.*, 1998). Studies in Tambopata, Peru (Hodgkinson, this volume) and in Manuel Antonio National Park, Costa Rica (Kauffman *et al.*, this volume) complete the scenario of what is known about the effects of tourism on Neotropical primates. All these studies present recommendations to improve tourism practices that are addressed in the next section.

Recommendations for better tourism practices

Some of the recommendations of the studies reviewed above are to improve the effectiveness of tourism control. Concentrating tourism activities in specific sites of a reserve, leaving areas free from tourism pressure, and limiting the number of tourists in heavily visited sites and seasons were three of the recommendations to mitigate the impacts found on primates in the Cuyabeno Reserve (de la Torre *et al.*, 2000; Mullner & Pfrommer, 2001). These recommendations have been slowly incorporated as management regulations of this and other reserves in the country, especially recommendations referring to the creation and reinforcement of tourism-free areas (AETS, 2009). It has been more difficult to control tourist numbers, both in Cuyabeno and in other private or public protected areas (e.g. Sacha Lodge Reserve and Yasuni National Park), given all the incentives at the national and local levels to increase tourism in the country.

A viable alternative to controlling the total number of visitors in an area is to control the number of tourists present at a specific time in a tourist site. Tourist group visits to the different attractions in an area, whether landscapes, trails, or primate territories, should be coordinated so that only one tourist group visits a given attraction at a time. Time of visit and between visits should be determined in advance and continuously evaluated to balance the interests of tourists with the need to minimize the potential negative impacts of visits (AETS, 2009; de la Torre *et al*, 2000). This strategy requires the implementation of monitoring programs on important bio-indicators in tourism areas, including several primate behaviors (de la Torre *et al.*, 2000).

It is also necessary to train tourists, tourist guides, and staff to reduce the disturbances they cause. Recommendations include: requiring that they make no attempts to interact or physically contact the monkeys, increasing their awareness of the health risks to the monkeys imposed by sick tourists, and improving waste disposal in tourism facilities (de la Torre *et al.*, 2000; Treves & Brandon, 2005; Wallis & Lee, 1999).

Currently, some protected areas in Ecuador have regulations that include tourism-free sites and controls on the number of tourists and groups that visit a site at the same time. However, we do not know if these strategies have been effective in mitigating the negative effects of tourism on the behavior of individual primates and on primate population dynamics. It is also possible that at least in some areas these regulations have not been enforced. Previous studies on tourism activities in the Cuyabeno Reserve, for example, suggested that the enforcement of regulations was weak, partly because the Ministry of the Environment did not have sufficient budget and personnel to control tourist operators (Yépez & de la Torre, 2000). Furthermore, even in areas where regulations have been enforced, no continuous monitoring programs have been implemented to evaluate tourism impacts on animal populations and to take corrective actions if needed.

With my research team I have made several attempts to develop such monitoring programs in different tourism areas in Amazonian Ecuador, recommending the behavioral variables that could be measured in primates (e.g. use of strata, play rates, or vocalization rates) and simple data analyses for the evaluations. In the aims of developing responsible primate tourism, we have made recommendations about the behavior that tourists should maintain to reduce the stress and health risks on the primate groups visited. We have also prepared relevant and attractive information about the primate species in the areas to increase tourists' interest in them. Until now, we have not been able to achieve continuity in monitoring programs. Funding limitations and personal and institutional interests are important factors that have affected the participation of tourist operators, local communities, and governmental agencies.

One possible solution to some of these participation problems could be to create programs of primate tourism in areas with flagship species or with highly diverse primate communities that could be particularly attractive to tourists (Kruger, 2005). In these areas, tourism should be restricted to a minimum of well-habituated groups, leaving the remaining population free from tourism stress. If such a program were to be developed in Ecuador, the economic incentives might increase people's interest and participation in monitoring programs and their acceptance of all the regulations mentioned above. It is also possible, however, that these same economic incentives could cause an increase in the number of tourists, relaxing the necessary controls so that more pressure is put on primate populations. These problems could be overcome with appropriate incentives and penalties developed and applied by control agencies, and with enhanced education programs for all stakeholders, including tourists. All these strategies should be part of a national action plan for Ecuadorian primate conservation. Such an action plan needs to be developed and implemented in the near future by the Ecuadorian government in collaboration with universities, environmental organizations, and local communities.

The alternative is to leave things as they are now. The fact that the areas where ecotourism and nature-based tourism are being implemented are among the few that are still forested shows that tourism is already a solution to habitat loss, the most important threat affecting Ecuadorian primates (de la Torre, 2010; Tirira, 2011). However, unless tourist operators consider primates as their main attractions and become interested in their conservation, it will be difficult to enforce regulations to reduce the negative effects of tourism on primate populations.

Responsible primate tourism in Ecuador has the potential to contribute positively to primate conservation if relatively simple regulations, such as the ones mentioned above, are applied and enforced. This will require close collaboration among tourist operators, local communities, universities, local governments, and environmental agencies. All stakeholders need to be aware of the importance of primate conservation for economic, cultural, biological, and ethical reasons. Such a combination of interests is not always easy to find, so responsible primate tourism may not be possible in all forests with primates, at least in present times. We are currently developing a proposal for a pilot program in a private reserve, owned by a tourist operator, with seven sympatric primate species in north-eastern Ecuador. Once implemented, this program should be continuously evaluated and modified according to the circumstances specific to the area to assure its contribution to primate conservation. If this private initiative is successful, we could use it as a model to replicate in other regions of the country.

References

AETS. (2009). *Reporte de logros 2007–2009*. Quito: Alianza Ecuatoriana para el Turismo Sostenible.

Albuja, L. and Arcos, R. (2007). Evaluación de las Poblaciones de *Cebus albifrons* cf. *aequatorialis* en los bosques suroccidentales ecuatorianos. *Revista Politécnica Biología*, 7: 59–69.

Brenowitz, E. A. (1986). Environmental influences on acoustic and electric animal communication. *Brain, Behavior and Evolution*, 28: 32–42.

de la Torre, S. (2010). Los primates ecuatorianos, estudios y perspectivas. *Avances en Ciencias e Ingenierías*, 2: B27–B35.

de la Torre, S., Snowdon, C. T., and Bejarano, M. (1999). Preliminary survey of the effects of ecotourism and human traffic on the howling behaviour of red howler monkeys, *Alouatta seniculus*, in Ecuadorian Amazon. *Neotropical Primates*, 7: 84–86.

de la Torre, S., Snowdon, C. T., and Bejarano, M. (2000). Effects of human activities on pygmy marmosets in Ecuadorian Amazon. *Biological Conservation*, 94: 153–163.

de la Torre, S., Utreras, V., and Campos, F. (1995). An overview of primatological studies in Ecuador: primates of the Cuyabeno Reserve. *Neotropical Primates*, 3: 169–171.

Kruger, O. (2005). The role of ecotourism in conservation: panacea or Pandora's box? *Biodiversity and Conservation*, 14: 579–600.

Marsh, L. (2004). Primate species at the Tiputini Biodiversity Station, Ecuador. *Neotropical Primates*, 12: 75–78.

Mindo Cloudforest Foundation. (2006). *Estrategia Nacional de Aviturismo*. Quito: CORPEI.

Ministerio de Turismo del Ecuador. (2010). *Estadísticas del Turismo en el Ecuador*. www.turismo.gob.ec (accessed February 20, 2011).

Ministerio del Ambiente, EcoCiencia & UICN. (2001). *La Biodiversidad del Ecuador, Informe 2000*. Quito: Ministerio del Ambiente/EcoCiencia/UICN.

Mittermeier, R. A., Robles, P., and Goettsch-Mittermeier, C. (1997). *Megadiversidad, los Países Biológicamente Más Ricos del Mundo*. México: CEMEX S.A. & Agrupación Sierra Madre.

Mullner, A. and Pfrommer, A. (2001). *Turismo de bosque húmedo y su impacto en especies seleccionadas de la fauna silvestre del Río Cuyabeno, Ecuador*. Eschborn: TÖB F- IV/8s. GTZ.

Sekulic, R. (1982). The function of howling in red howler monkeys (*Alouatta seniculus*). *Behaviour*, 81: 38–54.

SENPLADES. (2007). *Plan Nacional de Desarrollo 2007–2010, Turismo*. Quito: SENPLADES.

Stuart, M., Pendergast, V., Rumfelt, S. *et al.* (1998). Parasites of wild howlers (*Alouatta* spp.). *International Journal of Primatology*, 19: 493–512.

Tirira, D. G. (ed.). (2011). *Libro Rojo de los mamíferos del Ecuador*. 2ª. edición. Publicación especial sobre los mamíferos del Ecuador 8. Quito: Fundación Mamíferos y Conservación. Pontificia Universidad Católica del Ecuador and Ministerio del Ambiente del Ecuador.

Treves, A. and Brandon, K. (2005). Tourism impacts on the behavior of black howler monkeys (*Alouatta pigra*) at Lamanai, Belize. In: J. Paterson and J. Wallis (eds.), *Commensalism and Conflict: The Primate-human Interface*. Winnipeg, Manitoba: Hignell Printing, pp. 147–167.

Wallis, J. and Lee, D. R. (1999). Primate conservation: The prevention of disease transmission. *International Journal of Primatology*, 20: 803–826.

Whiteman, J. (1996). Ecotourism promotes, protects environment. *Forum for Applied Research and Public Policy*, 11: 96–101.

Yépez, P. and de la Torre, L. (2000). *Diagnóstico y Mapeo de la Actividad Turística en la Reserva de Producción Faunística Cuyabeno*. Quito: Proyecto Petramaz-Unión Europea.

Yu-Douglas, W., Hendrickson, T. and Castillo, A. (1997). Ecotourism and conservation in Amazonian Peru: Short-term and long-term challenges. *Environmental Conservation*, 24: 130–138.

Part V

Broader issues

15 Economic aspects of primate tourism associated with primate conservation

Glen T. Hvenegaard

Introduction

This chapter considers primate tourism as a form of wildlife tourism, that is tourism based on encounters with non-domesticated animal species (Higginbottom, 2004), stressing encounters in their natural environment and intended to be non-consumptive. The rapid growth of wildlife tourism around the world is influenced by many different groups, including tourists, tour operators, local communities, conservation organizations, and governments. These groups are involved for a variety of motives, including recreational enjoyment, business development, community development, protection of wildlife and their habitats, and tax revenues.

All these motivations have economic aspects (Lindberg, 2001). Consider a few examples. Recreation enjoyment is substantial and can be valued monetarily; such benefits can rival or exceed those of other types of land uses. Tourist expenditures can stimulate local development, such as transportation or communications infrastructure, that can benefit local residents and tourists. Tourism revenues can raise funds for wildlife conservation projects, provide local residents with alternatives to less sustainable resource uses, and support governmental and non-governmental educational goals. Tourism-related businesses can support various levels of government by generating tax revenues, and businesses that receive tourist expenditures will, in turn, re-spend some of that money in the local region. Other economic benefits from wildlife tourism include local employment, industry stimulation, economic diversification, and infrastructure improvements (McNeely et al., 1991). On the other hand, economic costs result from wildlife tourism, notably in establishing and controlling tourist facilities and services. They may grow if, for example, tourist expenditures increase inflation or tourist activities harm the wildlife, natural habitats, or regions visited.

Wildlife tourism focused on primates is growing steadily in terms of participants and economic impact (Andersson et al., 2005; Butynski & Kalina, 1998), although there are concerns about the health of the primates visited (Wallis & Lee, 1999; in this volume see chapters by Goldsmith, Muehlenbein & Wallis, Sapolsky, Wright et al.), the welfare of local people, the provision of satisfying experiences,

Primate Tourism: A Tool for Conservation?, ed. Anne E. Russon and Janette Wallis. Published by Cambridge University Press. © Cambridge University Press 2014.

and revenue generation (Greer & Cipolletta, 2006). This chapter aims to examine some of the theoretical and practical links between the economic aspects of primate tourism and primate conservation. I emphasize ecological economics because the valuation of biodiversity allows for a direct comparison with alternative uses of biodiversity (Nunes *et al.*, 2001). Similarly, local economic benefits, participation, and support for wildlife tourism and conservation are crucial for long-term sustainability (Fiallo & Jacobson, 1995; Strum & Manzolillo Nightingale, this volume). I use examples of successes and failures from a variety of wildlife tourism situations, including primate tourism wherever possible.

Before proceeding, I highlight the distinctive characteristics of tourism with primates versus other wildlife and the limits to the economic information available. In the context of wildlife tourism, primates are highly sought-after and create an ongoing tourist demand. Most primates live in the tropics, primarily in developing countries, which influences proximity to tourists' origins, chances of disease transmission, level of infrastructure development, local residents' ability to participate in tourism ventures, and levels of political unrest. Each characteristic has implications for economic aspects, such as travel costs, predictable tourism attractions, economic leakage, community development, and stable tourist flows. Recognizing and understanding these distinctive characteristics will provide a better basis for managing primate tourism in a sustainable manner.

Economic values of primate tourism

The first way to consider economic issues in primate tourism is in terms of *economic value*, or the benefit gained by tourists, measured by what they would be willing to spend in total for an experience such as primate watching (Wells, 1997). The focus is often consumer surplus, the willingness to pay beyond existing tourist expenditures, such as food, accommodation, travel, and souvenirs. Consumer surplus is usually estimated by surveying tourists about their willingness to pay directly or extrapolating from their current travel costs. While it can be difficult to determine and differentiate the various components of economic value, a considerable literature exists that conceptualizes and measures economic value for various situations involving wildlife and environmental services (Tisdell & Wilson, 2004). This economic value may take different forms (Bergstrom *et al.*, 1990; Ward & Beal, 2000); these are sketched below in terms of a person's willingness to pay for specific situations or services.

- *Direct use value:* for an experience (e.g. watch primates);
- *Indirect use value:* to maintain ecological services or experience nature vicariously (e.g. have primates as an essential part of the ecosystem, watch primate documentaries);
- *Option value:* to preserve the possibility of a future direct or indirect use, above any expected consumer surplus of that experience (e.g. see primates in the future);

- *Quasi-option value*: for some new kind of purpose in the future (e.g. possible use of primate viewing for an unanticipated use);
- *Bequest value:* to protect nature for the benefit of future generations (e.g. primate viewing for a person's grandchildren);
- *Existence value:* for the knowledge that nature will continue to exist, regardless of any expectation of use (e.g. knowing that healthy primate populations will exist, without expecting that the respondent or their descendants will use them).

Most attempts to measure these values for primate tourism have focused on gorillas and on direct use value. For the mountain gorilla viewing areas of Bwindi and the Virunga Massif protected forest parks in eastern/central Africa, the annual consumer surplus for international visitors from gorilla viewing in 2000–2001 was estimated at $5.89 million US (Hatfield, 2005). In Uganda alone, the annual consumer surplus for tourists from gorilla viewing in 1999 was estimated at $1.7 million US (Moyini & Uwimbabazi, 2000). Assuming this annual consumer surplus continues in perpetuity, the total value of mountain gorilla tourism from 1999 onwards, adjusted to 1999 dollars, was conservatively estimated at $7–34 million US (Moyini & Uwimbabazi, 2000).

Much controversy surrounds any assessment of the economic value or impact of tourism. Concerns about economic valuation relate to selecting proper valuation methods, choosing an appropriate time frame and discount rate, and categorizing components of a resource or activity (Nunes *et al.*, 2001). Concerns about economic impact assessments relate to narrowing the factors considered, such as the range of tourism activities, consistency and size of region, time frame, methods of measurement, the range of costs and benefits, and perception of those impacts (Hvenegaard & Manaloor, 2007).

Economic impacts of primate tourism

Aside from economic value, the most common way to consider economic issues in primate tourism is in terms of its impact, the effects that tourist expenditures generate on a local economy. A local economy includes all residents, businesses, employees, government agencies, conservation projects, and so on within a defined area, usually a political jurisdiction for which economic data are available. Economic impact studies can assist wildlife conservation by assessing the full range of economic benefits that could be lost or opportunity costs that could be gained if wildlife tourism activities were discontinued. Economic impacts can be categorized as direct, indirect, and induced (Lindberg, 1998; Wells, 1997).

Positive economic impacts

Most attention has been given to the positive side of economic impacts, when economic benefits accrue to tourists, businesses, tour operators, local residents, conservation, and governments in the form of enjoyment, revenue, nature protection, and taxes. Many people suggest that the economic impacts of primate tourism can serve as a positive force for primate conservation by, for example, generating fees to fund the protection of primate habitat.

Gorilla tourism revenue (entrance and viewing fees) throughout Africa produced more than $1.5 million US annually between 1985 and 1996 (Varty *et al.*, 2005; Weber, 1993). At a local level, it averaged $525 000 US (1985–1989) in Volcanoes, Rwanda; $250 000 US (1986–1990) in Virunga, Democratic Republic of Congo; and $450 000 US (1994–1996) in Bwindi, Uganda (Adams & Infield, 2003; Butynski & Kalina, 1998; Wilkie & Carpenter, 1999). Annual gorilla tourism revenues in Virunga, Volcanoes, and Bwindi were an estimated $7.75 million US in 2000–2001 (Hatfield, 2005). In Bwindi alone, gorilla tourism has the potential to produce $36 000 US per day (Goldsmith, this volume).

In some cases, primate tourism represents a significant proportion of total tourism spending in a country. In Uganda, gorilla tracking fees provided 80% of all foreign tourist revenues in 2010 (Tumusiime & Svarstad, 2011). Tourism involving other primates can create significant economic impacts. Annually, for example, over 800 000 tourists visit the Upper Rock Nature Reserve in Gibraltar to view Barbary macaques (*Macaca sylvanus*) and 100 000–200 000 visit Padangtegal in Bali, Indonesia to view long-tail macaques (*Macaca fascicularis*); entrance fees are $12.67 US and $1.00 US per person, respectively (Fuentes *et al.*, 2007).

Direct impacts also result when tourists spend money in places such as tour areas, parks, hotels, restaurants, and shops. In 1989, beyond the $1 million US in revenues from gorilla viewing at Volcanoes National Park, Rwanda, tourists spent another $3–5 million US in the country (Weber, 1993). Between 1994 and 1999, gorilla viewing in Mgahinga and Bwindi, Uganda, generated an estimated $11.86 million US annually in direct sales (Moyini & Uwimbabazi, 2000).

Additional indirect impacts result when those businesses re-spend that money on various goods and services to run their operations or pay government taxes. In 2000–2001, for Virunga and Bwindi gorilla tourism areas, indirect impacts generated within the Ugandan, Rwandan, and Democratic Republic of Congo economies included $4.48 million US in income and $3.10 million US in tax revenues (Hatfield, 2005). In Mgahinga and Bwindi, indirect impacts of gorilla viewing between 1994 and 1999 generated $3.67 million US annually in government taxes (Moyini & Uwimbabazi, 2000). Further induced impacts result when employees of those businesses spend their wages locally. Park guards and gorilla viewing staff spend money on various goods and services, such as accommodation and food. This can be substantial and influential but depends on the availability, desirability, and choice of goods and services. Annually between 1994 and 1999, for gorilla viewing in Mgahinga and Bwindi, estimates for indirect and induced impacts were $2.02 million US in sales, $626 586 US in government revenue, and 218 person-years of jobs (Moyini & Uwimbabazi, 2000).

Economic multipliers, the number of times money is re-spent in a defined economic region, help determine the overall economic impact of an activity in that region;

they are calculated from tracking the flow of money through a region (Bergstrom *et al.*, 1990). Since many wildlife tourism sites are located in remote areas of a country, multiplier effects are often low (Healy, 1988). However, in one of the few studies related to primates, Moyini and Uwimbabazi (2000) calculated a multiplier of 2.0 for gorilla tourism within Uganda; that is, for every $1 US tourists spent, businesses and their employees spent another $2 US in the country. This is greater than the average multiplier of 1.72 at the country scale (van Leeuwen *et al.*, 2011).

Leakage occurs when money flows out of the region to purchase imported goods (e.g. materials, capital, consumables, insurance, advertising, foreign employees), thus lowering the economic multiplier. The economic contribution to a region from primate tourism therefore depends on how much of the initial tourist spending stays in the region. Ignoring issues of commodification, lower leakage rates can increase the local justification to maintain primate populations and their habitats for continued economic benefits from tourism. For the late 1980s and 1990s, Butynski and Kalina (1998) estimated that only 5% of income from gorilla tourism in Rwanda, Uganda, or the Democratic Republic of Congo was spent on gorilla conservation or earned by people living near the protected areas. In Tangkoko Dua Saudara Nature Reserve, Indonesia, well known for macaque viewing, as little as 2% of total tourist expenditures remained in the reserve (Kinnaird & O'Brien, 1996). For comparison, on average 55% of gross tourism revenues leaks out of the host countries (Butynski & Kalina, 1998). Since almost all primate tourism occurs in developing countries, leakage seriously undermines the economic rationale for primate tourism as a conservation measure and reducing leakage from the region offering primate tourism and from the host country represents a considerable challenge.

Negative economic impacts

Less attention has been given to the negative side of economic impacts. Tourism could raise local rates of inflation, result in inequitable revenue distribution (especially to local residents and conservation projects) and unstable revenues (e.g. due to seasonality, political sensitivity, competing attractions), or undermine the welfare and conservation of the wildlife or areas visited (see various chapters in this volume on such costs). Civil unrest in Rwanda in the early 1990s, for instance, caused gorilla tourism to drop off dramatically in Volcanoes National Park (Weber, 1993). During the resulting decline in tourism income and park surveillance, one gorilla died from direct military action and illegal activities increased within the park (Butynski & Kalina, 1998).

Several types of costs are incurred in establishing and maintaining wildlife tourism sites (Dixon & Sherman, 1990; in this volume see Strum & Manzolillo Nightingale, Wright *et al.*). These costs can be measured but few attempts have been made for primate tourism. Direct costs are relatively easy to measure, and include costs incurred by various agencies to allocate land, develop facilities and skills, and prepare and implement management plans. For primate tourism, they also include trail construction, primate habituation, hiring and training tourist guides and park rangers, and permits to use primate habitat. These costs can be high. Blom (2001a) estimated that the habituation of one group of gorillas over two years in Dzanga-Sangha, Central African Republic, cost $250 000 US.

Indirect costs occur as a result of maintaining a wildlife tourism site. For primate tourism, they include responding to tourist-habituated primates' crop-raiding, harming or threatening people, or disrupting research (Feistner *et al.*, 2006; Matheson *et al.*, 2006; Naughton-Treves *et al.*, 1998; in this volume see chapters by Berman *et al.*, Dellatore *et al.*, Goldsmith, Kauffman) and to tourists harming the area's wildlife and habitats (e.g. injuries, disease transmission, habituation, behavioral change, disturbance, increased garbage, and degradation of heavily visited areas) (Berman *et al.*, 2007; Dellatore, 2007; Grossberg *et al.*, 2003; Sandbrook & Semple, 2006; Treves & Brandon, 2005; Woodford *et al.*, 2002; in this volume see Berman *et al.*, Dellatore *et al.*, de la Torre, Goldsmith, Hodgkinson *et al.*, Kauffman, Kurita, Leasor & MacGregor, Sapolsky, Wright *et al.*).

Opportunity costs refer to the benefits to the region that are lost to protect a site (e.g. foregone harvesting rights and alternative land uses) so that primate tourism can occur. In Mgahinga, Uganda, the main opportunity costs of establishing the park included losses of productive farmland and property (with some compensation provided), grazing areas, and commonly used resources (bamboo, water, thatching grass, medicinal plants, sites for beehives) (Adams & Infield, 1998). These authors estimated the annual value of lost agricultural production in 1995 at $855 000 US. These costs are generally poorly estimated, however, and potentially ignore habitat degradation associated with alternative land uses.

Interactions among multiple economic impacts can be difficult to manage, as seen in Hatfield's (2005) assessment of gorilla viewing in the Virunga Volcanoes and Bwindi. Overall, gorilla tourism generated $20.6 million US in annual use value benefits, with 53%, 41%, and 6% accruing to national, international, and local levels, respectively. Good-quality data were not available for costs associated with wildlife damage, such as gorilla damage to property and crops adjacent to their protected forest habitat. The opportunity cost of foregone agricultural productivity due to the existence of those forests was $13.4–15.0 million US, assuming 50% of the park areas were cultivable. Overall, considering both benefits and costs among stakeholders, Hatfield estimated gains attributable to gorilla tourism of $12.2 million US and $195.94 million US at national and international levels, respectively, but losses of $11.7 million US to local communities. Given this distribution of benefits and costs, the international community gained a disproportionate share of benefits while local communities bore a disproportionate share of costs. Thus, the international community should contribute substantially to the conservation of primates and their habitats required for long-term provision of primate tourism opportunities.

Inequitable situations may also occur within the region when economic benefits do not accrue to those bearing the costs of providing wildlife tourism (Groom *et al.*, 1991). For example, communities near primate tourism sites in Rwanda and Uganda did not receive sufficient compensation for the opportunity costs of conserving those habitats for primates (Ahebwa *et al.*, 2012, Tumusiime & Svarstad, 2011; Weber, 1993). Similarly, benefits may not be shared equitably among members of a local community. Most often, community-based tourism projects produce (at best) modest cash benefits. Although even small amounts of extra income are very welcome in rural areas, they are often captured by a relatively small proportion of the community (Alexander, 2000; Kiss, 2004).

Other concerns about local economic benefits include tourism instabilities. Political tensions or economic volatility within local areas, the country visited, nearby countries, or around the world may be particularly relevant for primate tourism, since many host countries have experienced such instabilities in the recent past (Butynski & Kalina, 1998; Wright *et al.*, this volume). The economic crisis that started in 2008 was predicted to slow international travel globally (World Tourism Organization, 2009), raising concerns that it would undermine conservation progress made at destinations benefiting from primate tourism (Block, 2009). Nevertheless, such setbacks are usually, but not always, temporary (Groom *et al.*, 1991). Further, wildlife tourism is often seasonal, thus generating local employment for only portions of the year and potentially disrupting normal patterns of local life. Other negative economic and social impacts occur when inflation at tourism locations causes prices to rise for local residents. This disproportionately affects local residents, who are less able to bear that cost.

Linking economics and conservation

Wildlife ecotourism has been considered one of the least damaging ways of using the environment and a basis for supporting biodiversity conservation economically and politically (Dabrowski, 1994; Tisdell, 1995). Many wildlife and protected area tourism ventures have been promoted on these beliefs. Links promoted between the economics of wildlife tourism and conservation are sketched below.

1. Wildlife tourism may provide much-needed revenue for the management and protection of primate habitat (Wilkie & Carpenter, 1999). The most common of many mechanisms for raising revenues through tourism is user fees. Across mountain gorilla viewing locations, foreign visitor user fees have increased steadily; a typical one-day gorilla viewing permit costs $500 US per foreign non-resident (Ahebwa *et al.*, 2012). At Kahuzi-Biega National Park, Democratic Republic of Congo, 40% of gorilla viewing fees have been allocated to park management and development in nearby communities (Butynski & Kalina, 1998). In Mgahinga, Uganda, about 20% of the gorilla tourism revenue has been used to manage the park (Archabald & Naughton-Treves, 2001).

   However, revenue from primate tourism areas may not be sufficient to manage and protect them adequately (Blom, 2001b). In the 1990s, tourist revenues in Tangkoko Dua Saudara Nature Reserve, Indonesia, where macaque tourism plays a significant role, did not generate enough money to implement appropriate reserve management. In 1993, only 2% of total entry fee revenues were allocated to the Department of Forestry and only a fraction of this was re-allocated to the reserve (Kinnaird & O'Brien, 1996). In the Dzanga-Sangha region, Central African Republic, tourism revenues covered only about 4% of the park management budget (Blom, 2000). User fees can also be used to justify reduced budget allocations to manage or protect wildlife or habitat areas (Varty *et al.*, 2005) and their collection and distribution increases the potential for bribery and corruption (Weber, 1993). At present, financing conservation with tourism revenues is probably viable in only a minority of cases.

2. Wildlife tourism activities can offer a preferred financial alternative over more environmentally damaging activities (Gould, 1999). For example, in 1984, gorilla tourism in Rwanda was estimated to earn about $200 US/ha, compared with $15 US/ha for cattle grazing, a land use option in the surrounding landscape (Harcourt, 1986). Potential income from gorilla tourism in Cameroon was estimated to far exceed income from gorilla hunting or forest harvesting (Djoh & van derWal, 2001). Similarly, in 1991–1992, guiding fees for viewing colonies of ex-captive stumptail macaques (*Macaca arctoides*) in Veracruz, Mexico, generated more income per hectare than renting land for cattle grazing and more income per family than other occupations such as working on farms (Serio-Silva, 2006). Such assessments can give local residents and authorities reason to support primate conservation by changing or reducing local extractive resource uses, and did so in Cuyabeno Wildlife Reserve in Amazonian Ecuador (Wunder, 2000).

3. Direct economic benefits from wildlife tourism activities can raise the welfare of local residents and communities in ways that benefit conservation, commonly through local employment, industry stimulation, economic diversification, and infrastructure improvements (McNeely *et al.*, 1991). By the late 1980s, for example, gorilla tourism in Volcanoes National Park, Rwanda, employed about seventy guards and eight to ten guides; guides' incomes were more than four times greater than the national average (Shackley, 1995; Weber, 1993). In Uganda, from 1993 to 1998, a program for sharing revenues from national parks hosting substantial primate tourism with surrounding communities resulted in $83 000 US spent on schools, clinics, a bridge, and a road (Archabald & Naughton-Treves, 2001). Following increases in gorilla tourism in Dzanga-Sangha from 1993 to 1998, law enforcement in the area, funded by gorilla tourism fees, improved (Blom, 2000). Fixed and ongoing economic costs also occur, some of which will increase in proportion to the level of tourism activity.

4. For any wildlife tourism project, the benefits and costs accrued and shared among local communities will influence how local people assess prospects for conservation (Alexander, 2000). If it benefits local communities, it can encourage conservation among local residents by creating positive attitudes toward local wildlife and their habitats and by promoting conservation activities (Butynski & Kalina, 1998; Hvenegaard & Manaloor, 2007; Treves & Brandon, 2005). Positive local attitudes to wildlife and their habitats are considered important in shaping wildlife-friendly behavior (Kaiser *et al.*, 1999).

One method of generating positive local attitudes to primate conservation is by sharing revenues generated by primate tourism with local communities. This may provide local incentives to resist environmental threats from outside forces, such as poaching and road development in primate tourism sites (Archabald & Naughton-Treves, 2001; Wunder, 2000). Programs of tourism revenue sharing have been established in some primate tourism areas. In gorilla tourism parks in Uganda, revenue-sharing programs have existed since 1994, with several variations. In 1995–1996, 12% of the park entry fees were allocated to local community projects (8%), local district administrations (2%), and a national pool (2%) (Adams & Infield, 2003). Communities near Bwindi are supposed to receive 20% of all park entry fees; starting in 2006, an additional $5 US per gorilla viewing permit has raised more funds for local communities (Ahebwa *et al.*, 2012). Revenue sharing from park entrance fees in Dzanga-Ndoki allocated 40% to a local NGO for local projects (e.g. road maintenance, health care, and drinking water), 50% to park management, and 10% to a forestry and tourism fund that finances conservation projects elsewhere in the Central African Republic (Blom, 2000). In 1995, local community benefits from tourism fees, income, and employment amounted to $18 478 US (Blom, 2000).

Following the introduction of a revenue-sharing program in 1994 in western Uganda's protected areas that offer primate tourism, attitudes toward the protected areas improved (Archabald & Naughton-Treves, 2001). Surveys of local residents showed their attitudes to gorilla tourism for Bwindi improved from 1997 (when a gorilla tourism program began) to 1999 (Lepp, 2002). For example, fewer local residents felt that the costs of living near the park were greater than the benefits (68% in 1997 to 46% in 1999). Local people nonetheless maintained some concern about the opportunity costs of lost agricultural land (Adams & Infield, 2003) and many are ambivalent about gorilla tourism and the national park (Tumusiime & Svarstad, 2011). Local acceptance of Dzanga-Sangha increased with growing revenue from gorilla tourism (Blom, 2000). Local residents' conservation attitudes toward gorilla tourism and Volcanoes National Park became more favorable from 1979 (soon after gorilla viewing began) to 1984 (Weber, 1987). At the Community Baboon Sanctuary in Belize, which protects black howling monkeys (*Alouatta pigra*), local landowners reported receiving benefits from monkey tourism and supporting the sanctuary's management efforts, primarily for reasons of monkey protection, jobs, and tourism income (Alexander, 2000; Hartup, 1994). Several chapters in this volume report similar local views toward primate tourism (e.g. Strum & Manzolillo Nightingale, Wright *et al.*).

Local economic benefits from tourism have also enhanced conservation activities. In the Virunga Volcanoes, Rwanda, gorilla poaching decreased dramatically between 1979 (soon after gorilla viewing began) and 1984 (Weber, 1987). Densities of snares and poachers' tracks in research and tourism areas were also 25–50% lower than in areas without research and tourism (McNeilage, 1996), presumably resulting in fewer wildlife deaths from poachers in tourist areas.

Efforts by local Rwandans played a role in maintaining gorilla populations during the civil strife of the 1990s (Salopek, 1995).

Importantly, economic incentives from tourism and regulatory safeguards need to work together to promote gorilla conservation (Brown *et al.*, 2008). Many factors can affect the tourism–income connection, including the availability of labor, land tenure, market factors, and knowledge. Income from tourism may not be used as intended. In the Uganda tourism revenue-sharing program, only 60% of this income was spent on the intended projects (e.g. school classrooms) (Adams & Infield, 2003). In Bwindi, despite the revenue-sharing program, some local communities did not feel adequately compensated for lost opportunities due to conservation (Ahebwa *et al.*, 2012). For gorilla tourism revenue-sharing programs to succeed, they must have long-term institutional support of implementing agencies, identify the appropriate target community, and be transparent and accountable (Archabald & Naughton-Treves, 2001; Tumusiime & Vedeld, 2012).

These links between the economic benefits of primate tourism and enhanced primate conservation should be interpreted cautiously. An important question is whether expectations are normally or ever realized. It has often been assumed that attitudes will translate into action, although this is far from guaranteed (Russon & Russell, 2005). The extent and quality of evidence currently available on this issue remains limited. Some of the studies reported provide anecdotal or non-systematic evidence, and others need more contextual information to evaluate the conservation contributions thoroughly. The importance of tourism-generated economic incentives for conservation, for instance, should be assessed relative to total tourism spending or economic activity. Some of the relevant studies are also older and whether their conclusions remain relevant today is uncertain (e.g. Goldsmith, this volume). Finally, studies have turned up negative as well as positive findings, notably realities that fail to meet expectations or promises.

Optimizing economic impacts

Concluding that maximizing wildlife tourism's economic impact would best promote the links with conservation is hasty for several reasons. The links are preliminary and need further research to examine their subtle dynamics. The links vary from situation to situation, depending on many variables, and both the benefits and the costs of primate tourism must be considered in the context of long-term conservation (Harcourt, 2001). Maximizing economic impact promotes some conservation objectives at the expense of others, impacts can be perceived as good or bad depending on one's position (Lindberg, 2001), and frequently, other considerations of economic value are ignored.

Proponents of primate tourism are better advised to define their objectives for primate tourism relative to primate conservation carefully rather than to focus on its economic impacts alone, and then to assess thoroughly the positive and negative impacts of any proposed or existing primate tourism activity on primate conservation. The optimal level of wildlife tourism should be determined, with a focus on pricing policies and visitor numbers but also considering many other biological and social variables. Also important is critical analysis of proposed and realized economic benefits and costs, before and after implementation of a primate tourism program. Such information and analyses are necessary for effective decision making and planning regarding primate tourism development, local development, and primate conservation.

There are many ways to increase the economic benefits from wildlife tourism activities to local communities that are applicable to primate tourism (Hvenegaard & Manaloor, 2001). Many protected areas have multi-tiered user fee structures in which foreigners pay substantially more than nationals, many primate tourism sites included (Andersson et al., 2005). Monteverde Cloud Forest Preserve, Costa Rica, uses such a system and foreigners generated 97% of its 1994 entry fee revenues although they represented only 80% of visits (Aylward et al., 1996). Entrance fees might be raised considerably for some primate tourism destinations, given experiences with mountain gorilla tourism in Uganda (Andersson et al., 2005; Moyini & Uwimbabazi, 2000) and wildlife tourism in the Galapagos (Edwards, 1991).

Incoming primate tourist numbers could be increased if increases are realistic given the tourism market and the conservation, ecological, and tourist-carrying capacities of local sites and facilities. The risk is that expanding primate tourism to maximize financial returns may be irresistible (Butynski & Kalina, 1998), to the potential detriment of the primate populations visited. There have been pressures to increase mountain gorilla tourist numbers; at present, 44% of the global mountain gorilla population has been habituated for tourism (Goldsmith, this volume). However, Virunga and Bwindi gorilla tourism areas were recently operating at 41% of full capacity (Hatfield, 2005). Revenues could be increased by simply filling the available capacity.

Leakage could be reduced by encouraging tourists to spend more money locally, so that more benefits accrue to local residents. Offering additional attractions could encourage lengthening their stay or returning. The Nyungwe Forest, Rwanda has used this approach: the original tourist attraction was viewing chimpanzees (*Pan troglodytes*) but black-and-white colobus monkeys (*Colobus guereza*), blue monkeys (*Cercopithecus mitis*), grey-cheeked mangabeys (*Lophocebus albigena*), and a variety of birds could be added (Weber, 1993). Economic participation by local residents may be increased (Uddhammar, 2006; Wunder, 2000) and failing to do so can be a missed opportunity. Various economic incentives can encourage locals to invest in community-based wildlife tourism (Strum & Manzolillo Nightingale, this volume; Victurine, 2000). Locals can sell and sometimes supply, rather than import, items that tourists like to purchase such as handicrafts, literature, souvenirs, apparel, equipment, and food (Hvenegaard et al., 1989). Local people might also be hired instead of outsiders to guide wildlife viewing.

Finally, tourists can be encouraged to donate to local conservation or social projects; indeed many are willing to do so (Hvenegaard & Dearden, 1998; Kangas *et al.*, 1995). Moyini and Uwimbabazi (2000) found that 46% of gorilla tourists would be willing to donate to a fund that was established to support gorilla conservation, law enforcement, and community development. Non-governmental and international organizations might also be persuaded to provide donor support because of the emphasis on primate tourism (e.g. Greer & Cipolletta, 2006).

Numerous complexities must be considered. Local people often lack the necessary expertise, experience, and capital, so developing local handicrafts for sale, training local guides with appropriate language and natural history expertise, and encouraging local investment in hotels can entail considerable costs (Lindberg, 2001; Weber, 1993). Local inequities resulting from the concentration of revenues within a few businesses or households, due to patterns of ownership in tourism-related services, proximity to the tourism site, gender, and availability of time to participate in tourism services, can create considerable tension within a community. Thus, equitable governance structures are needed to allocate opportunities (Brown *et al.*, 2008; Uddhammar, 2006). Some communities with primate tourism, such as those near the Community Baboon Sanctuary, Belize, have created such structures; for instance, managers and members of the sanctuary rotate opportunities to provide accommodation to tourists among households (Alexander, 1999). However, because of management inconsistencies, results have been mixed. Even though many members of the sanctuary had concerns about how the benefits of tourism are shared equitably, most wanted to maintain the sanctuary, primarily for jobs, habitat protection, and tourism benefits (Alexander, 2000).

Discussion

It is recognized that wildlife tourism is not a panacea for nature conservation, it should not be relied on to meet the conservation objectives of a particular site, and its contribution to conservation is limited both by its ability to contribute to long-term protection and its tendency to degrade the environment (Butynski, 1998; Isaacs, 2000; Kinnaird & O'Brien, 1996). It is just one of many tools available (Forsyth *et al.*, 1995) and it should be implemented in a coordinated and integrated fashion.

The ultimate test of whether the economic impact of wildlife tourism benefits conservation is its effect on the wildlife populations and habitats visited (Newsome *et al.*, 2005). In Rwanda, prior to the civil unrest, Weber (1993) argued that the economic impact of mountain gorilla tourism was positive. In its first ten years (1979–1989), when it charged tourism fees to fund park security, conservation education, and the gorilla tourism program, the mountain gorilla population increased from 260 to 320. Conversely, unlimited primate tourism can exceed an area's capacity to manage environmental impacts and can reduce its primate populations. In Tangkoko Dua Saudara Nature Reserve, Indonesia, the basis for primate tourism – the macaque population – declined by 75% from 1978 to 1993 (Kinnaird & O'Brien, 1996). Other chapters in this volume report outcomes of both types.

Sustainability of primate tourism is critical, many factors affect it, and one cannot escape the pervasive role of economic issues in the many critical social, environmental, and political decisions involved. The long-term success of primate tourism sites depends on the financial viability of the primate tourism operations (Kiss, 2004). Adams and Infield (2003) provide a valuable matrix of interests in the economic aspects of primate tourism, from local to international scale, and cross-cut with historical, conservation, employment, national, and ethical issues. Sustainable primate tourism entails providing tourists with satisfying experiences for which they will pay high prices and which they will promote to others, providing the means to support the visitor and management infrastructures and the local communities, and protecting the wildlife and areas visited. Primate tourism that is less profitable than competing alternatives could result in fewer funds for conservation, governments, or local people and could encourage developers or local people to reduce conservation efforts (Tisdell, 1995).

Thus, a number of recommendations regarding the economic links between primate tourism and conservation are suggested. First, economic issues are important but should not be the driving or only consideration in primate tourism. Many other factors should be considered as well, especially conservation goals, ethics, and local community welfare (Hill, 2002), so that economic and conservation issues can be coordinated in decision making and management.

Second, careful attention should be paid to community-based primate ecotourism, that is tourism ventures that aim to promote environmental conservation and sustain the well-being of local people through the active involvement of local communities (Kiss, 2004). Ideally, legitimate ecotourism would directly link protecting primates and their habitat with protecting local income. Sustaining community-based ecotourism depends on three major sources (Kiss, 2004): incentives derived from income dependent on biodiversity; reinvestment of some of that income to sustain the tourism business and protect the biodiversity on which the tourism activities depend; and securing private sector involvement once tourism initiatives are established (community support and infrastructure). Creating unrealistic expectations about the economic benefits from primate tourism could cause people to abandon their involvement because of under-performance. Before tourism income reaches incentive levels, external incentives may also be needed. Stable operations with consistent economic benefits will more likely attract tourists, tourism investors, donations from international conservation organizations, and support from local and national governments. Diversification in tourism products can also promote long-term stability.

Third, much more research is needed to document and evaluate the economic impacts of primate tourism. Full and fair comparisons of all net benefits and costs of primate tourism to local residents and to the conservation of the primates and habitat visited versus no primate tourism at a particular site are rare, but necessary to make sound decisions. For example, an economic cost–benefit analysis of a wildlife tourism operation at Possum Point Biological Station, Belize showed a net profit only in the latter two years of a three-year pilot study (Kangas *et al.*, 1995). While it remains difficult to identify the actual impacts of primate tourism on primate conservation, since primate tourism is often just one of several activities within larger-scale conservation projects (Butynski & Kalina, 1998), it is important to determine the amount of conservation funding that wildlife tourism enterprises are absorbing or generating and doing so entails more rigorous assessment and analysis of existing projects.

It is equally important to study links between primate tourism, economic impacts and values, conservation goals, and community conservation incentives, to assess the contributions of tourism to primate conservation, local people's level of participation in tourism, and the extent to which local people link economic prosperity from tourism with primate conservation (Kiss, 2004; Tumusiime & Svarstad, 2011). Answers to the following and related questions will help create stronger links with conservation. How does primate tourism change long-term attitudes and behavior regarding primate conservation? What are the influencing variables and dynamics involved? To what extent are all benefits and costs (local, national, international) considered in setting up and operating primate tourism operations, and how does this affect economic expectations? How can agencies share revenues and costs from primate tourism equitably among stakeholders? How can leakage be reduced to improve local and primate conservation benefits?

In conclusion, for primate tourism to contribute to conservation, conservation must take priority over economic or political concerns, decisions should be based on sound science, regulations need to be enforced rigorously, and economic benefits to conservation need to increase substantially (Butynski & Kalina, 1998; Macfie & Williamson, 2010; Williamson & Macfie, this volume). Primate tourism needs to embrace and implement policies like those found in Bwindi's development plan: tourist activity must support conservation, local participation in tourism activities must be encouraged, and tourism development must promote environmental awareness. Such policies need also to be bolstered by careful design, support, and regulation.

Acknowledgments

Thanks to Anne E. Russon, Connie Russell, and Janette Wallis for their helpful comments. The University of Alberta and the Centre d'Écologie Fonctionnelle et Evolutive (of the Centre Nationale de la Recherche Scientifique) provided support during the writing of this chapter.

References

Adams, W. M. and Infield, M. (1998). *Community conservation at Mgahinga Gorilla National Park, Uganda.* Working Paper, No. 10. Manchester, UK: Institute for Development Policy and Management, University of Manchester.

Adams, W. M. and Infield, M. (2003). Who is on the gorilla's payroll? Claims on tourist revenue from a Ugandan National Park. *World Development*, 31: 177–190.

Ahebwa, W. M., van der Duim, R., and Sandbrook, C. (2012). Tourism revenue sharing policy at Bwindi Impenetrable National Park, Uganda: A policy arrangements approach. *Journal of Sustainable Tourism*, 20: 377–394.

Alexander, S. E. (1999). The role of Belize residents in the struggle to define ecotourism opportunities in monkey sanctuaries. *Cultural Survival Quarterly*, (summer): 21–23.

Alexander, S. E. (2000). Resident attitudes towards conservation and black howler monkeys in Belize: The Community Baboon Sanctuary. *Environmental Conservation*, 27: 341–350.

Andersson, P., Croné, S., Stage, J., and Stage, J. (2005). Potential monopoly rents from international wildlife tourism: An example from Uganda's gorilla tourism. *Eastern Africa Social Science Research Review*, 21 (1): 1–18.

Archabald, K. and Naughton-Treves, L. (2001). Tourism revenue-sharing around national parks in Western Uganda: Early efforts to identify and reward local communities. *Environmental Conservation*, 28: 135–149.

Aylward, B., Allen, K., Echeverria, J., and Tosi, J. (1996). Sustainable ecotourism in Costa Rica: The Monteverde Cloud Forest Reserve. *Biodiversity and Conservation*, 5: 315–343.

Bergstrom, J. C., Stoll, J. R., Titre, J. P., and Wright, V. L. (1990). Economic value of wetlands-based recreation. *Ecological Economics*, 2: 129–147.

Berman, C. M., Li, J. H., Ogawa, H., Ionica, C., and Yin, H. (2007). Primate tourism, range restriction, and infant risk among *Macaca thibetana* at Mt. Huangshan, China. *International Journal of Primatology*, 28: 1123–1141.

Block, B. (2009). Recession may hinder sustainable tourism. *Worldwatch Institute Eye on Earth* article. www.worldwatch.org/node/6028 (accessed Mar. 18, 2009).

Blom, A. (2000). The monetary impact of tourism on protected area management and the local economy in Dzanga-Sangha (Central African Republic). *Journal of Sustainable Tourism*, 8: 175–189.

Blom, A. (2001a). Potentials and pitfalls of tourism in Dzanga-Sangha. *Gorilla Journal*, 22: 40–41.

Blom, A. (2001b). *Ecological and Economic Impacts of Gorilla-based Tourism in Dzanga-Sangha, Central African Republic*. Unpublished PhD Thesis. Wageningen, The Netherlands: Wageningen University.

Brown, M., Bonis-Charancle, J. M., Mogba, Z., Sundararajan, R., and Warne, R. (2008). Linking the Community Options Analysis and Investment Toolkit (COAIT), Consensys® and Payment for Environmental Services (PES): A model to promote sustainability in African gorilla conservation. In: Stoinski, T. S., Steklis, H.D., and Mehlman, P. T. (eds.) *Conservation in the 21st Century*. New York, NY: Springer, pp. 205–227.

Butynski, T. M. (1998). Is gorilla tourism sustainable? *Gorilla Journal*, 16: 15–19.

Butynski, T. M. and Kalina, J. (1998). Gorilla tourism: A critical look. In: E. J Milner-Gullard and R. Mace (eds.), *Conservation of Biological Resources*. Oxford, UK: Blackwell Science, pp. 294–313.

Dabrowski, P. (1994). Tourism for conservation, conservation for tourism. *Unasylva*, 45: 42–44.

Dellatore, D. F. (2007). *Behavioural Health of Reintroduced Orangutans (Pongo abelii) in Bukit Lawang, Sumatra Indonesia*. UK: MSc thesis, Oxford Brookes.

Dixon, J. A. and Sherman, P. B. (1990). *Economics of Protected Areas: A New Look at Benefits and Costs*. Washington, DC: Island Press.

Djoh, E. and van derWal, M. (2001). Gorilla-based tourism: A realistic source of community income in Cameroon? Case study of the villages of Koungoulou and Karagoua. *Rural Development Forestry Network Paper*, 25e (III): 31–37.

Edwards, S. F. (1991). The demand for Galapagos vacations: Estimation and application to wilderness preservation. *Coastal Management*, 19: 155–169.

Feistner, A. T., Razafiarimalala, A., and Wright, P. C. (2006). Primate research and ecotourism: Conflict or collaboration? *International Journal of Primatology*, 27 (Supplement 1): 222–223.

Fiallo, E. A. and Jacobson, S. K. (1995). Local communities and protected areas: Attitudes of rural residents towards conservation and Machalilla National Park, Ecuador. *Environmental Conservation*, 22: 241–249.

Forsyth, P., Dwyer, L., and Clarke, H. (1995). Problems in use of economic instruments to reduce adverse environmental impacts of tourism. *Tourism Economics*, 1: 265–282.

Fuentes, A., Shaw, E., and Cortes, J. (2007). Qualitative assessment of Macaque tourist sites in Padangtegal, Bali, Indonesia, and the Upper Rock Nature Reserve, Gibraltar. *International Journal of Primatology*, 28: 1143–1158.

Gould, K. A. (1999). Tactical tourism: A comparative analysis of rainforest development in Ecuador and Belize. *Organization & Environment*, 12: 245–262.

Greer, D. and Cipolletta, C. (2006). Western gorilla tourism: Lessons learned from Dzanga-Sangha. *Gorilla Journal*, 33: 16–19.

Groom, M. J., Podolsky, R. D., and Munn, C. A. (1991). Tourism as a sustained use of wildlife: A case study of Madre de Dios, southeastern Peru. In: J. G. Robinson and K. H. Redford (eds.), *Neotropical Wildlife Use and Conservation*. University of Chicago Press, pp. 393–412.

Grossberg, R., Treves, A., and Naughton-Treves, L. (2003). The incidental ecotourist: Measuring visitor impacts on endangered howler monkeys at a Belizean archaeological site. *Environmental Conservation*, 30: 40–51.

Harcourt, A. H. (1986). Gorilla conservation: Anatomy of a campaign. In: K. Benirschke (ed.), *Primates: The Road to Self-sustaining Populations*. New York, NY: Springer-Verlag, pp. 31–46.

Harcourt, A. (2001). The benefits of mountain gorilla tourism. *Gorilla Journal*, 22: 36–37.

Hartup, B. K. (1994). Community conservation in Belize: Demography, resource uses, and attitudes of participating landowners. *Biological Conservation*, 69: 235–241.

Hatfield, R. (2005). *The Economic Value of the Bwindi and Virunga Gorilla Mountain Forests*. Washington, DC: African Wildlife Foundation.

Healy, R. G. (1988). *Economic Considerations in Nature-oriented Tourism: The Case of Tropical Forest Tourism*. FPEI Working Paper No. 39. Research Triangle Park, NC: Southeastern Center for Forest Economics Research.

Higginbottom, K. (2004). Wildlife tourism: An introduction. In: K. Higginbottom (ed.) *Wildlife Tourism: Impacts, Management and Planning*. Altona, Australia: Common Ground Publishing, pp. 1–14.

Hill, C. M. (2002). Primate conservation and local communities – Ethical issues and debates. *American Anthropologist*, 104: 1184–1194.

Hvenegaard, G. T., Butler, J. R., and Krystofiak, D. K. (1989). The economic values of bird watching at Point Pelee National Park, Canada. *Wildlife Society Bulletin*, 17: 526–531.

Hvenegaard, G. T. and Dearden, P. (1998). Linking ecotourism and biodiversity conservation: A case study of Doi Inthanon National Park, Thailand. *Singapore Journal of Tropical Geography*, 19: 193–211.

Hvenegaard, G. T. and Manaloor, V. (2001). Snow Goose Festival generates economic benefits for Tofield, Alberta. *Edmonton Naturalist*, 29 (2): 28–31.

Hvenegaard, G. T. and Manaloor, V. (2007). A comparative approach to analyzing local expenditures and visitor profiles of two wildlife festivals. *Event Management*, 10: 231–239.

Isaacs, J. C. (2000). The limited potential of ecotourism to contribute to wildlife conservation. *Wildlife Society Bulletin*, 28: 61–69.

Kaiser, F. G., Wölfing, S., and Fuhrer, U. (1999). Environmental attitude and ecological behaviour. *Journal of Environmental Psychology*, 19: 1–19.

Kangas, P., Shave, M., and Shave, P. (1995). Economics of an ecotourism operation in Belize. *Environmental Management*, 19: 669–673.

Kinnaird, M. F. and O'Brien, T. G. (1996). Ecotourism in the Tangkoko Dua Saudara Nature Reserve: Opening Pandora's Box? *Oryx*, 30: 65–73.

Kiss, A. (2004). Is community-based ecotourism a good use of biodiversity conservation funds? *Trends in Ecology and Evolution*, 19: 232–237.

Lepp, A. (2002). Uganda's Bwindi Impenetrable National Park: Meeting the challenges of conservation and community development through sustainable tourism. In: R. Harris, T. Griffin, and P. Williams (eds.), *Sustainable Tourism: A Global Perspective*. Amsterdam, Netherlands: Elsevier Butterworth-Heinemann. pp. 211–220.

Lindberg, K. (1998). Economic aspects of ecotourism. In: K. Lindberg and M. E. Wood (eds.), *Ecotourism: A Guide for Planners and Managers*, vol. 2. North Bennington, VT: The Ecotourism Society, pp. 87–117.

Lindberg, K. (2001). Economic impacts. In: D. B. Weaver (ed.), *The Encyclopedia of Ecotourism*. Oxon, UK: CABI, pp. 363–377.

Macfie, E. J. and Williamson, E. A. (2010). *Best Practice Guidelines for Great Ape Tourism*. Gland, Switzerland: IUCN/SSC Primate Specialist Group.

Matheson, M. D., Sheeran, L. K., Li, J. H., and Wagner, R. S. (2006). Tourism impact on Tibetan macaques. *Anthrozoös*, 19: 158–168.

McNeely, J. A., Thorsell, J. W., and Ceballos-Lascurain, H. (1991). *Guidelines for Development of Terrestrial and Marine National Parks and Protected Areas for Tourism*. Gland, Switzerland: International Union for Conservation of Nature and Natural Resources.

McNeilage, A. (1996). Ecotourism and mountain gorillas in the Virunga Volcanoes. In: V. J. Taylor and N. Dunstone (eds.), *The Exploitation of Mammal Populations*. London, UK: Chapman & Hall, pp. 334–344.

Moyini, Y. and Uwimbabazi, B. (2000). *Analysis of the Economic Significance of Gorilla Tourism in Uganda*. Kampala, Uganda: Environmental Monitoring Associates, Ltd.

Naughton-Treves, L., Treves, A., Chapman, C., and Wrangham, R. (1998). Temporal patterns of crop-raiding by primates: Linking food availability in croplands and adjacent forest. *Journal of Applied Ecology*, 35: 596–606.

Newsome, D., Dowling, R., and Moore, S. (2005). *Wildlife Tourism*. Aspects of Tourism 24, Clevedon, UK: Channel View Publications.

Nunes, P. A. L. D. and van den Bergh, J. C. J. M. (2001). Economic valuation of biodiversity: Sense or nonsense? *Ecological Economics*, 39: 203–222.

Russon, A. E. and Russell, C. L. (2005). Orangutan tourism. In: J. Caldecott and L. Miles (eds.), *World Atlas of Great Apes and their Conservation*. Berkeley, CA: University of California Press. pp. 264–265.

Salopek, P. F. (1995). Gorillas and humans: An uneasy truce. *National Geographic*, 188 (4): 72–83.

Sandbrook, C. and Semple, S. (2006). The rules and the reality of mountain gorilla *Gorilla beringei beringei* tracking: How close do tourists get? *Oryx*, 49: 428–433.

Serio-Silva, J. C. (2006). Las Islas de los Changos (the Monkey Islands): The economic impact of ecotourism in the region of Los Tuxtlas, Veracruz, Mexico. *American Journal of Primatology*, 68: 499–506.

Shackley, M. (1995). The future of gorilla tourism in Rwanda. *Journal of Sustainable Tourism*, 3 (2): 61–72.

Tisdell, C. (1995). Investment in ecotourism: Assessing its economics. *Tourism Economics*, 1: 375–387.

Tisdell, C. and Wilson, C. (2004). Economics of wildlife tourism. In: K. Higginbottom (ed.), *Wildlife Tourism: Impacts, Management and Planning*. Altona, Australia: Common Ground Publishing, pp. 167–186.

Treves, A. and Brandon, K. (2005). Tourist impacts on the behaviour of black howler monkeys (*Alouatta pigra*) at Lamanai, Belize. In: J. D. Paterson and J. Wallis (eds.), *Commensalism and Conflict: The Human-primate Interface*. Norman, OK: American Society of Primatologists, pp. 146–167.

Tumusiime, D. M. and Svarstad, H. (2011). A local counter-narrative on the conservation of Mountain Gorillas. *Forum for Development Studies*, 38: 239–265.

Tumusiime, D. M. and Vedeld, P. (2012). False promise or false premise? Using tourism revenue sharing to promote conservation and poverty reduction in Uganda. *Conservation and Society*, 10: 15–28.

Uddhammar, E. (2006). Development, conservation and tourism: Conflict or symbiosis? *Review of International Political Economy*, 13: 656–678.

Van Leeuwen, E. S., Nijkamp, P., and Rietveld, P. (2011). A meta-analytic comparison of regional output multipliers at different spatial levels: Economic impacts of tourism. In: A. Matias, P. Nijkamp, and M. Sarmento (eds.), *Advances in Tourism Economics: New Developments*. New York: Physica-Verlag/Springer, pp. 13–33.

Varty, N., Ferriss, S., Carroll, B., and Caldecott, J. (2005). Conservation measures in play. In: J. Caldecott and L. Miles (eds.), *World Atlas of Great Apes and their Conservation*. Berkeley, CA: University of California Press, pp. 242–275.

Victurine, R. (2000). Building tourism excellence at the community level: Capacity building for community-based entrepreneurs in Uganda. *Journal of Travel Research*, 38: 221–229.

Wallis, J. and Lee, D. R. (1999). Primate conservation: The prevention of disease transmission. *International Journal of Primatology*, 20: 803–826.

Ward, F. A. and Beal, D. (2000). *Valuing Nature with Travel Cost Models: A Manual*. Cheltenham, UK: Edward Elgar.

Weber, A. W. (1987). *Ruhengeri and its Resources: An Environmental Profile of the Ruhengeri Prefecture, Rwanda*. Kigali, Rwanda: Ruhengeri Resource Analysis and Management Project.

Weber, A. W. (1993). Primate conservation and ecotourism in Africa. In: C. S. Potter, J. I. Cohen, and D. Janczewski (eds.), *Perspectives on Biodiversity: Case Studies of Genetic Resource Conservation and Development*. Washington, DC: American Association for the Advancement of Science, pp. 129–150.

Wells, M. P. (1997). *Economic Perspectives on Nature Tourism, Conservation and Development*. Environment Department Paper. No. 55. Washington, DC: World Bank.

Wilkie, D. S. and Carpenter, J. F. (1999). Can nature tourism help finance protected areas in the Congo Basin? *Oryx*, 33: 332–338.

Woodford, M. H., Butynski, T. M., and Karesh, W. B. (2002). Habituating the great apes: The disease risks. *Oryx*, 36: 153–160.

World Tourism Organization. (2009). International tourism challenged by deteriorating global economy. *UNWTO World Tourism Barometer*, 7 (1): 1, 5–8.

Wunder, S. (2000). Ecotourism and economic incentives – An empirical approach. *Ecological Economics*, 32: 465–479.

16 Considering risks of pathogen transmission associated with primate-based tourism

Michael P. Muehlenbein and Janette Wallis

Introduction

Sustainable, nature-based tourism should attempt to educate visitors about wildlife and the environment, while minimizing modification or degradation of the natural resources in the sites they visit. When possible, these activities should broadly benefit the social and natural environments by involving the participation of local communities (Ceballos-Lascuráin, 1996). And, if managed well, nature-based tourism should facilitate species conservation by raising the needed funds for wildlife and habitat conservation, while increasing public awareness of conservation issues. Unfortunately, rapid and unmonitored development of tourism projects in protected areas can produce deleterious effects on the very species we wish to conserve. Such risks may include habitat degradation caused by pollution and environmentally damaging development of infrastructure; animal crowding into restricted areas; and the introduction of invasive species.

These issues take on special importance when the species in question are (non-human) primates. Most wild populations of primates are relatively small and their reproductive cycles are protracted (with low reproductive rates relative to most mammals of similar size), so they are particularly vulnerable to population decreases, including those that result from human activities. Humans have contributed to significant population declines of wild primates through hunting and bushmeat consumption, habitat loss and fragmentation, and illegal capture of live primates for entertainment or other purposes. Tourism activities involving primates in their native habitat (including free-living rehabilitants as well as their wild counterparts) could benefit primate conservation but we must remain diligent about monitoring potential negative consequences, especially since primate tourism has been increasing in popularity over the past few decades. For example, the habituation to human presence essential to primate tourism may increase the likelihood that these animals will raid crops, invade garbage pits, and break into vehicles or lodgings for food. They may also become more vulnerable to poaching as a result of their loss of fear, their natural diet and ranging may become permanently altered, and their normal social behaviors may become altered (e.g. the appearance of a group of tourists

Primate Tourism: A Tool for Conservation?, ed. Anne E. Russon and Janette Wallis. Published by Cambridge University Press. © Cambridge University Press 2014.

may interrupt mating behaviors). Many of the chapters in this volume document these tourism impacts on the primates visited. Habituation could even lead to alterations in animal stress responses, possibly leading to immunosuppression with decreased reproductive success and increased susceptibility to infectious diseases (Muehlenbein, 2009; Muehlenbein *et al.,* 2012).

We focus this chapter on describing human pathogen transmission to primates, positioning the potential role of tourists in the spread of these pathogens, and discussing important steps toward minimizing the impact of tourist-borne diseases in the face of many roadblocks to success. A key feature of a successful primate tourism experience is visitor access to these popular species, and rewarding sightings often depend upon relatively close encounters, that is outside a vehicle. This makes zoonotic (nonhuman animal to human) and anthropozoonotic (human to nonhuman animal) pathogen transmission of vital concern, particularly given the increasing demand from tourists to visit free-ranging primates.

Pathogen transmission to nonhuman primates

Pathogen transmission from humans to nonhuman primates is arguably one of the most dangerous outcomes of human–wildlife interactions (see chapters by Sapolsky, Williamson & Macfie in this volume for additional details). Infectious organisms include thousands of species of viruses (and bacteriophages), bacteria (including rickettsiae), parasitic protozoa and helminthes (nematodes, cestodes, and trematodes), and fungi. These parasitic organisms live all or part of their lives in or on a host from which biological needs are derived. This state of metabolic dependence usually results in host energy loss, lowered survival, and reduced reproductive potential.

Several pathogen transmission events from human to nonhuman primate populations have been either suspected or confirmed to date. These outbreaks have affected nearly all major long-term chimpanzee and gorilla study populations, such as Bwindi, Mahale, Gombe, and others. Confirmed cases that have spread through some type of fecal–oral transmission include *Giardia duodenalis* in gorillas (Graczyk *et al.,* 2002a; Johnston *et al.,* 2010; Salzer *et al.,* 2007) and *E. coli* in chimpanzees (Goldberg *et al.,* 2007) and gorillas (Rwego *et al.,* 2008). Other highly suspected cases of intestinal pathogen transmission include several helminthes and protozoa in gorillas (Ashford *et al.,* 1990), *Schistosoma mansoni* in olive baboons (Mueller-Graf *et al.,* 1997), *Campylobacter* and *Salmonella* in gorillas (Nizeyi *et al.,* 2001), *Encephalitozoon* in gorillas (Graczyk *et al.,* 2002b), a variety of gram-negative bacteria in yellow baboons (Rolland *et al.,* 1985), and polio in chimpanzees (Goodall, 1986; Kortlandt, 1996).

Respiratory infections have long been suspected to be major causes of mortality in wild primates. Confirmed cases include: respiratory syncytial virus in chimpanzees (Köndgen et al., 2008), metapneumovirus in chimpanzees (Kaur *et al.,* 2008; Köndgen et al., 2008), and metapneumovirus in gorillas (Palacios *et al.,* 2011). Other suspected cases include *Streptococcus pneumoniae* in chimpanzees (Chi *et al.,* 2007), *Pasteurella multocida* in chimpanzees (Chi *et al.,* 2007), influenza in chimpanzees (Hanamura *et al.,* 2008; Hosaka, 1995a, 1995b; Lukasik-Braum & Spelman, 2008; Nishida *et al.,* 2003; Wallis & Lee, 1999; Williams *et al.,* 2008), influenza in bonobos (Sakamaki *et al.,* 2009) and gorillas (Macfie, 1991; Sholley & Hastings, 1989), and measles in gorillas (Byers & Hastings, 1991; Ferber, 2000; Sholley, 1989). We also know from studies of captive primates that they can be very susceptible to tuberculosis (Burgos-Rodriquez, 2011).

Wild primates are also susceptible to a number of other pathogens. For example, Ebola virus infection has devastated several gorilla and chimpanzee populations in Gabon, Cameroon, and Democratic Republic of Congo over the past several decades (Huijbregts *et al.,* 2003; Leroy *et al.,* 2004). Scabies can cause significant morbidity in gorillas (Kalema-Zikusoka *et al.,* 2002; Macfie, 1996) and *Bacillus anthracis* has been particularly deadly in chimpanzees (Klee *et al.,* 2006; Leendertz *et al.,* 2006). Both herpesvirus type 1 and yellow fever can be very deadly in New World monkeys (Almeida *et al.,* 2012; Costa *et al.,* 2011). Tuberculosis can be fatal in wild primates, as evidenced by death of baboons that fed on contaminated meat at a tourist lodge garbage dump (Sapolsky, this volume).

Although human populations (particularly through environmental contamination) are suspected as the primary source of most of the above-mentioned cases in wild primates, the precise reservoirs or points of transmission for these pathogen "spill-overs" are unknown. The most likely sources of these transmission events have been local populations (including park personnel) or researchers (Muehlenbein & Ancrenaz, 2009; Wallis & Lee, 1999). There are no documented cases of pathogen transmission from tourists to wildlife, and this is quite understandable. It is difficult to confirm with complete certainty the origin of pathogen transmission events, particularly those involving indirect human contact. Pathogens spread during short-term visits by tourists would likely reveal themselves only after the tourist is gone, making it impossible to trace the exact source of infection. However, we contend that tourists should be considered a health risk to wildlife, particularly to primates whose immune systems are usually naïve to new human pathogens and who, because of the genetic similarity to humans, are particularly sensitive to many of our pathogens (Brack, 1987).

The potential role of tourists in pathogen transmission

The relative contribution of tourists to the spread of pathogens to wildlife is unknown, but the number of tourists visiting wildlife areas worldwide has increased steadily over recent years. A major shortcoming of international travelers in general is their poor knowledge, attitudes, and practices about travel health (Hamer & Conner, 2004; Wilder-Smith *et al.,* 2004). Many travelers do not use pre-travel preventive health strategies, including physician advice and chemoprophylaxes (Crockett & Keystone, 2005; Van Herck *et al.,* 2003). Moreover, traveler compliance to physician advice is surprisingly low, even with regard to avoiding dangerous food items such as salads, shellfish, and non-treated water (Steffen *et al.,* 2004). Many travelers do not understand the basic risks of infection, including their sources and causes (Van Herck *et al.,* 2004; Wilder-Smith *et al.,* 2004; Zuckerman & Steffen, 2000).

A significant proportion of travelers to tropical regions (where most primate-based tourism takes place) are not protected against vaccine-preventable illnesses (Lopez-Velez & Bayas, 2007; Prazuck *et al.,* 1998; Schunk *et al.,* 2001; Van Herck *et al.,* 2004). A majority of these travelers demonstrate poor recall of their actual vaccination status (Hamer & Connor, 2004), as verified by reference to vaccination certificates or serological testing (Hilton *et al.,* 1991; Toovey *et al.,* 2004; Van Herck *et al.,* 2004). Long-distance travelers may also be stressed due to sleep dysregulation, unfamiliar diets and climate, and exposure to novel pathogens, and this stress may make them even more susceptible to health problems. Although some aircraft are designed to circulate cabin airflow to minimize disease transmission, travelers may contract respiratory disease during long flights. Moreover, once in a country, tourists may find themselves traveling in close proximity to others on buses, trains, boats, and other local transport. Consequently, illness during travel is very common, particularly gastrointestinal and respiratory infections (Rack *et al.*, 2005).

Like most other travelers, the majority of tourists that visit primates in their native habitat (which may include both wild and rehabilitated/released but not zoo-living individuals) probably underestimate their own risk of infection, as well as their potential contribution to the spread of diseases. Even conservation-oriented tourists who select travel itineraries that take them to view endangered species may be largely unaware of their potential impact on the health of the wildlife they visit.

Risk of pathogen transmission at primate-based tourism destinations varies with a number of factors, including the species visited, frequency of sightings, degree of proximity, possibility of contact, use of legal or illegal provisioning, and the behavior and infection status of tourists. Thus, as proximity, contact, and food sharing between humans and nonhuman primates vary by location (Fuentes, 2006), risk of pathogen transmission also varies from site to site.

Case study: Tourists at the Sepilok Orangutan Rehabilitation Centre

Here we describe a study by M. Muehlenbein and colleagues that illustrates tourist attitudes and behaviors at a large primate tourism destination, the Sepilok Orangutan Rehabilitation Centre (SORC) in the Malaysian state of Sabah in northern Borneo. Operated by the Sabah Wildlife Department, SORC was established in 1964 as a center for the rehabilitation of orphaned, injured, and/or confiscated orangutans (*Pongo pygmaeus morio*) and other endangered species. Following a six-month quarantine period, orangutans are helped to learn how to navigate the forest, forage for food, and live within an orangutan community. Following rehabilitation and health inspections, the orangutans are eventually released into the adjacent Kabili Virgin Jungle Reserve, the Tabin Wildlife Reserve, or other locations.

To facilitate public education and generate operational funds, the public is allowed to view two daily feedings of the free-ranging orangutans. A multilingual information sign indicates that smoking, eating, and spitting are not allowed; that visitors should keep their distance from the orangutans and macaques; that visitors should not bring medications, bags, or insect repellant; and other miscellaneous information. Park rangers are present during orangutan feedings, and the visitor viewing area is separated from the actual feeding platforms by approximately 10 m. Although tourist visits tend to be relatively short (approximately 30 minutes), the number of tourists at any given feeding is not restricted. In recent years, approximately 100 000 visitors attended these feedings annually (Ambu, 2007). The orangutans and macaques that surround Sepilok exhibit much terrestrial activity and proximity to humans, and direct contact between tourists and orangutan or macaque populations does happen occasionally.

To understand better the risks of anthropozoonotic pathogen transmission from tourists to Sepilok's orangutans and macaques, a team led by M. Muehlenbein began surveying tourists at SORC. In 2007, 633 visitors at SORC completed a detailed survey regarding travel health and nature-based tourism. A little more than half of the respondents reported having current vaccinations against tuberculosis, hepatitis A, hepatitis B, polio, and measles (Muehlenbein *et al.*, 2008). Despite the fact that the majority of visitors to SORC are from temperate regions where influenza is more prevalent, 67.1% of those surveyed *with medical-related occupations* (and so some formal training in infection risks) reported not being currently vaccinated for influenza (Muehlenbein *et al.*, 2008). Such results lend support to the recommendation of requiring all visitors to present standardized vaccination certificates at wildlife tourism locations, particularly those with primates. However, routine, required, and recommended vaccines vary between countries and by age and health status of the recipient. And because vaccination certifications are not usually standardized between countries, it is difficult to accurately ascertain current immune status for most travelers. Despite these difficulties, travelers should be urged to examine their actual vaccination status prior to traveling, and update it in consideration of the wildlife species they plan to visit.

In 2007, 15% of tourists surveyed at SORC self-reported at least one of the following current symptoms: cough, sore throat, congestion, fever, diarrhea, and vomiting (Muehlenbein *et al.*, 2010). Those participants reporting recent animal contact (e.g. livestock, wildlife at other sanctuaries, unfamiliar domestic pets) were more likely to report current respiratory symptoms compared with individuals with no such animal contact. Such results highlight the fact that currently ill and potentially infectious tourists still visit wildlife sanctuaries, creating a risk of direct pathogen transmission to the primates they come to visit as well as to local human inhabitants and staff, which the latter could in turn spread to the primates. Some tourists may ignore such risks, whereas most (we suspect) are unaware or uninformed about such risks prior to travel. They not only underestimate their potential contribution to the spread of disease, but also underestimate their own risk of acquiring infection.

An additional 650 tourist surveys were obtained in 2009 at SORC. Of the respondents, 48% had visited other countries to specifically view monkeys or apes. Only 11% of these were made aware of health regulations at their destinations, and only 5.7% thought that such health regulations were enforced (Muehlenbein *et al.*, unpublished data). Despite the fact that 96% of respondents believed humans can give diseases to wild animals, 35% of these respondents reported that they would try to touch a wild primate if they had the opportunity. If tourism and conservation professionals are to prevent disease transmission, then we must find effective ways to educate tourists about their risky behaviors.

Recommendations for disease prevention and future research

We agree with the recently published IUCN Best Practice Guidelines for Great Ape Tourism (Macfie & Williamson, 2010). Some of their key recommendations for preventing disease transmission to great apes through tourism include minimum participant ages and viewing distances, with maximum visit durations and group sizes. Visitors are usually required to report voluntarily any current illnesses. Human feces must be adequately buried, and littering, smoking, eating, flash photography, feeding or touching great apes, coughing, spitting or nose blowing are not permitted. Other recommendations include the use of disposable face masks (see Figure 16.1) and gloves; mandatory hand washing and shoe disinfection before and after visiting great apes; medical screening of tourists; and required current vaccinations. Park personnel and surrounding local populations should be educated about disease risks, and efforts should be made to provide for adequate healthcare for park personnel, including vaccinations.

There is no doubt that adoption of many of these recommendations will require extensive resources for materials, personnel, and infrastructure. Risk assessment at different locations must be increased if we are to identify what precautions are most effective in preventing pathogen transmission from tourists to primates. Costs of implementing such monitoring and intensification of regulations may reduce immediate revenue, but may increase the long-term availability of healthy primates. It is understandable that many tourists will be inconvenienced by such regulations, having spent a large amount of money to travel to exotic destinations to view wildlife. Yet tourist lack of knowledge over issues such as disease prevention cannot be justified. We cannot risk the lives of the last remaining wild primates when we know that preventive measures could protect them. Of course, additional research is necessary to quantify and document actual disease risks, but this cannot stop us from implementing more strict regulations at primate tourism destinations now. Experts have been discussing such issues for years, yet little has changed since the seminal report by Homsy (1999) detailing unacceptably high disease risks.

Central to minimizing the costs of primate-based tourism is increased education of tourists. While several organizations (e.g. World Conservation Union, The International Ecotourism Society, Convention on Biological Diversity, Conservation

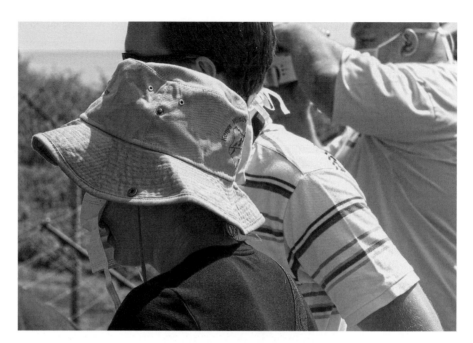

Figure 16.1   Face masks are provided to visitors at several wildlife sanctuaries, such as Ngamba Island Chimpanzee Sanctuary, Entebbe, Uganda. However, tourists may not always wear masks properly (incorrect use observed here). (Photograph by Michael Muehlenbein.)

International, the Cooperative Research Centre for Sustainable Tourism, the World Tourism Organization, the United Nations Environment Programme) provide some basic guidelines for nature tour operators, including how to minimize visitor impacts on the environment through proper behavior in the field, specific health-related behaviors are not emphasized. Such information is also not usually available on commercial travel websites (Horvath *et al.*, 2003). Tourists must be informed better about health risks before they enter wildlife parks. Improved tourist brochures and informational videos will be useful, and risks of zoonoses and anthropozoonoses should be part of information readily available to tourists before they begin traveling. The relative absence of detailed, yet accessible information regarding the justification for such regulations may be one reason why tourists often violate the rules at some primate tourism destinations (Sandbrook & Semple, 2006). Park personnel, tour operators, hoteliers, and surrounding community members must also be informed about the justifications for rules, be empowered to enforce the rules, and be rewarded when they do so.

Throughout this process, we must be cautious not to give the impression that wildlife health is more important than human health (Wallis & Lee, 1999). Such an attitude can lead to a bad relationship between researchers and local populations, which could lead to negative consequences for primate conservation in general. In addition, we must avoid causing alarm. Engendering fear of wildlife and their potential zoonotic diseases will not facilitate the much-needed support from local

Figure 16.2   Visitors are allowed to feed wild monkeys at several places throughout the world. Despite the fact that feeding Japanese macaques (*Macaca fuscata*) is only permitted through a wire barrier at Arashiyama, Kyoto, Japan, this practice perpetuates the view that close proximity and direct contact with primates are acceptable, even encouraged, and thereby increases risks of disease transmission. (Photograph by Michael Muehlenbein.)

human populations. Still the public must be educated about the risks of anthropo-zoonoses as well as zoonoses. Over half of all human infections are zoonotic in origin (Woolhouse & Gaunt, 2007) and several pathogens have been transmitted from nonhuman primates to humans. These include malaria (*Plasmodium knowlesi*: Cox-Singh & Singh, 2008), Cercopithecine herpesvirus 1 (B virus) (Huff & Barry, 2003), and several retroviruses like simian foamy virus, simian T-lymphotropic virus, and simian immunodeficiency virus (Gao *et al.* 1999; Jones-Engel *et al.* 2008; Santiago *et al.*, 2005; Van Heuverswyn *et al.*, 2006).

Despite the fact that many people may realize the potential for pathogen transmission between humans and primates, our affinity for gaining close proximity with primates often makes us behave irresponsibly at tourist destinations. What makes a tourist participate in risky behaviors? One likely factor is their experience viewing others doing the same thing. We see hundreds of photographs on the internet showing people holding or feeding wild primates. Unfortunately, many

Figure 16.3   Up-close visits between tourists and monkeys (*Macaca sylvanus* pictured here) are even available in Europe, for example at Trentham Monkey Forest, Staffordshire, England. In this situation, tourist and monkey behaviors are monitored by staff. While such experiences may foster our affection for these monkeys, allowing affection to take priority over visitor and wildlife health effectively undermines the behavioral precautions established to protect against zoonoses and anthropozoonoses. (Photograph by Michael Muehlenbein.)

professional primatologists are guilty of this as well, often with their own websites or books picturing themselves making physical contact with these animals (including embracing, holding hands, and allowing primates to crawl on them). This is irresponsible behavior by scientists who, of all people, should know the risks of disease transmission and the harm in setting a bad example. Famous actors are portrayed in documentaries caring for orphaned primates, further suggesting that close contact is acceptable and something to admire. Indeed, there are several sites in Mexico, Belize, Costa Rica, Honduras, Panama, and other locations where a tourist can hold a monkey, and close proximity is permissible at many locations in Europe, Indonesia, China, Uganda, and South Africa, among others (see Figures 16.2 and 16.3).

Additional factors that may influence tourists to exhibit risky behaviors include the playful behavior of primates. We can easily see behaviors in these animals that mirror our own; immature primates can be very cute, and even adults of some species can be quite neotenous. The question remains, why do such notions outweigh the use of common sense and healthy behaviors?

As responsible primatologists, we need to increase awareness of the health issues described in this chapter. Perhaps future research could focus on cross-cultural analysis of tourists' environmental values and attitudes, assessing ways to foster ecologically sensitive behaviors in travelers. We must educate tourists through various means of communication, providing understandable reasons for our regulations, and be willing to enforce the rules against violations. Of course, in most sites, the management of tourists is not under the control of researchers but is part of the site personnel's responsibility. Therefore, primatologists need to form collaborative relationships with tourist site staff and management to help raise awareness about these issues.

We must also work to learn why tourists and others in the tourism industry may violate safety and health regulations, and what information – or compensation – would prevent them from doing so. Such analyses must be conducted at a variety of locations that vary by tourist health knowledge, attitudes, and practices. In the interim it is critical that we proceed at once with intensive evaluation of current regulations at primate tourism destinations.

Acknowledgments

MPM wishes to thank (in alphabetical order) the following individuals and institutions for intellectual, logistical, and/or financial support of his primate-based tourism research: Sylvia Alsisto, Laurentius Ambu, Marc Ancrenaz, Patrick Andau, Jessieca Audrine, Carol Berman, Tony Blignaut, Cecilia Boklin, Garth Brink, Fany Brotcorne, Ben Chapman, Cleveland Metroparks Zoo, Hidemi DeHays, Lee Dekker, Agustin Fuentes, Shiho Fujita, Grace Fuller, Tony Goldberg, Benoit Goossens, Kym Gopp, Hideo Hasegawa, Michael Huffman, Indiana University-Bloomington, Sail Jamaludin, Vijver Jonck, Clayrina Julianus, Kinabatangan Orangutan Conservation Programme, Hiroyuki Kurita, Isabelle Lackman, Andrea Lemke, Cari Lewis, Sahdin Lias, Kristen Lukas, Liz Macfie, Andrew MacIntosh, Joanna Malukiewicz, Leigh Ann Martinez, Marc Mendelson, Lara Mostert, Sen Nathan, Sean Prall, Mary Ann Raghanti, Diana Ramirez, Red Ape Encounters, Ian Redmond, Sabah Wildlife Department, Rosman Sakong, Milena Salgado-Lynn, Sepilok Orangutan Rehabilitation Centre, Robert Steffen, University of Wisconsin-Milwaukee, Liz Williamson, Dominik Winkel, Dan Wittwer, Tim Wright, and Toni Ziegler. All primary research conducted by the authors and described in this chapter was ethically approved by their respective institutions.

References

Almeida, M. A. B., Santos, E. D., Cardoso, J. D. C. *et al.* (2012). Yellow fever outbreak affecting *Alouatta* populations in southern Brazil (Rio Grande do Sul State), 2008–2009. *American Journal of Primatology*, 74: 68–76.

Ambu, L. (2007). Strategy of the Sabah Wildlife Department for wildlife conservation in Sabah. *First International Conservation Conference in Sabah: The Quest for Gold Standards.* Kota Kinabalu, Malaysia: Sabah Wildlife Department.

Ashford, R. W., Reid, G. D. F., and Butynski, T. M. (1990). The intestinal faunas of man and mountain gorillas in a shared habitat. *Annals of Tropical Medicine and Parasitology*, 84: 337–340.

Brack, M. (1987). *Agents Transmissible from Simians to Man*. Berlin: Springer-Verlag.

Burgos-Rodriquez, A. G. (2011). Zoonotic diseases of primates. *Veterinary Clinics of North America: Exotic Animal Practice*, 14: 557–575.

Byers, A. C. and Hastings, B. (1991). Mountain gorilla mortality and climatic factors in the Parc National des Volcans, Ruhengeri Prefecture, Rwanda, 1988. *Mountain Research and Development*, 2: 145–151.

Ceballos-Lascuráin, H. (1996). *Tourism, Ecotourism, and Protected areas: The State of Nature-based Tourism Around the World and Guidelines for its Development*. Gland, Switzerland: World Conservation Union.

Chi, F., Leider, M., Leenandertz, F. *et al.* (2007). New *Streptococcus pneumoniae* clones in deceased wild chimpanzees. *Journal of Bacteriology*, 189: 6085–6088.

Costa, E. A., Luppi, M. M., Malta, M. C. C. *et al.* (2011). Outbreak of human Herpesvirus Type 1 infection in nonhuman primates (*Callithrix penincillata*). *Journal of Wildlife Diseases*, 47: 690–693.

Cox-Singh, J. and Singh, B. (2008). Knowlesi malaria: Newly emergent and of public health importance? *Trends in Parasitology*, 24: 406–410.

Crockett, M. and Keystone, J. (2005). "I hate needles" and other factors impacting on travel vaccine uptake. *Journal of Travel Medicine*, 12: S41–S46.

Ferber, D. (2000). Primatology. Human diseases threaten great apes. *Science*, 289: 1277–1278.

Fuentes, A. (2006). Human culture and monkey behavior: Assessing the contexts of potential pathogen transmission between macaques and humans. *American Journal of Primatology*, 68: 880–896.

Gao, F., Bailes, E., Robertson, D. L. *et al.* (1999). Origin of HIV-1 in the chimpanzee *Pan troglodytes troglodytes*. *Nature*, 397: 436–441.

Goldberg, T. L., Gillespie, T. R., Rwego, I. B. *et al.* (2007). Patterns of gastrointestinal bacterial exchange between chimpanzees and humans involved in research and tourism in western Uganda. *Biological Conservation*, 135: 511–517.

Goodall, J. (1986). *The Chimpanzees of Gombe: Patterns of Behavior*. Cambridge, MA: Harvard University Press.

Graczyk, T. K, Nizeyi, J. B., Ssebide, B. *et al.* (2002a). Anthropozoonotic *Giardia duodenalis* genotype (assemblage) A infections in habitats of free-ranging human habituated gorillas, Uganda. *Journal of Parasitology*, 88: 905–909.

Graczyk, T. K., Nizeyi, J. B., da Silva, A. J. *et al.* (2002b). A single genotype of *Encephalitozoon intestinalis* infects free-ranging gorillas and people sharing their habitats in Uganda. *Parasitology Research*, 88: 926–931.

Hamer, D. H. and Connor, B. A. (2004). Travel health knowledge, attitudes and practices among United States travellers. *Journal of Travel Medicine*, 11: 23–26.

Hanamura, S., Kiyono, M., Lukasik-Braum, M. *et al.* (2008). Chimpanzee deaths at Mahale caused by a flu-like disease. *Primates*, 49: 77–80.

Hilton, E., Singer, C., Kozarsky, P. *et al.* (1991). Status of immunity to tetanus, measles, mumps, rubella, and polio among US travellers. *Annals of Internal Medicine*, 115: 32–33.

Homsy, J. (1999). *Ape Tourism and Human Diseases: How Close Should we Get?* Report for the International Gorilla Conservation Programme Regional Meeting, Rwanda.

Horvath, L. L., Murray, C. K., and DuPont, H. L. (2003). Travel health information at commercial travel websites. *Journal of Travel Medicine*, 10: 272–279.

Hosaka, K. (1995a). Epidemics and wild chimpanzee study groups. *Pan Africa News*, 2: 1–2.

Hosaka, K. (1995b). Mahale: A single flu epidemic killed at least 11 chimps. *Pan Africa News*, 2: 3–4.

Huff, J. L. and Barry, P. A. (2003). B-virus (Cercopithecine herpesvirus 1) infection in humans and macaques: potential for zoonotic disease. *Emerging Infectious Diseases*, 9: 246–250.

Huijbregts, B., De Wachter, P., Ndong, L. S., Akou, O., and Akou, M. E. (2003). Ebola and the decline of gorilla *Gorilla gorilla* and chimpanzee *Pan troglodytes* populations in Minkebe Forest, north-eastern Gabon. *Oryx*, 37: 437–443.

Johnston, A. R., Gillespie, T. R., Rwego, I. B. *et al.* (2010). Molecular epidemiology of cross-species *Giardia duodenalis* transmission in western Uganda. *PLoS Neglected Tropical Diseases*, 4: e683.

Jones-Engel, L., C. May, C., Engel, G. A. *et al.* (2008). Diverse contexts of zoonotic transmission of simian foamy viruses in Asia. *Emerging Infectious Diseases*, 14: 1200–1208.

Kalema-Zikusoka, G., Kock, R. A., and Macfie, E. J. (2002). Scabies in free-ranging mountain gorillas (*Gorilla beringei beringei*) in Bwindi Impenetrable National Park, Uganda. *Veterinary Research*, 150: 12–15.

Kaur, T., Singh, J., Tong, S. *et al.* (2008). Descriptive epidemiology of fatal respiratory outbreaks and detection of a human-related metapneumovirus in wild chimpanzees (*Pan troglodytes*) at Mahale Mountains National Park, western Tanzania. *American Journal of Primatology*, 70: 755–765.

Klee, S. R., Ozel, M., Appel, B. *et al.* (2006). Characterization of *Bacillus anthracis*-like bacteria isolated from wild great apes from Côte d'Ivoire and Cameroon. *Journal of Bacteriology*, 188: 5333–5344.

Köndgen, S., Kühl, H., N'Goran, P. K. *et al.* (2008). Pandemic human viruses cause decline of endangered great apes. *Current Biology*, 18: 1–5.

Kortlandt, A. (1996). An epidemic of limb paresis (polio?) among the chimpanzee population at Beni (Zaire) in 1964, possibly transmitted by humans. *Pan Africa News*, 3: 9–10.

Leendertz, F. H., Ellerbrok, H., Boesch, C. *et al.* (2006). Anthrax kills wild chimpanzees in a tropical rainforest. *Nature*, 430: 451–452.

Leroy, E. M., Rouquet, P., Formenty, P. *et al.* (2004). Multiple Ebola virus transmission events and rapid decline of Central African wildlife. *Science*, 303: 387–390.

Lopez-Velez, R. and Bayas, J. M. (2007). Spanish travellers to high-risk areas in the tropics: Airport survey of travel health knowledge, attitudes, and practices in vaccination and malaria prevention. *Journal of Travel Medicine*, 14: 297–305.

Lukasik-Braum, M. and Spelman, L. (2008). Chimpanzee respiratory disease and visitation rules at Mahale and Gombe National Parks in Tanzania. *American Journal of Primatology*, 70: 734–737.

Macfie, E. (1991). The Volcano Veterinary Centre update. *Gorilla Conservation News*, 5: 20.

Macfie, E. (1996). Case report on scabies infection in Bwindi gorillas. *Gorilla Journal*, 13: 19–20.

Macfie, E. J. and Williamson, E. A. (2010). *Best Practice Guidelines for Wild Great Ape Tourism*. Gland, Switzerland. IUCN/SSC Primate Specialist Group.

Muehlenbein, M. P. (2009). The application of endocrine measures in primate parasite ecology. M. Huffman and C. Chapman (eds.), *Primate Parasite Ecology: The Dynamics of Host-Parasite Relationships*. New York: Cambridge University Press, pp. 63–81.

Muehlenbein, M. P. and Ancrenaz, M. (2009). Minimizing pathogen transmission at primate ecotourism destinations: The need for input from travel medicine. *Journal of Travel Medicine*, 16: 229–232.

Muehlenbein, M. P., Ancrenaz, M., Sakong, R. *et al.* (2012). Ape conservation physiology: Fecal glucocorticoid responses in wild *Pongo pygmaeus morio* following human visitation. *PLoS ONE*, 7 (3): e33357.

Muehlenbein, M. P., Martinez, L. A., Lemke, A. A. *et al.* (2008). Perceived vaccination status in ecotourists and risks of anthropozoonoses. *EcoHealth*, 5: 371–378.

Muehlenbein, M. P., Martinez, L. A., Lemke, A. A. *et al.* (2010). Unhealthy travellers present challenges to sustainable ecotourism. *Travel Medicine and Infectious Disease*, 8: 169–175.

Mueller-Graf, C. D. M., Collins, D. A., Packer, C., and Woolhouse M. E. J. (1997). *Schistosoma mansoni* infection in a natural population of olive baboons (*Papio cynocephalus anubis*) in Gombe Stream National Park, Tanzania. *Parasitology*, 115: 621–627.

Nishida, T., Corp, N., Hamai, M. *et al.* (2003). Demography, female life history, and reproductive profiles among the chimpanzees of Mahale. *American Journal of Primatology*, 59: 99–121.

Nizeyi, J. B., Mwebe, R., Nanteza, A. *et al.* (2001). Campylobacteriosis, salmonellosis, and shigellosis in free-ranging human-habituated mountain gorillas of Uganda. *Journal of Wildlife Diseases*, 37: 239–244.

Palacios, G., Lowenstine, L., Cranfield, M. *et al.* (2011). Human metapneumovirus infection in wild mountain gorillas, Rwanda. *Emerging Infectious Diseases*, 17: 711–713.

Prazuck, T., Semaille, C., Defayolle, M. *et al.* (1998). Immunization status of French and European tropical travellers: Study of 9,156 subjects departing from Paris to 12 tropical destinations. *Revue d'Epidemiologie et de Santé Publique*, 46: 64–72.

Rack, J., Wichmann, O., Kamara, B. *et al.* (2005). Risk and spectrum of diseases in travellers to popular tourist destinations. *Journal of Travel Medicine*, 12: 248–253.

Rolland, R. M., Hausfater, G., Marshall, B., and Levy, S. B. (1985). Antibiotic resistant bacteria in wild primates: Increased prevalence in baboons feeding on human refuse. *Applied Environmental Microbiology*, 49: 791–794.

Rwego, I. B., Isabirye-Basuta, G., Gillespie, T. R., and Goldberg, T. L. (2008). Gastrointestinal bacterial transmission among humans, mountain gorillas, and livestock in Bwindi Impenetrable National Park, Uganda. *Conservation Biology*, 22: 1600–1607.

Sakamaki, T., Mulavwa, M., and Furuichi, T. (2009). Flu-like epidemics in wild bonobos (*Pan paniscus*) at Wamba, the Luo Scientific Reserve, Democratic Republic of Congo. *Pan Africa News*, 16: 1–4.

Salzer, J. S., Rwego, I. B., Golderg, T. L., Kuhlenschmidt, M. S., and Gillespie, T. K. (2007). *Giardia* sp. and *Cryptosporidium* sp. infections in primates in fragmented and undisturbed forest in Western Uganda. *Journal of Parasitology*, 93: 439–440.

Sandbrook, C. and Semple, S. (2006). The rules and reality of mountain gorilla *Gorilla beringei beringei* tracking: How close do tourists get? *Oryx*, 40: 428–433.

Santiago, M. L., Range, F., Keele, B. F. *et al.* (2005). Simian immunodeficiency virus infection in free-ranging sooty mangabeys (*Cercocebus atye atys*) from the Tai Forest, Côte d'Ivoire: Implications for the origin of epidemic human immunodeficiency virus type 2. *Journal of Virology*, 79: 12515–12527.

Schunk, M., Wachinger, W., and Nothdurft, H. D.. (2001). Vaccination status and prophylactic measures of travellers from Germany to subtropical and tropical areas: results of an airport survey. *Journal of Travel Medicine*, 8: 260–262.

Sholley, C. (1989). Mountain gorilla update. *Oryx*, 23: 57–58.

Sholley, C. and Hastings, B. (1989). Outbreak of illness among Rwanda's gorillas. *Gorilla Conservation News*, 3: 7.

Steffen, R., Tornieporth, N., Costa Clemens, S. A. *et al*. (2004). Epidemiology of travellers' diarrhea: Details of a global survey. *Journal of Travel Medicine*, 11: 231–238.

Toovey, S., Jamieson, A., and Holloway. M. (2004). Travellers' knowledge, attitudes and practices on the prevention of infectious diseases: results from a study at Johannesburg International Airport. *Journal of Travel Medicine*, 11: 16–22.

Van Herck, K., Castelli, F., Zuckerman, J. *et al*. (2004). Knowledge, attitudes and practices in travel-related infectious diseases: The European airport survey. *Journal of Travel Medicine*, 11: 3–8.

Van Herck, K., Zuckerman, J., Castelli, F. *et al*. (2003). Travellers' knowledge, attitudes, and practices on prevention of infectious diseases: Results from a pilot study. *Journal of Travel Medicine*, 10: 75–78.

Van Heuverswyn, F., Li, Y., Neel, C. *et al*. (2006). SIV infection in wild gorillas. *Nature*, 444: 164.

Wallis, J., and Lee, D. R. (1999). Primate conservation: The prevention of disease transmission. *International Journal of Primatology*, 20: 803–826.

Wilder-Smith, A., Khairullah, N. S., Song, J. H., Chen, C. Y., and Torresi, J. (2004). Travel health knowledge, attitudes and practices among Australasian travellers. *Journal of Travel Medicine*, 11: 9–15.

Williams, J. M., Londsdorf, E. V., Wilson, M. L. *et al*. (2008). Causes of death in the Kasekela chimpanzees of Gombe National Park, Tanzania. *American Journal of Primatology*, 70: 766–777.

Woolhouse, M. and Gaunt, E. (2007). Ecological origins of novel human pathogens. *Critical Reviews in Microbiology*, 33: 1–12.

Zuckerman, J. N. and Steffen, R. (2000). Risks of Hepatitis B in travellers as compared to immunization status. *Journal of Travel Medicine*, 7: 170–174.

17 Guidelines for best practice in great ape tourism

Elizabeth A. Williamson and Elizabeth J. Macfie

Introduction

Tourism based on the viewing of great apes is increasingly promoted as a means of generating revenue for range states, local communities, and the private sector (e.g. GRASP, 2006). This is despite known risks from tourism, including disease transmission, which have caused concern among conservationists and prompted the International Union for Conservation of Nature to publish guidelines on best practices for great ape tourism (Macfie & Williamson, 2010). IUCN is one of the world's most respected authorities on species conservation, and brings together governments, UN agencies, and NGOs to conserve biodiversity and to ensure that any use of natural resources is equitable and ecologically sustainable.

Great apes are of high conservation concern because all species and subspecies are listed as Endangered or Critically Endangered (IUCN, 2013) and are protected throughout their range by both national and international laws. They are particularly appealing to human observers because they are behaviorally and physically so similar to people. However, their genetic closeness also makes great apes susceptible to human diseases for which they have no immunity (in this volume, see chapters by Dellatore et al.; Desmond & Desmond; Goldsmith; Muehlenbein & Wallis; Russon et al.).

Tourism regulations specific to great apes were first developed with the advent of mountain gorilla tourism in the 1970s (Weber, 1993). These were initially based on intuition and common sense, but have been adapted and revised over time on the basis of field experience and impact studies, and subjected to scientific review (Homsy, 1999). Mountain gorilla tourism has enabled improved monitoring and protection of habituated gorilla groups, enhanced the profile of great apes at both national and international levels, and helped improve the livelihoods of local communities (Blomley et al., 2010; Gray & Rutagarama, 2011; Plumptre & Williamson, 2001; Robbins et al., 2011; Williamson & Fawcett, 2008). However, tourism also creates risks to the great apes visited, notably stress and disease transfer (Palacios et al., 2011), making it important to institute measures to minimize these risks by controlling the conditions of visits, such as their frequency and duration, the distance to be maintained between apes and observers, and the health and number

Primate Tourism: A Tool for Conservation?, ed. Anne E. Russon and Janette Wallis. Published by Cambridge University Press. © Cambridge University Press 2014.

of visitors per group (Homsy, 1999; Ryan & Walsh, 2011). To this end, IUCN has developed best practice guidelines for great ape tourism through the collaborative and consensual input of many experts. These guidelines aim to establish and promote international standards, increase awareness of appropriate conservation practices, decrease likelihood of practices and projects that are wrong or harmful, and inform and support sound policy decisions relating to the protection and management of great apes and their habitat.

For great ape tourism to contribute to the conservation of great apes and their habitats, these guidelines must be rigorously applied and tourism activities strictly controlled. In this chapter we summarize IUCN guidelines, focusing on their recommendations for tourism implementation, monitoring and evaluation, and visitation "rules." We also present nine guiding principles for using tourism as a great ape conservation tool. The IUCN publication additionally includes a history of 30 years of great ape tourism, a review of lessons learned, potential impacts (key positive and critical negative), clear recommendations as to when tourism with apes is and is not appropriate, guidance in the planning and development of tourism initiatives, and great ape species-specific guidelines.

Visit regulations

Individual great ape tourism sites should develop detailed regulations incorporating lessons learned from other sites, and should monitor, reinforce, and improve these regulations throughout the lifespan of their program. Site-specific regulations should be developed in consultation with specialists in medicine, ecology, and behavior, as well as travel and tourism practitioners (Muehlenbein & Ancrenaz, 2009). However, good plans are meaningless without effective enforcement, and poor enforcement has been a perennial problem for great ape tourism (e.g. Sandbrook & Semple, 2006). Therefore, it is critical that conservation managers have the authority to institute tourism regulations, to exercise authority once tourism is underway, and to maintain that authority over the long term. This will help to foster compliance by both staff and tourists.

Ideally, all visitors should be informed of the rationale behind the measures instituted to minimize disease risks and other negative impacts of tourism, both during the booking process and again prior to their arrival at a great ape tourism site. Printed regulations should be sent to tour operators, marketing, or booking agents and, if possible, posted on a website.

The general regulations given below are relevant to most great ape tourism sites.

Presentation of tourism impacts and safety issues upon arrival

Site authorities should provide appropriate information on the various impacts of tourism on great apes when the tourists arrive. Presentation should be thorough and consist of both active discussion of the regulations that minimize risks and passive information transfer (such as written materials in accommodation facilities and displays and signage in check-in areas). To better prepare all visitors, this can be reinforced with demonstrations of the required safe distance, and role play with guides on how to respond to an approaching ape. Tourists will be more likely to remember and enact what they have been taught if they have practiced acting it out. Safety precautions should also be explained at this time and, if required by local regulations, all visitors should sign liability waivers.

Immunization requirements

Many great ape sites require that tourists present proof of vaccination and/or a current negative test for a number of diseases. Vaccination requirements may include polio, tetanus, measles, mumps, rubella, hepatitis A and B, yellow fever, meningococcal meningitis, typhoid, and tuberculosis. For tuberculosis, proof of a negative skin test within the last six months may be acceptable. The immunization requirement has a number of advantages beyond preventing the spread of these diseases. It reinforces the visitor's perception that tourism poses a risk to the apes. Awareness of this risk should also stimulate any responsible tourist's willingness to adhere to guidelines for their visit, including self-reporting potential medical conditions and volunteering not to visit.

While vaccination requirements may stimulate awareness of the disease risks and control some potential infections, proof of vaccination or a negative test alone will not control all infections of concern, such as the common cold and influenza, for which there is either no vaccine or vaccines for certain strains only. Neither will it guarantee tourists' compliance with health regulations. Vaccinated tourists may develop a false sense of security and feel that they can violate other regulations because they are immunized. In addition, lead-times for vaccination mean that vaccination requirements may not be easy to manage when tourists arrive (e.g. vaccinating only one day before a visit is generally not protective, and a modified live vaccine can theoretically infect other contacts, apes included). To avoid visitor disappointment and enhance compliance, vaccination and health regulations should be provided at the time of booking and should be widely available on websites providing booking information, so that tourists can arrange for any immunizations or tests required and obtain the necessary documentation before traveling.

Guided health evaluation prior to departure

During final check-in for a tourist visit, staff should inspect vaccination certificates rather than rely on self-reporting (Muehlenbein *et al.*, 2008). Self-evaluation will identify those willing to decline a visit on health grounds and facilitate the process of refunding tourists who self-report illness, but does not ensure compliance because some tourists will try to conceal symptoms.

Tourists should then be guided through a self-evaluation designed to highlight whether they might be infectious or otherwise unable to participate in the visit. This should include a checklist of symptoms such as sneezing, coughing, fever, or diarrhea within the previous 48 hours, and exposure to any significant risks (e.g. disease, bat caves). Bats are thought to carry Marburg virus (Timen *et al.*, 2009) and several species of fruit bat have been implicated in the complex transmission cycles of Ebola virus (Leroy *et al.*, 2005); both of these hemorrhagic diseases are highly lethal to great apes and humans, so tourist visits to bat caves or roosts should be scheduled *after* viewing great apes or avoided altogether in countries with a history of Marburg or Ebola.

Professional health evaluation

An on-site health professional could perform routine health checks, such as measuring body temperature, heart rate, and respiratory rate. This will not be possible at all sites, but large tourism programs should consider having a nurse or doctor on staff, along with employee health programs. Health professionals will also be able to advise on local and global disease patterns and propose additional precautions as needed. Guides should also be trained to recognize tourists who are unwell and given authority to exclude them from great ape tourism activities.

"The rules"

Despite strict enforcement of vaccination and other preventive health measures outlined above, tourists who have traveled long distances (usually at great expense) may try to hide illness, and some could be infectious without knowing it or carry infections that health checks fail to detect. Consequently, everyone who approaches great apes poses a potential disease risk and must be required to behave accordingly. Strict regulations are also important to minimize the behavioral impacts of tourist visits. Any site claiming that they adhere to best practice in great ape tourism must implement the following:

No visits by people who are sick

People who are unwell will not be allowed to visit the apes. This must be made very clear at the time of booking. It is critical to encourage tourists to self-report their illnesses and to offer them incentives to refrain from visiting if necessary. Incentives should not include a postponed visit (it is probable that the person would continue to be infectious for a few days), but could be a refund on-site or vouchers for other tourism services (e.g. accommodation, hiking). Similarly, staff members who are ill must not participate in ape visits and must be given incentives to remain away from apes, such as guaranteed "sick days" and a policy of non-discrimination if they cannot work because of illness.

Children younger than 15 years old prohibited from visiting

Children under 15 years old must not be allowed to visit great apes. This safeguard is primarily for health reasons. Young people are more likely to be infected with common childhood diseases, even when properly vaccinated, and therefore pose a much greater health risk to habituated apes.

One tourist visit per day of limited duration

There should be no more than one visit per day to each group of apes (or individual/party/forest area in the case of chimpanzee and orangutan tourism) with a maximum duration of one hour (see "One-hour time limit" below). Any existing site that has been operating more than one visit per day should reduce the schedule to one visit a day per group or individual. This can be done by closing second-visit bookings over time, or in some cases by habituating a new group. New groups should be habituated only if a full impact assessment indicates the conservation benefits will outweigh the costs. In addition, tourism accommodation located in or near ape habitat must control visitor movements away from the facility to prevent uncontrolled ape viewing.

Maximum number of tourists per group

To minimize behavioral disturbance and disease risk, strict limits on the number of tourists allowed to visit each day must be set and adhered to. In dense forest where visibility is poor, any sudden noise or movement could cause alarm and unpredictable reactions. In addition, finding a good viewing spot for each tourist can be challenging. Tourists must stay together and avoid encircling the apes being viewed. To facilitate the control of visitors, minimize danger, and enhance visitor satisfaction, the number of people per party should be no more than four tourists accompanied by two guides/trackers. This should achieve a reasonable balance between apes and humans, and reduce behavioral disturbance, stress, and their associated effects. Small numbers also favor high permit prices, as tourists tend to value being part of a small and exclusive group of visitors.

This general guideline should be implemented at all new sites. Species-specific recommendations on tourist numbers are discussed in Macfie and Williamson (2010). A few sites currently operate with fewer than four tourists per visiting group and their success suggests that numbers can and should remain low at these sites. Mountain gorilla sites and some chimpanzee sites currently operate with more than four tourists, and these sites should assess whether reducing tourist numbers *toward* this recommended maximum of four could be feasible in future. Any new ape groups opened for tourism should be visited by a smaller number of tourists.

N95 respirator masks

All tourists and staff who are likely to approach habituated apes to within 10 m should wear a surgical quality N95 respirator mask for the duration of their one-

Figure 17.1   Tourists wearing face masks while viewing chimpanzees in Mahale Mountains National Park. (© Toshisada Nishida, Tanzania.)

hour visit (Figure 17.1). Respirators that filter out higher percentages of aerosolized particles are also acceptable (i.e. N99 or N100). Masks should be carried by trackers/guides in appropriate waterproof containers so that they are not damaged and rendered less effective during transport. They should be distributed to tourists just before they begin actually viewing the apes and collected for appropriate disposal afterwards.

A surgical mask could give the wearer a false sense of security regarding the risks they pose while in proximity to apes; therefore all other regulations (concerning hygiene, distance from the apes, visit duration) must be enforced alongside mask use. Appropriate education in hygiene must be given to staff and tourists alike. Tourists feeling the urge to sneeze or cough while in proximity to the apes should turn their head away even when wearing a surgical mask, but should not remove the mask; if the mask worn becomes soiled or damp, however, staff should offer a replacement mask.

Masks are disposable and should not be re-used. They should be collected by the trackers/guides immediately after the visit and disposed of appropriately after the visit, as they pose a disease risk to apes and other wildlife if accidentally dropped or otherwise disposed of in the forest. Masks must be burned upon return to tourism administration or accommodation facilities, away from areas where apes range.

Staff must receive training in mask management, including proper fit-testing, wear, use, and disposal. Trained staff should demonstrate appropriate mask use in full to tourists at the visit departure point and review this before they reach the 10-m distance from any apes, so that masks are not put on incorrectly in a rush to see the apes. A mask that becomes damp or wet is less effective at blocking pathogens and should be exchanged for a new one.

Mask management should be monitored as part of a broader tourism monitoring program, and results used to inform and improve procedures. Tourist compliance and feedback should also be taken into consideration when reviewing mask management procedures.

Procurement systems must ensure a reliable supply of appropriate masks on site. If N95 respirator masks are not available, surgical quality multi-layer masks may be used until N95 respirators are procured because they provide a barrier to large droplets. Their use should be temporary because they are less effective in preventing disease transmission than N95 respirators.

Minimum distance to habituated great apes

The minimum distance to which visitors wearing N95 surgical masks are permitted to approach great apes is 7 m. For visitors not wearing N95 masks, the minimum distance permitted is 10 m (see also Klailova *et al.*, 2010).

One-hour time limit

Tourist visits must be restricted to no more than one hour. This limit, combined with the restriction of one visit per day, should ensure that no ape is visited by tourists for more than one hour on any single day. If apes are not easily visible when first approached, staff should escort tourists away to a distance of 200 m to await a time when the apes are resting or have moved into more open vegetation, and then begin the permitted hour.

Non-essential personnel to remain at a distance from great apes

Non-essential personnel such as military escorts or porters must stay as far away from the apes as feasible during tourist visits, out of sight *and* earshot. Such personnel should remain in contact with guides via walkie-talkie radios so that they can be instructed to move if the apes start moving in their direction.

Hand-washing and hygiene

Basin facilities and soap should be provided at departure points and tourists encouraged to wash their hands before departure. Additionally, guides should carry hand disinfectant spray (such as chlorhexidine), gel, or wipes for all visitors and staff to use before approaching apes.

Latrines must be provided at departure points and tourists encouraged to use them before departure. Latrines should be constructed at appropriate distances from watercourses (at least 30 m). If tourists or staff need to urinate or defecate while in the forest, they should do so at least 500 m away from apes and watercourses, and feces must be buried in a 30 cm-deep hole.

Smoking is not permitted in ape habitat due to the risks of fire and of disease transmission via contaminated cigarette butts. The smell of smoke may also scare wildlife.

Nose blowing and spitting on the ground are not allowed. Staff and tourists should use handkerchiefs or tissues as needed and ensure these are disposed of appropriately, as with masks, and away from the apes.

The same boots and clothing should not be worn to visit more than one group of apes, by staff or visitors, unless they have been washed and dried between visits.

Prevent contamination of the habitat with food waste

Eating and drinking are not allowed during visits. Food and drink must not be visible while observing great apes, and should be left with porters or other personnel who remain out of sensory range of the apes. Food must not be consumed within 500 m of apes. Food waste and all other rubbish must be stowed in backpacks and carried out of the forest to prevent deposition of infectious waste in the habitat. This will minimize accidental contaminated waste and prevent the apes from developing an association between humans and food.

Food must never be used to attract apes toward tourists.

Tipping policies and staff salaries

Tourists should be informed that tips and gifts cannot be used to encourage staff to break regulations, and staff must not view tips as justification to ignore regulations. Both infractions also reduce the professionalism of the operation. Tipping policies should be clearly displayed and explained so that tourists are aware of the issues before starting their activity. Tourists dislike having rules presented to them and then seeing them broken, and reduces respect for both staff and regulations. This message must be communicated to staff through education, training, and monitoring, to enhance their compliance. Regular monitoring and staff supervision should be used to reinforce tipping practices, and a no-tipping policy should be considered if tips are judged to be a prime factor in staff relaxing regulations.

All tourism staff, from check-in clerks to trackers and guides, should benefit from tips via a shared tip box with tips distributed equally among all staff each day. Policies specifying that pooled tips will be divided among all tourism staff will help prevent irregularities and should be posted where they are visible to visitors.

Tourism staff should be paid satisfactory salaries (at least a "living wage" and preferably higher) to minimize temptations to violate regulations for higher tips.

Monitoring and enforcement of rules

It is imperative that all staff understand the rules, can explain their rationale to visitors, and can enforce them. Therefore, tourism staff should be regularly monitored and evaluated on their conduct, and results should be discussed openly between evaluators and staff. Regular refresher courses reinforce staff understanding and adherence to tourism regulations and should include training on enforcement techniques.

A post-visit checklist provided to tourists and staff could help to reinforce staff compliance, and specific cases where staff had problems enforcing rules could be used in staff training exercises.

Site management

Infrastructure designed to minimize impact on apes and habitat

Environmental impact assessments should be carried out for all tourism-related infrastructure developments, in keeping with national environmental legislation. Tourism infrastructure, such as lodges, campsites, and visitor centers, should be constructed in areas where impacts on apes and their habitats are minimal. Any disruption to native vegetation, especially forest, should be kept to a minimum. If possible, tourism infrastructure should be located far enough outside ape habitat to prevent unauthorized access.

Tourism infrastructure should not be built in areas frequented by apes, due to risks of encountering people, food preparation areas, waste disposal, or sanitation facilities, and risk of injury from electrical cables or other hazards (Figure 17.2). If infrastructure on any scale is necessary in ape habitat, attention should be paid to reducing the impact of tree felling on the apes' feeding and ranging requirements (see Morgan & Sanz, 2007). Tourism infrastructure should not include installations that could attract apes, such as the planting of crops or fruit trees.

Tourism infrastructure must not introduce additional disease risks to ape populations. Attention to appropriate sanitation, hygiene, and waste disposal is critical in this regard.

Staff housing and administrative infrastructure

Staff and administrative buildings should be sited to maximize the oversight and control of tourism programs. These buildings must be located and designed to minimize impacts on apes and their habitat from noise and other hazards (e.g. fuel, power lines, toxins). Managers and law-enforcement teams should be posted on-site so that monitoring and protection activities can be carried out routinely.

Tourism accommodation should benefit local communities

Accommodation in lodges or campsites should be managed to maximize community benefits through community ownership, employment opportunities, or revenue-sharing schemes that provide income to members of the community or funding for social services.

Figure 17.2 Bwindi mountain gorillas ranging in the vicinity of tourist accommodation. (© Uwe Kribus, Uganda.)

Tourist accommodation that benefits local communities should be protected from external competition. This can be achieved through zoning that limits the number of facilities allowed to operate at preferred locations.

Special considerations for small and Critically Endangered populations

Particular caution is required before developing or expanding tourism with Critically Endangered taxa. This classification is given to three of the four gorilla subspecies (mountain, western lowland, and Cross River) and the Sumatran orangutan (IUCN, 2013). Although the three subspecies of Bornean orangutan are listed as Endangered, the northwestern subspecies and the East Kalimantan populations of the northeastern subspecies also merit special consideration because their remaining populations are similar in size to those of the Sumatran orangutan (Soehartono *et al.*, 2007). Special consideration is also merited for Grauer's gorillas, as they live in a volatile region and over the last 20 years their numbers have been drastically reduced (Maldonado *et al.*, 2012).

Risk-management programs

IUCN guidelines recommend that a number of impact-management measures accompany all great ape tourism programs. Additionally, funding for risk management must be guaranteed before any tourism activities are launched with small or Critically Endangered populations, to ensure that negative impacts are identified and immediately addressed.

Optimize before expanding

A number of sites with Critically Endangered great apes are already conducting tourism. In some of these, tourism has made a positive contribution, generating income for comprehensive conservation programs in and around ape habitat (Gray & Rutagarama, 2011; Plumptre & Williamson, 2001; Robbins *et al.*, 2011; Williamson & Fawcett, 2008). Income to national treasuries and a range of stakeholders has resulted in enhanced perceptions of great apes and has stimulated long-term support for great ape conservation (Plumptre & Williamson, 2001; Williamson & Fawcett, 2008). While keeping these successes in mind, it is important to step back and evaluate the future of tourism at these sites, to protect the programs from complacency and to prevent them from sliding toward over-exploitation of the apes. There has been a general tendency to expand tourism by habituating additional animals (Macfie, 2007), but for conservation to remain the primary objective, it is important to resist the temptation to expand for economic gain. Economic benefits can be achieved in ways that do not involve subjecting the apes to additional tourists or exposing more individuals to tourism. The recommendations below should be followed at all sites operating tourism with Critically Endangered apes.

Encourage income generation that does not involve great ape tourism expansion

Governments and conservation authorities should encourage alternative means of stimulating earnings by authorities, the private sector, and local economies, such as investment in national enterprise development, micro-credit schemes for local enterprises, and support for other business developments.

Do not increase the number of great ape groups habituated for tourism

Sites with Critically Endangered great apes should avoid increasing the number of habituated groups or individuals. It is important to maintain a balance of exposed and unexposed groups to better mitigate negative impacts of tourism.

Limit the number of individual great apes habituated for tourism

Habituation decisions should not be based on habituating the largest groups of apes, or the greatest number of individuals, for tourism. Habituation decisions must include consideration of maintaining a significant proportion of the population as unexposed to tourism, as the larger the proportion that is exposed to tourism, the greater the risk that disease could result in drastic reduction of the population.

Maximize revenue per tourism permit

If there is pressure to increase revenues from great ape tourism, the first measure to take should be to increase permit prices. Revenue per permit can also be maximized by diversifying tourism activities at each site and building ape tourism into national tourism circuits. Extending the average length of in-country stay of great ape tourists would increase the earnings associated with each permit at local, regional, and national levels.

Monitoring and evaluation

Tourism programs should be supported by independent impact assessments to inform and improve tourism policy and management systems. Formal mechanisms of review and incorporation of research results into management and policy will help ensure that conservation impacts are optimized.

Applied research

Tourism should stimulate the development of research programs to meet the requirements of tourism-impact monitoring and applied research. Research programs associated with great ape tourism should include the following:

Disease monitoring

Disease is the most serious risk associated with great ape tourism. Regular health monitoring is needed to detect patterns of disease, allow management to design prevention measures (e.g. quarantine, tourist vaccination regulations, community health projects), and respond to disease outbreaks. Routine observations of tourist-visited great apes by trained personnel and non-invasive screening should be supplemented by opportunistic sampling of animals that have had to be immobilized (IUCN guidelines are in preparation).

Behavioral monitoring

Tourism can have serious negative impacts on the behavior, physiology, and social dynamics of habituated great apes. Independent research on the behavior of great apes visited by tourists will highlight potential or incipient problems before they become severe and will allow corrective management (see Hodgkinson & Cipolletta, 2009; Muyambi, 2005).

Ecological monitoring

Heavy tourist traffic may cause soil compaction, erosion, trampling, and damage to vegetation. Controls to minimize degradation of the habitat should include prohibition of cutting or removal of seedlings and vegetation, walking off trails, and fire.

Regular ecological monitoring should be instituted to enable the improvement and enforcement of these controls.

Population monitoring

Population monitoring is an essential adjunct to tourism management; it serves as a longer-term indicator of the impacts of tourism and other conservation interventions on great ape populations.

Law-enforcement monitoring

The development and operation of tourism must not divert attention and resources away from the central goal of protecting great apes and their habitat. It is, therefore, important to monitor trends in illegal activities and assess the performance and results of law-enforcement activities. Law-enforcement monitoring will highlight areas for improvement or the need for increased surveillance, and can inform management when apes are ranging into areas of illegal activity, so that prevention and response to those activities can be enhanced.

Conflict monitoring

Human–great ape conflicts can be alleviated through the provision of tourism benefits to local communities, or exacerbated by tourism altering the apes' ranging behavior and bringing them into conflict situations more frequently. It is important that conflicts are systematically monitored and the success of mitigation efforts measured.

Economic assessments

The motivation for initiating great ape tourism is often the economic benefits anticipated by various institutional, local, and national stakeholders, in both the public and private sectors. However, as stated throughout this document, conservation must be the ultimate goal of great ape tourism and should be given priority over all other interests, especially economic ones. Therefore, it is important to monitor the economic impacts of great ape tourism to assess whether its continuing operation is justified and to inform management decisions, such as pricing structures and booking systems. Methodology can be adapted from previous studies (e.g. Hatfield & Malleret-King, 2006; WCS Gabon, 2008; Wilkie & Carpenter, 1999).

Program monitoring and evaluation

Financial monitoring and transparency

Insofar as tourism is a tool to provide funding for conservation, it is crucial that systems are in place to monitor tourism revenue generation and distribution. Financial controllers must be able to demonstrate that income is supporting great ape conservation, protected area management and operations, community projects, and revenue-sharing programs. Transparency will go a long way to reassuring critics of great ape tourism that this is an appropriate and effective conservation measure.

Program reporting

Progress reports and the results of tourism-impact monitoring and applied research should be produced at regular intervals (preferably quarterly, but at least annually) to stimulate internal review and timely identification of issues to be addressed.

Program evaluation

Regular medium-term (every two years) internal assessments of the performance, management, and impacts of great ape tourism programs must be carried out to accurately monitor progress and to allow for program review and improvement. The results of management-related research (see "Applied research") should be used to guide improvement and adaptation in tourism program management. In the longer term, external evaluations should take place every five years to ensure appropriate implementation and to foster learning and exchange with other great ape tourism sites.

Staff monitoring

Staff working in great ape tourism must be fully supported in their role as the prime defenders of great apes against the negative impacts of tourism. They need to be, and feel, able to discuss and enforce tourism rules and regulations. Their roles must be evaluated regularly to assess effectiveness and modify management, as needed. This can be achieved by regular supervision, including evaluation in the field, evaluation during tourism-impact research, and feedback from tourists.

Conclusions and guiding principles for using tourism as a great ape conservation tool

These guidelines were developed not only for conservation practitioners, but also for development organizations, donors, and private companies promoting tourism development in great ape habitat. The general principle that determined the orientation of these guidelines is that first and foremost, tourism must contribute to the conservation of great apes in their natural habitat. Additionally, tourism programs should be designed to support other activities such as law enforcement, habitat protection, and community involvement in conservation.

A number of biological, geographical, economic, and global factors will determine the success or failure of a great ape tourism project: for example, the failure of the tourism market to provide revenue sufficient to cover development and operating costs or the failure to protect the great apes from the negative impacts inherent in tourism. Once apes are habituated to human observers, they are permanently at increased risk from poaching and other forms of conflict with humans. Therefore, they must be protected in perpetuity even if tourism fails or ceases for any reason.

Great ape tourism should not be developed without conducting critical feasibility analyses to ensure that there is sufficient potential for success. Strict attention must be paid to the design of the enterprise, its implementation, and its continual management capacity in a manner that avoids, or at least minimizes, the negative impacts of tourism on local communities and on the apes themselves. Monitoring programs to track costs and impacts, as well as benefits, are essential to inform management on how to optimize tourism for conservation benefits.

Nine guiding principles

Tourism is not a panacea for great ape conservation or revenue generation

Tourism can contribute to great ape conservation but will not be viable at all sites: sites that do not meet the criteria listed in IUCN guidelines are not appropriate for great ape tourism. Sites that fail to generate the revenue anticipated may suffer a backlash against the conservation effort, so care must be taken to avoid raising false expectations among politicians, managers, and local communities.

Great ape tourism must be based on sound objective science

Great ape tourism can be controversial and not all conservationists agree that it is an acceptable activity. To defend great ape tourism as a sustainable component of a conservation strategy, in addition to the requirement that conservation must take priority over economic and political interests, decisions affecting tourism must be based on sound and objective science and regulations governing visits must be scientifically formulated and rigorously enforced.

Comprehensive understanding of impacts must guide tourism development

Great ape tourism has a number of advantages and disadvantages, all of which must be clearly understood by everyone involved in the planning and implementation. These issues should be kept in mind at all stages of the design, development, and management of great ape tourism. The best practice guidelines are founded on the principle of optimizing conservation impacts for great apes and their habitats. Any site that cannot sustain impact-optimizing activities, financially or institutionally, should not initiate a great ape tourism program.

Conservation benefits must significantly outweigh risks

Great ape tourism development proposals should undergo full feasibility and impact assessments, and should not be implemented unless the benefits anticipated outweigh the potential risks. Tourism and its associated impact mitigation measures must significantly improve the conservation outcome compared with a no-tourism scenario. Only programs that will enhance conservation efforts and improve protection of the resident great ape population should go ahead. While this is a general guideline for all great apes, it is crucial for Critically Endangered and small populations due to their precarious conservation status.

Conservation investment and action must be assured in perpetuity

Anti-poaching activities must be launched in parallel with habituation efforts, especially in areas where poaching of great apes is common (e.g. Central Africa where poaching for meat is at its highest levels, and Southeast Asia where poaching for the pet trade is high). Once habituated, great apes and their home ranges must be protected and monitored daily by law-enforcement teams with on-call veterinary expertise. These activities are necessary not only for conservation, but also to support tourism development and management, and must be continued in perpetuity to protect great apes that have been habituated, and to ensure the viability of a tourism enterprise. Financial contingency plans to ensure that conservation actions continue during periods of low tourism should be in place before tourism is developed.

Benefits and profit for local communities should be maximized

For great ape tourism to properly meet the criteria for *sustainable* tourism, it must maximize both direct and indirect benefits to adjacent communities that bear the costs of conservation, including opportunity costs (Grosspietsch, 2007). While conservation must take priority over other interests, tourism should strive to contribute to poverty reduction wherever possible and, at the very least, should do no harm to local communities (SGLCP, 2009). Direct benefits include local recruitment of tourism staff and sharing a percentage of tourism revenue with adjacent communities. Indirect benefits include marketing and support for services that earn additional income for communities (such as tourism infrastructure which is partially or wholly community owned and operated). Care should be taken to ensure that benefits are not focused on a small section of a community but are accessible to the majority. Full consultations should be conducted to ensure that benefits are provided in a manner both recognized and valued by local residents. Guidance on involving communities in tourism activities is available (e.g. Ancrenaz *et al.*, 2007; Gutierrez *et al.*, 2005; Rajaratnam *et al.*, 2008), as are lessons learned through the development and implementation of revenue-sharing and other community programs centered on great ape tourism (Adams & Infield, 2003; Archabald & Naughton-Treves, 2001; Blomley *et al.*, 2010).

Profit to private sector partners must not be a driving force

In the development of any great ape tourism activity, conservation principles must take precedence over profit generation for private sector stakeholders. While a successful tourism program will provide opportunities for income to accrue at various levels, the primary aim of developing and operating this revenue-generating mechanism is to support the cost of conservation efforts. The needs of communities living in or adjacent to ape habitats must also be a high priority concern. If the priorities become inverted, with profit to the private sector becoming the driving force behind great ape tourism, then stakeholders must analyze how the priorities could have gone astray and how to rebalance them.

Tourism can enhance long-term support for conservation

Tourism can be used to enhance the financial, esthetic, and cultural values of great apes and their habitats as perceived by local communities, policy-makers, and political leaders, thereby promoting long-term support for conservation of apes and their habitats.

Conservation must be the primary goal of great ape tourism

Conservation must be given priority over economic and political concerns at all great ape tourism sites. Any site that undertakes great ape tourism must place continued and enhanced emphasis on protection, law enforcement, environmental awareness-raising, and other conservation activities. The effort and resources required to develop and operate tourism should not divert resources and attention away from the conservation focus.

In closing, readers are urged to adhere to the guiding principles of best practice in great ape tourism, and to keep them in mind at all stages of planning, developing, implementing, and monitoring great ape tourism.

Acknowledgments

We wish to thank Anne E. Russon and Janette Wallis for the invitation to contribute to this volume. We also thank the following for contributing to the IUCN tourism guidelines: Marc Ancrenaz, Chloe Cipolletta, Debby Cox, Christina Ellis, David Greer, Chloe Hodgkinson, Anne E. Russon, and Ian Singleton.

References

Adams, W. M. and Infield, M. (2003). Who is on the gorillas' payroll? Claims on tourist revenue from a Ugandan national park. *World Development*, 31: 177–190.

Ancrenaz, M., Dabek, L., and O'Neil, S. (2007). The costs of exclusion: recognizing a role for local communities in biodiversity conservation. *PLoS Biology*, 5: e289.

Archabald, K. and Naughton-Treves, L. (2001). Tourism revenue-sharing around national parks in Western Uganda: early efforts to identify and reward local communities. *Environmental Conservation*, 28: 135–149.

Blomley, T., Namara, A., McNeilage, A., *et al.* (2010). *Development AND Gorillas? Assessing Fifteen Years of Integrated Conservation and Development in South-western Uganda. Natural Resource Series No. 23.* London and Edinburgh, UK: International Institute for Environment and Development (IIED).

GRASP (2006). *The Kinshasa Declaration*. Great Ape Survival Partnership, Nairobi. URL: www.unesco.org/mab/doc/grasp/E_KinshasaDeclaration.pdf (accessed Apr. 4, 2014).

Gray, M. and Rutagarama, E. (eds.) (2011). *20 Years of IGCP: Lessons Learned in Mountain Gorilla Conservation*. Kigali, Rwanda: International Gorilla Conservation Programme.

Grosspietsch, M. (2007). *Maximizing Tourism's Contribution to Poverty Reduction in Rwanda.* Doctoral dissertation, Westphalian Wilhelms-University, Münster, Germany.

Gutierrez, E., Lamoreux, K., Matus, S., and Sebunya, K. (2005). *Linking Communities, Tourism and Conservation: A Tourism Assessment Process.* Washington, DC: Conservation International and The George Washington University.

Hatfield, R. and Malleret-King, D. (2006). *The Economic Value of the Mountain Gorilla Protected Forests (The Virungas and Bwindi Impenetrable National Park).* Nairobi: International Gorilla Conservation Programme (IGCP).

Hodgkinson, C. and Cipolletta, C. (2009). Western lowland gorilla tourism: impact on gorilla behaviour. *Gorilla Journal*, 38: 29–32.

Homsy, J. (1999). *Ape Tourism and Human Diseases: How Close Should We Get? A Critical Review of Rules and Regulations Governing Park Management and Tourism for the Wild Mountain Gorilla, Gorilla gorilla beringei.* Nairobi: International Gorilla Conservation Programme (IGCP). www.igcp.org/wp-content/themes/igcp/docs/pdf/homsy_rev.pdf (accessed Apr. 4, 2014).

IUCN (2013). *The IUCN Red List of Threatened Species.* Version 2013.2. Gland, Switzerland and Cambridge, UK: International Union for Conservation of Nature. www.iucnredlist.org (accessed Apr. 4, 2014).

Klailova, M., Hodgkinson, C., and Lee, P. C. (2010). Behavioral responses of one western lowland gorilla (*Gorilla gorilla gorilla*) group at Bai Hokou, Central African Republic, to tourists, researchers and trackers. *American Journal of Primatology*, 72: 897–906.

Leroy, E. M., Kumulungui B., Pourrut X., *et al.* (2005). Fruit bats as reservoirs of Ebola virus. *Nature*, 438: 575–576.

Macfie, E. J. (2007). *Habituation Impact Assessment: A Tool for the Analysis of Costs and Benefits Related to the Potential Habituation of a Gorilla Group for Either Tourism or Research.* Unpublished report to the Virunga Bwindi Gorilla Management Technical Advisory Committee, International Gorilla Conservation Programme, Nairobi.

Macfie, E. J. and Williamson, E. A. (2010). *Best Practice Guidelines for Great Ape Tourism.* Gland, Switzerland: IUCN/SSC Primate Specialist Group. www.primate-sg.org/best_practice_tourism (accessed Apr. 4, 2014).

Maldonado, O., Aveling, C., Cox, D., *et al.* (2012). *Grauer's Gorillas and Chimpanzees in Eastern Democratic Republic of Congo (Kahuzi-Biega, Maiko, Tayna and Itombwe Landscape): Conservation Action Plan 2012–2022.* Gland, Switzerland: IUCN/SSC Primate Specialist Group, Ministry of Environment, Nature Conservation and Tourism, Institut Congolais pour la Conservation de la Nature, and the Jane Goodall Institute.

Morgan, D. and Sanz, C. (2007). *Best Practice Guidelines for Reducing the Impact of Commercial Logging on Great Apes in Western Equatorial Africa.* Gland, Switzerland: IUCN/SSC Primate Specialist Group. www.primate-sg.org/best_practice_logging (accessed Apr. 4, 2014).

Muehlenbein, M. P. and Ancrenaz, M. (2009). Minimizing pathogen transmission at primate ecotourism destinations: the need for input from travel medicine. *Journal of Travel Medicine*, 16: 229–232.

Muehlenbein, M. P., Martinez, L. A., Lemke, A. A., *et al.* (2008). Perceived vaccination status in ecotourists and risks of anthropozoonoses. *EcoHealth*, 5: 371–378.

Muyambi, F. (2005). The impact of tourism on the behaviour of mountain gorillas. *Gorilla Journal*, 30: 14–15.

Palacios, G., Lowenstine, L. J., Cranfield, M. R., *et al.* (2011). Human metapneumovirus infection in wild mountain gorillas, Rwanda. *Emerging Infectious Diseases*, 17: 711–713.

Plumptre, A. J. and Williamson, E. A. (2001). Conservation–oriented research in the Virunga region. In: M. M. Robbins, P. Sicotte, and K. J. Stewart (eds.), *Mountain Gorillas: Three Decades of Research at Karisoke*. Cambridge University Press, pp. 361–390.

Rajaratnam, R., Pang, C., and Lackman-Ancrenaz, I. (2008). Ecotourism and indigenous communities: the Lower Kinabatangan experience. In: J. Connell and B. Rugendyke (eds.), *Tourism at the Grassroots: Villagers and Visitors in the Asia Pacific*. London, UK: Routledge, pp. 236–255.

Robbins, M. M., Gray, M., Fawcett, K. A., *et al.* (2011). Extreme conservation leads to recovery of the Virunga mountain gorillas. *PLoS One*, 6: e19788. doi:10.1371/journal.pone.0019788.

Ryan, S. J. and Walsh, P. D. (2011). Consequences of non-intervention for infectious disease in African great apes. *PLoS One*, 6: e29030. doi:10.1371/journal.pone.0029030.

Sandbrook, C. G. and Semple, S. (2006). The rules and the reality of mountain gorilla (*Gorilla beringei beringei*) tracking: how close do tourists get? *Oryx*, 40: 428–433.

SGLCP. (2009). *Response to Notification: Updating or Revision of the Convention after 2010*. Convention on Biological Diversity (CBD) Steering Group on Linking Conservation and Poverty (SGLCP). www.cbd.int/doc/strategic-plan/revision-input/Germany.pdf (accessed Apr. 4, 2014).

Soehartono, T., Susilo, H. D., Andayani, N., *et al.* (2007). *Orangutan Indonesia: Conservation Strategies and Action Plan 2007–2107*. Jakarta, Indonesia: Directorate General of Forest Protection and Nature Conservation, Ministry of Forestry of the Republic of Indonesia. www.primate-sg.org/action_plans/ (accessed Apr. 4, 2014).

Timen, A., Koopmans, M. P. G., Vosson, A. C. T. M., *et al.* (2009). Response to imported case of Marburg hemorrhagic fever, the Netherlands. *Emerging Infectious Diseases*, 15: 1171–1175.

WCS Gabon. (2008). *Langoué Bai, Ivindo National Park: Review of the Pilot Tourism project 2001–June 2008*. Libreville, Gabon: Wildlife Conservation Society. http://en.calameo.com/books/00000278504447bd38612 (accessed Apr. 4, 2014).

Weber, A. W. (1993). Primate conservation and eco-tourism in Africa. In: C. S. Potter, J. I. Cohen, and D. Janczewski (eds.), *Perspectives on Biodiversity: Case Studies of Genetic Resource Conservation and Development*. Washington, DC: American Association for the Advancement of Science Press, pp. 129–150.

Wilkie, D. S. and Carpenter, J. F. (1999). Can tourism finance protected areas in the Congo Basin? *Oryx*, 33: 332–338.

Williamson, E. A. and Fawcett, K. A. (2008). Long-term research and conservation of the Virunga mountain gorillas. In: R. Wrangham and E. Ross (eds.), *Science and Conservation in African Forests: The Benefits of Longterm Research*. Cambridge University Press, pp. 213–229.

Part VI

Conclusion

18 Primate tourism as a conservation tool: a review of the evidence, implications, and recommendations

Anne E. Russon and Janette Wallis

Introduction

Our aim in this book was to encourage balanced, empirically based assessments of primate tourism's strengths and weaknesses as a primate conservation tool. Such assessments are increasingly called for, since nature tourism has not proven to be as low impact or as strong a conservation tool as previously hoped (Higham, 2007a; Knight & Cole, 1995). In this chapter we review primate tourism's positive and adverse effects on primate conservation, primarily from the perspective of our contributors' assessments of the species and sites they study. We focus on the primates and habitats visited and, where possible, weigh tourism's adverse against positive effects. Insofar as this approach improves understanding of primate tourism's impacts, it can serve as a basis for developing ways to operate primate tourism to the net benefit of primate conservation.

Assessing primate tourism's conservation benefits and costs

Assessing the benefits and costs of primate tourism for primate conservation is complicated by the many factors and the long timeframes typically involved. Important complexities are that primate tourism is often only one of several activities at a site; primates are not necessarily the main tourist attraction; humans other than tourists may visit a site's primates, such as researchers, conservationists, park employees, and local people; conservation may not be the main focus of primate tourism at a site; and patterns associated with primate tourism can also correlate with broader economic, socio-political, and natural events. It is not always possible to isolate which of these factors is responsible for changes relevant to primate conservation. An important case in point is disease transmission: human diseases have infected primates that tourists visit, but there is no evidence that tourists are more likely than non-tourists to have spread the diseases (Muehlenbein & Wallis, this volume).

Second, primate tourism's effects on primate conservation can change over time. For example, mountain gorilla tourism benefited mountain gorilla conservation in

Primate Tourism: A Tool for Conservation?, ed. Anne E. Russon and Janette Wallis. Published by Cambridge University Press. © Cambridge University Press 2014.

the twentieth century. However, its costs to the gorillas visited have become more apparent and some costs have grown over time, so many experts are concerned that the risks may now outweigh the benefits (Goldsmith, this volume). Primate tourism practices have also changed over time, either to counteract tourism problems or to cater to tourists (in this volume, see Desmond & Desmond, Goldsmith, Kurita). The net result is to change its effects on primate conservation.

Third, assessments of tourism's costs and benefits to primate conservation depend on who reaps the benefits and who bears the costs. Nature or primate conservation work may gain major benefits from primate tourism at the national level, for example, while the primates and habitats visited bear the bulk of its costs and may reap relatively few of its benefits (in this volume, see Goldsmith, Wright *et al.*).

In this chapter, we focus on sites where primates are the main tourist attraction because this increases the chances of identifying links between primate tourism and primate conservation. We considered the effects of primate tourism at sites that are not conservation oriented as well as at sites that are, to generate a broad picture of what can result. We discuss two conservation benefits expected from nature tourism, nature protection and conservation funding, plus primate tourism's effects on primates' health and behavior. We do not discuss conservation education, an important expected benefit, only because our contributors did not assess it systematically.

Protecting primates and their habitat

Primate habitat

Habitat loss is the most important threat to survival for many primate species, so securing legal protected status for primate habitat has been singled out as the most effective strategy for protecting them (Bruner *et al.*, 2001; Desmond & Desmond, this volume; Pusey *et al.*, 2008). Without legal protection, natural areas increasingly risk conversion to human use. Protected status has mitigated primate habitat loss in several countries (e.g. Tanzania, Kenya, Madagascar, Rwanda, and Uganda: in this volume, see Desmond & Desmond, Goldsmith, Strum & Manzolillo Nightingale, Wright *et al.*).

Nature tourism has helped protect primate habitat in some countries by providing the impetus for granting protected status, funding protection, and/or helping guard the area (e.g. Indonesia, Uganda, Madagascar; in this volume, see Dellatore *et al.*, Goldsmith, Wright *et al.*). Due to its popularity, primate tourism has often been an important contributor (Kinnaird & O'Brien, 1996; Wollenberg *et al.*, 2011). In Madagascar, Ecuador, Indonesia, Malaysia, and Tanzania, primate habitat is in better condition inside protected areas with primate tourism than outside these areas (Fuentes, 2010; Singleton & Aprianto, 2001; in this volume, see de la Torre, Desmond & Desmond, Wright *et al.*). Community support for protecting primate habitat has in some cases improved because of primate tourism, especially when tourism revenue-sharing programs are provided (Ahebwa *et al.*, 2012; Tumusiime

& Svarstad, 2011; in this volume, see Goldsmith, Strum & Manzolillo Nightingale, Wright *et al.*).

Nature tourism is also well known for damaging habitat, for example erosion, tourist or tourist vehicle damage to vegetation and soil, infrastructure construction, water pollution, waste accumulation, and wildlife loss (Higham, 2007a; McCarthy, 1999; Xiang *et al.,* 2011). Such damage has been reported for several primate tourism sites (Cochrane, 1998; McCarthy, 1999; in this volume, see Dellatore *et al.*, Hodgkinson *et al.*, Leasor & Macgregor, Sapolsky, Wright *et al.*). Primate tourism sites where annual visitor numbers have reached the hundreds of thousands almost certainly face substantial damage (Ambu, 2007; Corpuz, 2004; Fuentes *et al.*, 2007; Kurita, this volume). The primates visited can damage habitat as a result of habituation or provisioning for tourism; crop-raiding and damaging natural areas due to over-population or concentrating activities around provisioning sites have both been reported (in this volume, see Dellatore *et al.*, Goldsmith, Kurita). Increasing primate tourism can also increase outsider influx and local community growth, aggravating pressures on the primate habitat visited (in this volume, see Hodgkinson *et al.*, Leasor & Macgregor, Wright *et al.*).

Primate survival

We consider the effects of primate tourism on the safety and population size of the primates visited, as most directly affecting their survival. We do not consider health or birth rates here: we discuss health separately, as an issue of major concern for primate conservation, and birth rates may offer biased estimates of population change. Rehabilitant orangutans visited by tourists show increases in both birth rates and infant mortality (Kuze *et al.*, 2012). Berman *et al.* (this volume) found no change in Tibetan macaque birth rates attributable to tourism, suggesting no effect on population size, but increased infant mortality and wounding that probably reflected infanticide linked with tourism.

Primate tourism has helped improve the safety of the primates visited by guarding, patrolling, and/or improving local people's willingness to protect or tolerate them. Such measures have reduced killings or injuries in mountain gorillas, chimpanzees, olive baboons, and Japanese macaques (in the volume see Desmond & Desmond, Goldsmith, Kurita, Strum & Manzolillo Nightingale).

Some primate populations have maintained or increased their size after tourism was introduced (e.g. chimpanzees, olive baboons, Japanese macaques, long-tailed macaques, mountain gorillas, golden bamboo lemurs: Fuentes, 2010; in this volume, see Desmond & Desmond, Goldsmith, Kurita, Strum & Manzolillo Nightingale, Wright *et al.*). In some primates, the improved nutritional condition resulting from long-term provisioning can increase birth rates by reducing the age at first birth or interbirth intervals and/or by improving infant survival (Cheney *et al.*, 2004; Kuze *et al.*, 2012; in this volume, see Dellatore *et al.*, Kurita). It is often difficult to determine whether these changes are due to tourism, co-occurring conditions (e.g. securing habitat protection status, researchers, predator change, and climate change), species differences, or interactions among these factors (in this volume, see Desmond & Desmond, Wright *et al.*).

Primate tourism can compromise the primates' safety by facilitating poaching and improving access to their habitat, especially previously inaccessible areas. Tourism workers captured marmosets from tourist-visited groups for the illegal pet market (de la Torre, this volume). Gorillas' reduced vigilance and wariness of humans, caused by habituation for tourism and research, has facilitated poachers' killing and capturing them (Goldsmith, this volume; Ilambu, 1998; Pole Pole Foundation, 2012). Tourist staff and guides have allowed or promoted tourist proximity, contact and hand feeding with orangutans, Tibetan macaques, and Japanese macaques; and tourism site staff have herded Tibetan and Japanese macaques for tourism; as a result of these activities, some of the primates were seriously injured or killed (Zhao, 2005; Zhao & Deng, 1992; in this volume, see Berman *et al.*, Dellatore *et al.*, Kurita, Russon & Susilo). Primates have been injured or killed because of conflicts with local communities that may have resulted from provisioning or habituation, both undertaken largely for tourism (Goldsmith, this volume; Russon, pers. obs.).

Long-term provisioning of Japanese macaques for tourism caused overpopulation relative to the available natural habitat, which in turn increased their crop-raiding, mortality, and conspecific competition (Kurita, this volume). Tourism correlated with decreases in population or group size in pygmy marmosets (de la Torre *et al.*, 2000), black howlers (Treves & Brandon, 2005), crested black macaques (Kinnaird & O'Brien, 1996), the endangered Milne-Edwards' sifaka, and the critically endangered greater bamboo lemur (Wright *et al.*, this volume). For both lemurs, tourism also correlated with decreased body weight (Wright *et al.*, this volume).

Funding primate conservation

A major motive for primate tourism has been funding primate conservation. Most primates inhabit developing tropical countries, where other sources of conservation funds are limited. Hvenegaard's chapter discusses economic issues in detail, so we summarize the main achievements and problems in using primate tourism to fund primate conservation.

Some primate tourism has generated hundreds of millions of dollars annually, a substantial portion of national tourism revenues. Notable examples are tourism to Rwanda's and Uganda's mountain gorillas and Madagascar's lemurs (Lanyero, 2011; Maekawa *et al.*, 2013; in this volume, see Goldsmith, Wright *et al.*). This funding support can be meaningful and sometimes substantial (Archabald & Naughton-Treves, 2001; Butynski & Kalina, 1998; in this volume, see Goldsmith, Hvenegaard, Wright *et al.*). Where primate tourism operates in protected areas and visitor fees revert to governments (e.g. Madagascar, Uganda, Indonesia), an area's tourism income has also subsidized other protected areas. Funds from primate tourism have improved law enforcement, reduced primate poaching, or improved local attitudes to primates and conservation-friendly behavior in some areas (Alexander,

2000; Archabald & Naughton-Treves, 2001; Blom, 2000; Hartup, 1994; Lepp, 2002; MacKenzie, 2012; McNeilage, 1996; Weber, 1987; in this volume, see Strum & Manzolillo Nightingale, Wright *et al.*). Local Rwandans helped protect mountain gorillas during the 1990s civil strife (Salopek, 1995), for example, and Ugandan villagers near Kibale National Park reduced poaching and other disturbances in protected areas, thereby reducing disease and snare injury risks to resident chimpanzees (MacKenzie, 2012). In Rwanda, such changes in local attitudes and staff commitment to mountain gorillas have been maintained (Maekawa *et al.*, 2013).

Funding primate conservation from primate tourism revenues faces important limitations, notably instabilities in tourism revenues and leakage out of the region that generated it. Primate tourism, like tourism in general, is a notoriously unstable source of income because it is sensitive to political, economic, and environmental instabilities and tourist trends (Butynski & Kalina, 1998; Maekawa *et al.*, 2013; Tran, 2013; in this volume, see Hvenegaard, Kurita, Strum & Manzolillo Nightingale, Wright *et al.*). Instabilities are usually temporary but have led to dramatic drops in primate tourism, tourism income, and area surveillance and to increased illegal activities; Rwanda's mountain gorilla tourism took about 10 years to recover (Maekawa *et al.*, 2013; Weber, 1993; in this volume, see Hvenegaard, Strum & Manzolillo Nightingale, Wright *et al.*).

Leakage of wildlife tourism revenues out of the host country or host area is typically high (in the volume see Hvenegaard, Russon & Susilo). For primate tourism, the portion of the host area's tourism revenues that funds primate conservation has typically been low. Conservation funds are most often allocated from visitor (user or entry) fees, a very small percentage of tourism revenues and typically very low for sites that are formally protected or offer "ready-to-view," unexciting, or elusive primates (Cochrane, 1997; Fuentes *et al.*, 2007; in this volume, see Dellatore *et al.*, Goldsmith, Hvenegaard, Russon & Susilo, Wright *et al.*). Only an estimated 5% of gorilla tourism income in Rwanda, Uganda, or DRC was spent on gorilla conservation or local people in the 1980s and 1990s (Butynski & Kalina, 1998). At one Sulawesi reserve known for wild macaque tourism, only 2% of visitor fees reached the Forestry Ministry and the fraction of that allocated to the reserve did not fund appropriate management (Kinnaird & O'Brien, 1996). The expensive tracking permits that some primate tourism sites require (e.g. daily, $500–$750 US for mountain gorillas, $50–150 US for chimpanzees) do not fund conservation in the same way as visitor fees (Goldsmith, this volume; Maekawa *et al.*, 2013; Uganda Wildlife Authority, 2013). Where visitor fees are low, income can be increased by high tourist numbers (e.g. annually, nearly 1 000 000 at several primate sites and 2 000 000 at one: Fuentes *et al.*, 2007; Kurita, this volume) but this easily causes adverse effects for primate conservation (Corpuz, 2004). Finally, the funds allocated to the host area's conservation are not necessarily commensurate with the tourism fees it generated (in this volume, see Goldsmith, Wright *et al.*). If the conservation funds that an area's primate tourism generates serve nature or primate conservation elsewhere, its own primates and habitat are thereby commodified and their own protection compromised.

Funding to local communities from primate tourism can support primate conservation; examples are revenue-sharing programs and employment in tourism. It, too, has typically been modest (Maekawa *et al.*, 2013; in this volume, see Goldsmith, Hvenegaard). Revenue-sharing programs have not been entirely successful: in some cases primate poaching continued and some local people have been dissatisfied with their share of revenues (Adams & Infield, 2003; Ahebwa *et al.,* 2012; Goldsmith, this volume; Hodgkinson, 2009; Maekawa *et al.*, 2013; Tumusiime & Svarstad, 2011; Weber, 1993). Local people have also complained of less involvement in tourism than desired, tourism employment mostly in few, low-paying tourism jobs, and not necessarily being better off because of primate tourism (Leasor & Macgregor, this volume; Wyman & Stein, 2010). Programs that improved local conditions have attracted new people to move in, which can aggravate pressures on the area's primates, habitat, and residents (Wright *et al.*, this volume).

In sum, primate tourism's funding contribution to primate conservation has not always been substantial. The conservation funding generated by mountain gorilla tourism, arguably the most lucrative primate tourism, is probably unrealistic for most other primates. At some times and sites, these funds have not clearly outweighed the costs to primate conservation that tourism generates. Two cases suggest current uncertainties. Mountain gorilla tourism generated funds that helped save dwindling populations in the twentieth century but it has become increasingly exploitive, for economic reasons, and may no longer weigh in their favor (Goldsmith, this volume). Primate tourism may contribute significantly to protecting Madagascar's biodiversity by helping fund nature protection: all its ecosystems are now within protected areas visited by tourists (Garbutt, 2009; Mittermeier *et al.*, 2010).

Primate tourism's effects on primate health

We consider primate tourism's effects on the health of the primates visited in terms of nutrition, stress, and disease. Health effects can be slow to appear so they are most evident in long-term studies. Assessments have focused on three contributing factors: provisioning, behavioral stressors, and human pathogens.

Nutrition

Provisioning linked to primate tourism has two common forms: formal, regulated daily staff provisioning and tourist or guide provisioning (authorized to uncontrolled: human foods offered or stolen, garbage eating) (Fuentes *et al.*, 2007; in this volume, see Berman *et al.*, Dellatore *et al.*, Kauffman, Kurita, Russon & Susilo, Sapolsky).

Formal provisioning for tourism has in some cases improved or had no detectable effects on the primates' nutritional condition, where improvement is indexed by higher body weight, birth rates, or population size (in this volume, see Berman *et al.*, Kurita). It has provided a substantial proportion of the primates' diet, for example at Takasakiyama monkey park, up to 50–75% of female Japanese macaques' daily ingested energy for several years (Kurita, this volume). Reproduction is accelerated in Japanese macaques and rehabilitant orangutans provisioned for tourism, compared with non-provisioned wild conspecifics, and regular provisioning is a probable cause (Cheney *et al.*, 2004; Kuze *et al.*, 2012; Kurita, this volume).

If provisioning provides a high-energy diet, too much food, or contaminated food, it can have adverse effects (in this volume, see Kurita, Sapolsky). Formal provisioning at Takasakiyama led to Japanese macaque overpopulation, which in turn increased their crop-raiding, damaging natural habitat, and competition with conspecifics (Kurita, this volume). Accelerated reproduction has been associated with increased infant mortality (Berman *et al.*, this volume; Kuze *et al.*, 2012). Garbage eating at tourist lodges caused obesity and pre-diabetic syndrome in savanna baboons (Sapolsky, this volume). Other primates that eat tourist garbage are also susceptible to similar problems, for example capuchins and common marmosets in a Costa Rican and a Brazilian national park (Cunha, 2010; Cunha & Vieira, 2004; Cunha *et al.*, 2006; Kauffman, this volume).

Stress

Tourists may be able to visit some primates without stressing them. Red howling monkeys ignored tourist presence (Hodgkinson *et al.*, this volume) and two highly habituated wild East Bornean orangutans showed immediate but not chronic physiological stress responses to it (Muehlenbein *et al.*, 2012). Caveats are that "ignore" does not always mean no response (Higham & Shelton, 2011) and sensitivity to tourists can vary taxonomically, ontogenetically, and experientially. Graell's tamarins showed no change in their flight response to tourists even after years of exposure (de la Torre, this volume), for example, but lemurs habituated for research since early in life appeared tolerant of human observers (Wright *et al.*, this volume).

In other circumstances, tourist presence, behavior, and numbers can stress the primates visited, which can increase susceptibility to disease by compromising immune function (Muehlenbein & Wallis, this volume). Tourists more often qualify as intrusive than benign: they are typically unfamiliar to the primates, may arrive in large numbers, and behave disruptively (in this volume see Berman *et al.*, Dellatore *et al.*, Leasor & Macgregor, Russon & Susilo, Wright *et al.*). Common behavioral indices of stress in the primates visited have been fear or threat responses, increased vigilance, social grooming, and self-directed behaviors (in this volume, see Berman *et al.*, de la Torre, Hodgkinson *et al.*, Wright *et al.*). Tourist, guide, and staff behaviors that elicit such stress responses include close proximity, making loud or sudden noises, flash photos, shaking trees, herding, and throwing rocks (in this volume, see Berman *et al.*, Dellatore *et al.*, Leasor & Macgregor, Russon & Susilo, Wright *et al.*). Regular tourism stress on primates has resulted in distorted ranging and energetics, decreased body mass, less cohesive groups, and increased infant mortality (Treves & Brandon, 2005; in this volume, see Berman *et al.*, Wright *et al.*). Avoiding tourist areas may explain the higher emigration than immigration that may have caused sifakas' group size to drop (Wright *et al.*, this volume).

Figure 18.1   Safer tourist viewing practices. A group of tourists views chimpanzees at Ngamba Island
Chimpanzee Sanctuary, Uganda, from a viewing platform. The viewing area is designed to
enforce a safe minimum distance between visitors and chimpanzees. Visitors are wearing
face masks, an additional measure to reduce the risks of transmitting human diseases to
the chimpanzees visited. (© J. Wallis.)

Disease

Introducing diseases to nonhuman primates, especially infectious human diseases,
is considered one of the most serious threats to their survival (in this volume, see
Desmond & Desmond, Goldsmith, Muehlenbein & Wallis, Williamson & Macfie).

Tourism has helped improve medical care for primates at some sites. Regular
health monitoring and illness prevention programs have been instituted for mountain gorillas and chimpanzees at several tourist sites (in this volume, see Desmond
& Desmond, Goldsmith). Prevention measures include health and vaccination
checks, visitor viewing rules to control against airborne and contact transmission
of human disease (e.g. maximum 1 hour visit, minimum 10 m distance, wearing
face masks), and stopping provisioning (in this volume, see Desmond & Desmond,
Goldsmith, Muehlenbein & Wallis) (see Figure 18.1). These measures were often
spurred by infectious disease outbreaks of probable human origin, however, to
which tourism could have contributed (Wallis & Lee, 1999).

Tourism presents disease risks for the primates visited. While current evidence
suggests that outbreaks of human disease in primates more likely originated in
local people or staff than in tourists, this could reflect the greater difficulty of
tracing pathogens to tourists than to humans regularly present (Muehlenbein &
Wallis, this volume). Tourism's risks stem from increasing the intensity of human

presence, the number of local people and staff included, and the range of human diseases introduced (Muehlenbein & Wallis, this volume). Infections of tourist origin are suspected; human nematodes were found only in sifaka groups that tourists constantly visited (Wright *et al.*, this volume). Provisioning linked with tourism facilitates disease transfer because human foods and garbage can carry human pathogens which can infect the primates that eat them and because provisioning promotes close proximity between humans and primates and between the primates themselves (Sapolsky, this volume). Eating infected meat from a tourist lodge garbage dump infected many baboons with tuberculosis and killed them (Sapolsky, this volume).

Tourist behavior increases disease risks. Tourists have concealed illnesses to continue their primate visit, even tourists informed of the health risks they pose (Adams *et al.*, 2001; Muehlenbein & Wallis, this volume). Close tourist–primate proximity (under 10 m) still occurs, physical contact included, and very few sites require face masks (in this volume, see Berman *et al.*, Dellatore *et al.*, Desmond & Desmond, Goldsmith, Kauffman, Kurita, Leasor & Macgregor, Russon & Susilo, Williamson & Macfie). Close proximity is difficult to counteract because it is very attractive to tourists, it is often promoted in wildlife tourism, the primates themselves can initiate it, and even where health controls are in place they are poorly enforced (Tapper, 2006; in this volume, see Berman *et al.*, Dellatore *et al.*, Kauffman, Russon & Susilo).

Primate tourism's effects on primate behavior: activity budgets

Our concern is primate tourism's effects on primate behaviors that may mediate its effects on survival or reproduction. We therefore focused on activity budgets (time spent daily on feeding, rest, travel, and sociality) as the most closely tied to energy management and reproduction.

Caveats are that primates' responses to tourists can vary between species (e.g. speed of habituation), within species (e.g. age, sex, reproductive or social status, learning, degree of habituation), and with context (e.g. season, time of day, tourist behavior). Common effects on behavior should therefore be relatively rare (in this volume, see Berman *et al.*, de la Torre, Goldsmith, Hodgkinson *et al.*, Kurita, Sapolsky, Wright *et al.*). As one example, four sympatric primate species at one site reacted differently to tourists: *Saimiri boliviensis* tended to flee, *Cebus apella* to flee or monitor, *Saguinus fuscicollis* to avoid or monitor, and *Alouatta seniculus* to monitor or ignore (Hodgkinson *et al.*, this volume).

Broadly, primates' responses to tourists tend to vary from engagement to avoidance, largely as a function of habituation, provisioning, and species. Engagement has been common in primates that are larger, more omnivorous (great apes, baboons, macaques, capuchins), well-habituated, and provisioned formally or informally (in this volume, see Berman *et al.*, Dellatore *et al.*, Goldsmith, Kauffman, Kurita, Russon & Susilo, Sapolsky). Avoidance has been common in primates that are smaller (marmosets, tamarins, squirrel monkeys, bamboo lemurs, proboscis monkeys), more folivorous, relatively unhabituated, and not provisioned (in this volume, see de la Torre, Hodgkinson *et al.*, Leasor & Macgregor, Wright *et al.*).

A few common patterns nonetheless stand out: notably, habituation and provisioning generate long-term learning-based changes in primate responses to tourists, and provisioning attracts primates to humans (Knight, 2009). Long-term changes are difficult to study and to link to tourism, but are especially important to conservation.

Activity budgets

Primate tourism has decreased primates' natural foraging time daily by orienting their behavior to tourists or by provisioning them with easily accessed, high-energy foods (in this volume, see de la Torre, Dellatore *et al.*, Goldsmith, Sapolsky, Wright *et al.*). Where tourism does not involve provisioning, primates typically move away from tourists so their daily feeding time decreases with this increase in travel time (in this volume, see de la Torre, Goldsmith, Hodgkinson *et al.*, Wright *et al.*). This could affect the primates' natural diet if avoiding tourists effectively restricts their access to natural feeding areas (in this volume, see de la Torre, Leasor & Macgregor, Wright *et al.*). Insofar as tourist-induced changes to the primates' feeding affect their diet and natural foraging skills, they can ultimately affect nutritional condition and health in the ways previously discussed.

Tourism is most often linked with reduced rest because primates often react to tourists with avoidance, hiding, or increased vigilance (in this volume, see de la Torre; Hodgkinson *et al.*; Wright *et al.*). When tourism involves provisioning, it can increase rest if primates wait for food at feeding sites and gain higher-energy foods (Dellatore *et al.*, this volume). Both can result in altered energy management and missed opportunities for natural behavior.

Primates visited by tourists tend to increase or decrease day travel time as a function of provisioning: they tend to remain near feeding sites and tourists at sites that provision but move away from tourists at sites that do not (in this volume, see de la Torre, Dellatore *et al.*, Goldsmith, Hodgkinson *et al.*, Kauffman, Wright *et al.*). Provisioning effectively restricts primate ranges by drawing them to feeding sites or tourist areas (Knight, 2009; in this volume, see Berman *et al.*, Dellatore *et al.*, Kauffman, Kurita, Sapolsky). Tourist presence has also altered primates' ranging by interfering with their normal travel (e.g. inhibiting proboscis monkeys' crossing rivers by swimming, causing some primates to move higher in the canopy, luring highly arboreal Sumatran orangutans to the ground: in this volume, see de la Torre, Dellatore *et al.*, Hodgkinson *et al.*, Leasor & Macgregor, Wright *et al.*). Such travel changes can affect primates' health, safety, and reproduction by altering their daily energy expenditure, increasing their exposure to predators or infections, and/or limiting access to natural resources (in this volume, see Dellatore *et al.*, Sapolsky, Wright *et al.*). Lemurs' avoiding tourist areas may have reduced their group size and reproductive success by favoring emigration at dispersal (Wright *et al.*, this volume).

Tourism's effects on primate social behavior include increased conspecific competition, aggression, infant wounding, group fission, and social grooming, decreased social play and vocal communication, and increased or decreased group cohesion (Maréchal *et al.*, 2011; Ram *et al.*, 2013; in this volume, see Berman *et al.*, de la Torre, Dellatore *et al.*, Goldsmith, Kurita, Leasor & Macgregor, Russon & Susilo, Wright *et al.*).

In summary, some of primate tourism's effects on primate behavior appear minor (e.g. monitoring, moving higher) and, on an occasional basis, may be. When frequent and chronic, however, those that affect activity budgets, stress, or social interaction can undermine primates' safety, health, growth, reproduction, and social structures (in this volume, see Berman *et al.*, de la Torre, Dellatore *et al.*, Goldsmith, Kurita, Leasor & Macgregor, Muehlenbein & Wallis, Russon & Susilo, Sapolsky, Wright *et al.*).

In the balance: weighing primate tourism's costs and benefits for primate conservation

Primate tourism generates conservation costs and benefits at multiple levels, from the local primates and habitats exposed to tourism to entire primate populations or species, host nations, and international agencies. We focus on weighing costs and benefits at the local level because most studies assess site-specific factors and harming the primates that tourists visit risks losing broader conservation and economic benefits, that is, killing the goose that laid the golden egg.

We know of no formal criteria for weighing tourism's conservation costs and benefits to the primates visited. Previous efforts often reflect judgments of knowledgeable researchers or conservationists (e.g. Butynski & Kalina, 1998; Goldsmith, this volume; Harcourt & Stewart, 2007; Macfie & Williamson, 2010; Rijksen & Meijaard, 1999). From the view that primate tourism should improve or at least maintain the viability of the primates visited, we used the following indicator of a net conservation benefit over a given period of operation: the primate populations or groups that tourists visited maintained their size or grew commensurate with their natural habitat's carrying capacity. Population size is not reported for all primate tourism sites and it can be slow to change, so we also used indicators of direction of change and/or reduction in threats to the primates visited (e.g. poaching or killing, better disease control). Change is typically assessed by comparing primates before and after experiencing tourism or over specific periods during tourism, or by comparing conspecifics that experienced tourism with those that did not.

With these criteria, primate tourism has operated to the net conservation benefit of the primates visited by fostering healthy population size in mountain gorillas (1979 – present), chimpanzees in Gombe National Park (1980s – present), lemurs in Ranomafana National Park (1996 – ca 2006), Japanese macaques at Takasakiyama (1952 – ca 1965, ca 1995 – present), and olive baboons in Kenya (1996 – present) (in this volume, see Desmond & Desmond, Goldsmith, Kurita, Strum & Manzolillo Nightingale, Wright *et al.*). At other sites and times, primate tourism may have generated net conservation loss. It has been linked with reduced group size in lemurs (Wright *et al.*, this volume). When it involves provisioning, it has caused overpopulation and perhaps increased birth rates and infant mortality (Berman *et al.*, 2007; Kuze *et al.*, 2012; Mallapur, 2013; in this volume, see Berman *et al.*, Dellatore *et al.*, Kurita). It has increased threats to the primates visited by exposing them to infectious human diseases and inducing metabolic disorders, deleterious range shifts, and loss of fear, reduced vigilance, or increased aggression toward humans that in turn increased their crop-raiding, stealing, or attacks (in this volume, see Berman *et al.,* Desmond & Desmond, Goldsmith, Kauffman, Kurita, Muehlenbein & Wallis, Sapolsky, Wright *et al.*). In sum, primate tourism has sometimes contributed to maintaining or improving the primates' natural habitat but has sometimes undermined it.

Corrective management measures have had some success in alleviating these costs, often at conservation-oriented sites. Japanese macaque overpopulation at Takasakiyama was brought under control but it took almost 35 years to reduce the population to its current size, still six times its pre-tourism size (Kurita, this volume). Health monitoring and disease prevention measures have been instituted for mountain gorilla and chimpanzee groups visited by tourists (in this volume, see Desmond & Desmond, Goldsmith, Muehlenbein & Wallis). Some sites have intensified their efforts to regulate and control tourist behavior (in this volume, see Dellatore *et al.*, Desmond & Desmond). How well these measures achieve and maintain net conservation benefits remains to be seen. Some problems, including primates' losing their fear of humans due to habituation or provisioning and contacting infectious human diseases, can be very hard or impossible to reverse (in this volume, see Goldsmith, Muehlenbein & Wallis). The best option is preventive measures instituted from the outset and systematically maintained.

The emerging pattern concerning the conservation value of primate tourism for the primates visited is an uncertain and unstable balance, even at some of the most conservation-oriented sites. On the one hand, tourism problems are increasingly understood and preventive and corrective practices are increasingly available and instituted. On the other hand, tourist demand for direct encounters with wildlife, especially primates, continues to increase (Muehlenbein & Wallis, this volume) and problems that are slow to develop or that depend on tourist volume are increasingly emerging and spreading (Goldsmith, this volume). Better tourism practices should reduce costs and more tourists should mean more funds for primate conservation, but this has not always resulted. Managing primate tourism to protect the primates visited is already difficult, however, some sites show few signs of altering their tourism practices for conservation, some problems may be irreversible, and increased tourism should increase management difficulties and risks to the primates visited. The cost–benefit balance could easily tip away from conservation at many primate tourism sites and probably already has at some (Goldsmith, this volume).

Discussion: lessons learned

Primate tourism is unlikely to disappear, given its popular and commercial success. Empirical studies like those presented in this book now offer considerable evidence on its positive and adverse effects on the primates visited and the causal factors involved. While these studies suggest it is probably impossible to eliminate all of primate tourism's costs to primate conservation, they enrich understanding of what is needed to manage primate tourism for net primate conservation benefit. Here we offer an overview of the lessons learned and broad recommendations. We focus on three problems in primate tourism that have major conservation costs: habituation, provisioning, and disease. Infecting primates with human diseases is among the most serious threats to their survival (Macfie & Williamson, 2010; Muehlenbein & Wallis, this volume). Habituation and provisioning are cornerstones of primate tourism and many of the problems it can generate, disease included (Goldsmith, this volume; Knight, 2010).

An important factor in primates' popularity with tourists is their biologically based close similarities to humans (Kinnaird & O'Brien, 1996), some of which predispose the constellation of tourism effects identified. Examples are primates' great behavioral plasticity, high potential for competition with humans over habitat and food resources, and high susceptibility to human diseases. Mixed with primate tourism, these traits easily lead to primates' approaching or attacking instead of ignoring or avoiding humans, crop- or garbage-raiding, and contracting human disease (in this volume, see Berman, Dellatore *et al.*, Goldsmith, Kauffman, Kurita, Russon & Susilo). Primates also show pronounced species differences, so species-specific traits further shape the effects that develop under primate tourism.

Primate tourism has served the conservation of the primates visited and their habitat relatively well where it has been undertaken in collaboration with researchers and conservationists who are knowledgeable about the primates' biology and behavior, who have long-term commitments to the area's biodiversity and human residents, and who maintain their involvement with and some influence over tourism practices and management. Some of the worst problems have been prevented or alleviated where monitoring and corrective measures have been used to maintain the balance of tourism effects in favor of conservation. More conservation costs are reported where primate tourism operates without primate specialists (e.g. government or commercial management) or as stand-alone ventures parachuted into an area. Primate specialists have certainly made primate tourism mistakes (e.g. Kurita, this volume), but non-specialists are more likely to take counter-productive actions because they lack the requisite specialized knowledge (primates, disease, behavior management) and to prioritize factors other than conservation, especially economic gain and satisfying tourists. Local communities can contribute positively to primate tourism and conservation, but not by fiat and probably not without long-term investments to integrate them.

Contextual factors may affect primate tourism's conservation impact. We mention two. First, the biological and socio-political landscape may enable or limit possibilities. Successes in Madagascar may be partly due to the facts that its relatively small national parks can sustain rich biodiversity and people living nearby are poor but have strong traditions and respect for laws (Wright *et al.*, this volume). This makes it relatively easy to operate primate tourism to benefit the primates visited, the national parks, and local communities (Wright & Andriamihaja, 2002). Second, tourism intensity makes a difference, that is tourist numbers and frequency of visits. High tourism intensity has correlated with primates' negative or aggressive responses to tourists, distorted activity budgets, avoiding tourist areas, and drops in group size (in this volume, see Berman *et al.*, Dellatore *et al.*, Hodgkinson *et al.*, Kauffman, Leasor & Macgregor, Wright *et al.*). Increasing tourist numbers often means habituating more primates; this aggravates many tourism impacts, risks of introducing human disease included (in this volume, see Goldsmith, Muehlenbein & Wallis). The take home message is that site-specific factors play a significant role in fostering or undermining primate tourism's benefits to the primates visited.

Primate tourism's effects change over time, as they do for wildlife tourism generally (Catlin *et al.*, 2011; Duffus & Dearden, 1990). Net conservation benefits for the primates visited are often reported in the early phases of primate tourism at a site. Conservation costs have tended to increase over time, as tourist numbers increase and problems emerge that are slow to develop or threshold-based. This is predictable, given that tourist area "life-cycles" typically show progressive deterioration over time as tourism shifts from specialized and responsible to general and mass, wildlife from naïve and wary to habituated and human-savvy, and control from conservation–local to economic–outsider parties (Butler, 1980; Catlin *et al.*, 2011; Duffus & Dearden, 1990). Tourism is recognized as fundamentally commercial, so making it serve conservation requires establishing mutually beneficial partnerships between conservation and economic interests. Such partnerships have rarely proven complementary, are hard to establish and harder to sustain, and entail important compromises that have often prioritized economics over conservation; some consider them fundamentally contradictory (Buckley, 2011; Higham, 2007b). The upshot is that the net conservation benefit is extremely difficult to maintain.

Recommendations

One of our aims is to contribute to recommendations for operating primate tourism to the net benefit of primate conservation. We work from the following position: primate tourism must operate to protect the primates and habitats visited, whether it aims to serve primates or humans. Otherwise it can easily destroy both, which serves neither. It is not impact free, its adverse effects on the primates and habitats visited can be significant, and it is unlikely to disappear. Largely, then, operating primate tourism to the net benefit of primate conservation comes down to keeping its adverse effects within acceptable levels. The design, implementation, and management of primate tourism are then critical determinants of how well it benefits the conservation of the primates and habitats visited.

Because many chapters in this volume provide detailed species- and site-specific recommendations for primate tourism, we refer the reader to those chapters most relevant to their interests. Here, we add the following general recommendations to those specific suggestions.

The adverse effects of greatest concern for primate tourism and the hurdles involved in limiting these concerns are sketched below. Many are rooted in the behavior of the primates, tourists, guides, site staff, or local people, so limiting them requires enforcing behavioral regulations. For adverse effects that look to be inevitable products of tourism, the best hope is alleviating them. For those that are irreversible once introduced, prevention is the only solution.

Habituation, which enables tourists to view primates by reducing primates' wariness and increasing their tolerance of human presence, has led to three major problems for the primates visited: increased vulnerability to poachers and conflicts with humans; long-term stress for primates that do not habituate easily; and disease, injury, or death for primates who became over-habituated due to excessive exposure to tourists. Habituation is progressive and very difficult to reverse in individuals and can spread socially to unhabituated conspecifics (Goldsmith, this volume), so its effects are not easily contained. Thus, we recommend limiting the degree of habituation for tourism, whenever possible. Maintaining a reasonable degree of wariness in the primates being visited offers tourists the experience of seeing genuinely wild animals. In most tourism settings, some level of proximity tolerance will develop over time without deliberate efforts on the part of humans to habituate the primates visited.

Provisioning is an avenue to habituation (Knight, 2010) so it generates many of the same effects. It generates additional ones because it attracts primates to humans and associates them with food (Knight, 2009). Known adverse effects include provoking conspecific competition and aggression, human–primate conflict over food (crop-raiding, stealing tourists' food), overpopulation, obesity, metabolic disorders, and contracting human diseases. Alleviating these problems is difficult. Stopping provisioning may be resisted for economic reasons, and stopping it abruptly has major negative effects on the primates. Enforcing rules against tourists' feeding primates and changing primates' behavior once they have become dependent on provisions or skilled in stealing food have both proven difficult and slow to achieve, but not impossible (Fuentes *et al.*, 2007; in this volume, see Berman *et al.*, Dellatore *et al.*, Kauffman, Kurita). Therefore, we recommend prevention of provisioning for any new primate tourism sites being developed and carefully controlled reduction of provisioning in existing primate tourism sites.

Infecting primates with human diseases has not, yet, been linked to tourists but it has been traced to humans in contact with primates (i.e. researchers, park personnel, or people involved in habituation for tourism). Tourists nonetheless present serious disease risks because of the close proximity they seek and the human foods and garbage they lead primates to consume. International tourists carry foreign diseases for which local humans have no immunity, and tourists have concealed compromised health to avoid having their primate visit cancelled. We therefore recommend strict enforcement of health regulations for all humans who may come into close proximity with primates, tourists included.

In broadest terms, our contributors indicate that the following are needed to achieve this.

1. Effective management is critical to success. An important reason it remains hard to succeed is ignoring lessons learned. Lax or non-existent enforcement is a chronic problem in primate tourism and recommended measures for preventing and alleviating known problems have often been ignored (in this volume, see Berman *et al.*, Dellatore *et al.*, Kauffman, Russon & Susilo). Management should apply and enforce its tourism regulations rigorously, ensure that all staff are suitably trained and all tourists informed of regulations, monitor tourism operations and impacts regularly to detect incipient and emerging problems and to assess the impact of its own practices and regulations, and develop and implement measures to counteract incipient or emerging problems.
2. Experts in the biology, behavior, and conservation of the primates involved need to be strongly involved in the planning, implementation, management, and evaluation of any primate tourism operation. They should be members of the tourism area's management committees along with representatives of local communities and other stakeholders, and should have the authority to enforce, review, and revise management practices
3. Regulations for primate tourism must be established that aim to ensure the protection of the primates visited and their habitat and that are in line with the species' behavior and biology and with currently recognized best practices. Our contributors offer several models, from the recent best practice guidelines for tourism with the great apes to detailed recommendations for particular primate species and sites.

References

Adams, W. M. and Infield, M. (2003). Who is on the gorilla's payroll? Claims on tourist revenue from a Ugandan National Park. *World Development*, 31: 177–190.

Adams, H. R., Sleeman, J. M., Rwego, I., and New, J. C. (2001). Self-reported medical history survey of humans as a measure of health risk to the chimpanzees (*Pan troglodytes schweinfurthii*) of Kibale National Park, Uganda. *Oryx*, 35 (4): 308–312.

Ahebwa, W. M., van der Duim R., and Sandbrook, C. (2012). Tourism revenue sharing policy at Bwindi Impenetrable National Park, Uganda: a policy arrangements approach. *Journal of Sustainable Tourism*, 20 (3): 377–394.

Alexander, S. E. (2000). Resident attitudes towards conservation and black howler monkeys in Belize: The Community Baboon Sanctuary. *Environmental Conservation*, 27: 341–350.

Ambu, L. (2007). Strategy of the Sabah Wildlife Department for Wildlife Conservation. In: *Sabah. First International Conservation Conference in Sabah: the Quest for Gold Standards.* Kota Kinabulu, Malaysia: Sabah Wildlife Department.

Anonymous (2006). Police hunt Leone 'killer chimps'. BBC News, Monday, 24 April 2006. http://news.bbc.co.uk/go/pr/fr/-/2/hi/africa/4938620.stm (accessed Feb. 7, 2013).

Archabald, K. and Naughton-Treves, L. (2001). Tourism revenue-sharing around national parks in Western Uganda: Early efforts to identify and reward local communities. *Environmental Conservation*, 28: 135–149.

Berman, C. M., Li, J. H., Ogawa, H., Ionica, C., and Yin, H. (2007). Primate tourism, range restriction and infant risk among *Macaca thibetana* at Mt. Huangshan, China. *International Journal of Primatology*, 28: 1123–1141.

Blom, A. (2000). The monetary impact of tourism on protected area management and the local economy in Dzanga-Sangha (Central African Republic). *Journal of Sustainable Tourism*, 8: 175–189.

Bruner, A. G., Gullison, R. E., Rice, R. E., and da Fonseca, G. A. B. (2001). Effectiveness of parks in protecting tropical biodiversity. *Science*, 291 (5501): 125–128.

Buckley, R. (2011). Tourism and environment. *Annual Review of Environment and Resources*, 36: 397–416. DOI: 10.1146/annurev-environ-041210–132637.

Butler, R. W. (1980). The concept of a tourist area cycle of evolution: implications for management of resources. *Canadian Geographer*, 24: 5–12.

Butynski, T. M. and Kalina, J. (1998). Gorilla tourism: A critical look. In: E. J. Milner-Gulland and R. Mace (eds.), *Conservation of Biological Resources*. Oxford: Blackwell Press, pp. 294–313.

Catlin, J., Jones, R., and Jones, T. (2011). Revisiting Duffus and Dearden's wildlife tourism framework. *Biological Conservation*, 144 (5): 1537–1544.

Cheney, D. L., Seyfarth, R. M., Fischer, J. *et al.* (2004). Factors affecting reproduction and mortality among baboons in the Okavango delta, Botswana. *International Journal of Primatology*, 25: 401–428.

Cochrane, J. (1997). *Factors Influencing Ecotourism Benefits to Small, Forest–Reliant Communities: A Case Study of Bromo Tengger Semeru National Park, East Java*. University of Hull, UK. www.recoftc.org/documents/Inter_Reps/Ecotourism/Cochrane.rtf (accessed May 22, 2005).

Cochrane, J. (1998). *Organisation of Ecotourism in the Leuser Ecosystem*. Unpublished report to the Leuser Management Unit.

Corpuz, R. (2004). *"Wild Borneo" – a perception: A Study of Visitor Perception and Experience of Nature Tourism in Sandakan, Sabah, Malaysian Borneo*. International Centre for Responsible Tourism, University of Greenwich: Unpublished MSc thesis.

Cunha, A. A. (2010). Negative effects of tourism in a Brazilian Atlantic Forest National Park. *Journal for Nature Conservation*, 18: 291–295.

Cunha, A. A. and Vieira, M. V. (2004). Present and past primate community composition of the Tijuca Forest, Rio de Janeiro, Brazil. *Neotropical Primates*, 12: 153–154.

Cunha, A. A., Grelle, C. E. V., and Vieira, M. V. (2006). Preliminary observations on diet, support and habitat use by two non-native primates in an urban fragment of Atlantic Forest: the capuchin monkey (*Cebus sp.*) and the common marmoset (*Callithrix jacchus*) in the Tijuca Forest, Rio de Janeiro, Brazil. *Urban Ecosystems*, 9: 351–359.

de la Torre, S., Snowdon, C. T., and Bejarano, M. (2000). Effects of human activities on pygmy marmosets in Ecuadorian Amazon. *Biological Conservation*, 94: 153–163.

Duffus, D. A. and Dearden, P. (1990). Non-consumptive wildlife-oriented recreation: A conceptual framework. *Biological Conservation*, 53: 213–231.

Fuentes, A. (2010). Natural cultural encounters in Bali: monkeys, temples, tourists, and ethnoprimatology. *Cultural Anthropology*, 24 (4): 600–624. doi: 10.1111/j.1548–1360.2010.01071.x.

Fuentes, A., Shaw, E., and Cortes, J. (2007). Qualitative assessment of macaque tourist sites in Padangtegal, Bali, Indonesia, and the Upper Rock Nature Reserve, Gibraltar. *International Journal of Primatology*, 28 (5): 1143–1158. doi: 10.1007/s10764-007-9184-y.

Garbutt, N. (2009). *Mammals of Madagascar*. New Haven, CT: Yale University Press.

Harcourt, A. H. and Stewart, K. J. (2007). *Gorilla Society: Conflict, Cooperation and Compromise Between the Sexes*. University of Chicago Press.

Hartup, B. K. (1994). Community conservation in Belize: Demography, resource uses, and attitudes of participating landowners. *Biological Conservation*, 69: 235–241.

Higham, J. E. S. (ed.) (2007a). *Critical Issues in Ecotourism: Understanding a Complex Tourism Phenomenon*. Oxford: Elsevier.

Higham, J. E. S. (2007b). Ecotourism: competing and conflicting schools of thought. In: J. E. S. Higham (ed.), *Critical Issues in Ecotourism: Understanding a Complex Tourism Phenomenon*. Oxford: Elsevier, pp. 1–19.

Higham, J. E. S. and Shelton, E. J. (2011). Tourism and wildlife habituation: Reduced population fitness or cessation of impact? *Tourism Management*, 32: 1290–1298.

Hodgkinson, C. (2009). *Tourists, Gorillas and Guns: Integrating Conservation and Development in the Central African Republic*. Unpublished PhD Dissertation, University College London.

Ilambu, O. (1998). *Impact de la guerre d'octobre 1996 sur la distribution spatiale de grands mammifères (gorilles et éléphants) dans le secteur de haute altitude du Parc National de Kahuzi-Biega*. Programme report of Wildlife Conservation Society in Kahuzi-Biega National Park.

Kinnaird, M. F. and O'Brien, T. G. (1996). Ecotourism in the Tangkoko Dua Saudara Nature Reserve: Opening Pandora's Box? *Oryx*, 30: 65–73.

Knight, J. (2009). Making wildlife viewable: habituation and attraction. *Society and Animals*, 17: 167–184.

Knight, J. (2010). The ready-to-view wild monkey: The convenience principle in Japanese wildlife tourism. *Annals of Tourism Research*, 37 (3): 744–762. doi:10.1016/j.annals.2010.01.003.

Knight, R. L. and Cole, D. N. (1995). Wildlife responses to recreationists. In: R. L. Knight and K. J. Gutzwiller (eds.), *Wildlife and Recreationists: Coexistence through Management and Research*. Washington, DC: Island Press, pp. 51–70.

Kuze, N., Dellatore, D., Banes, G. L. *et al.* (2012). Factors affecting reproduction in rehabilitant female orangutans: Young age at first birth and short inter-birth interval. *Primates*, 53 (2): 181–192.

Lanyero, F. (2011). UWA lowers gorilla tracking fees. *Daily Monitor*, www.monitor.co.ug/News/National/-/688334/1164758/-/c1i12wz/-/index.html. (accessed Sept. 19, 2011).

Lepp, A. (2002). Uganda's Bwindi Impenetrable National Park: Meeting the challenges of conservation and community development through sustainable tourism. In: R. Harris, T. Griffin, and P. Williams (eds.), *Sustainable Tourism: A Global Perspective*. Amsterdam, Netherlands: Elsevier Butterworth-Heinemann, pp. 211–220.

Macfie E. J. and Williamson, E. A. (2010). *Best Practice Guidelines for Great Ape Tourism*. Gland, Switzerland: International Union for Conservation of Nature and Natural Resources.

MacKenzie, C. A. (2012). Trenches like fences make good neighbours: Revenue sharing around Kibale National Park, Uganda. *Journal for Nature Conservation*, 20 (2): 92–100.

Maekawa, M., Lanjouw, A., Rutagarama, E., and Sharp, D. (2013). Mountain gorilla tourism generating wealth and peace in post-conflict Rwanda. *Natural Resources Forum*, 37: 127–137.

Mallapur, A. (2013). Macaque tourism: Implications for their management and conservation. In: S. Radhakrishna, M. A. Huffman, and A. Sinha (eds.), *The Macaque Connection: Cooperation and Conflict between Humans and Macaques*. New York: Springer, pp. 93–105.

Maréchal, L., Semple, S., Majolo, B. *et al.* (2011). Impacts of tourism on anxiety and physiological stress levels in wild male Barbary macaques. *Biological Conservation*, 144 (9): 2188–2193.

McCarthy, J. (1999). Nature based tourism. Case study: Gunung Leuser, Indonesia. In: *Institute for Sustainability and Technology 2000*. Perth, Australia: Murdoch University.

McNeilage, A. (1996). Ecotourism and mountain gorillas in the Virunga Volcanoes. In: V. J. Taylor and N. Dunstone (eds.), *The Exploitation of Mammal Populations*: London: Chapman & Hall, pp. 334–344.

Mittermeier, R. A., Louis, E. E. Jr., Richardson, M., *et al.* (2010). *Lemurs of Madagascar*. Washington, DC: Conservation International.

Muehlenbein, M. P., Ancrenaz, M., Sakong, R., *et al.* (2012). Ape conservation physiology: Fecal glucocorticoid responses in wild *Pongo pygmaeus morio* following human visitation. *PLoS ONE*, 7 (3): e33357.

Pole Pole Foundation. (2012). *History of Gorilla Tracking*. www.polepolefoundation.org/kbnp.php (accessed Apr. 6, 2011).

Pusey, A. E., Wilson, M. L., and Collins, D. A. (2008). Human impacts, disease risk, and population dynamics in the chimpanzees of Gombe National Park, Tanzania. *American Journal of Primatology*, 70 (8): 738–744.

Ram, S., Venkatachalam, S., and Sinha, A. (2013). Changing social strategies of wild female bonnet macaques during natural foraging and on provisioning. *Current Science*, 84: 781–790.

Rijksen, H. D. and Meijaard, E. (1999). *Our Vanishing Relative: The Status of Wild Orang-Utans at the Close of the Twentieth Century*. Kluwer Academic Publishers.

Salopek, P. F. (1995). Gorillas and humans: An uneasy truce. *National Geographic*, 188 (4): 72–83.

Singleton, I. and Aprianto, S. (2001). *The Semi-wild Orangutan Population at Bukit Lawang: A valuable 'ekowisata' resource and their requirements*. Unpublished paper presented at the Workshop on Eco-tourism Development at Bukit Lawang, Medan, Indonesia, April.

Tapper, R. (2006). *Wildlife Watching and Tourism: A study on the benefits and risks of a fast growing tourism activity and its impacts on species*. UNEP/CMS Secretariat, Bonn, Germany.

Tran, M. (2013). Conflict in DRC Congo threatens chimpanzee tourism programme. *The Guardian*, Feb. 8. www.guardian.co.uk/global-development/2013/feb/08/conflict-congo-drc-chimpanzee-tourism/print (accessed Feb. 19, 2013).

Treves, A. and Brandon, K. (2005). Tourism impacts on the behavior of black howler monkeys (*Alouatta pigra*) at Lamanai, Belize. In: J. Paterson and J. Wallis (eds.), *Commensalism and Conflict: The Primate-human Interface*. Winnipeg, Manitoba: Hignell Printing, pp. 147–167.

Tumusiime, D. M. and Svarstad, H. (2011). A local counter-narrative on the conservation of mountain gorillas. *Forum for Development Studies*, 38 (3), 239–265.

Uganda Wildlife Authority (2013). Visitor tariffs. www.ugandawildlife.org/about-uganda-master/visitor-tariffs (accessed Oct. 15, 2013).

Wallis, J. and Lee, D. R. (1999). Primate conservation: the prevention of disease transmission. *International Journal of Primatology*, 20: 803–825.

Weber, A. W. (1987). *Ruhengeri and its Resources: An Environmental Profile of the Ruhengeri Prefecture, Rwanda*. Kigali, Rwanda: Ruhengeri Resource Analysis and Management Project.

Weber, A. W. (1993). Primate conservation and ecotourism in Africa. In: C. S. Potter, J. I. Cohen, and D. Janczewski (eds.), *Perspectives on Biodiversity: Case Studies of Genetic Resource Conservation and Development*. Washington, DC: American Association for the Advancement of Science, pp. 129–150.

Wollenberg, K. C., Jenkins, R. K. B., Randrianavelona, R., *et al.* (2011). On the shoulders of lemurs: pinpointing the ecotouristic potential of Madagascar's unique herpetofauna. *Journal of Ecotourism*, 10 (2): 101–117. doi: 10.1080/14724049.2010.511229.

Wright, P. C. and Andriamihaja, B. A. (2002). Making a rain forest national park work in Madagascar: Ranomafana National Park and its long-term research commitment. In: J. Terborgh, C. van Schaik, M. Rao, and L. Davenport (eds.), *Making Parks Work: Strategies for Preserving Tropical Nature*. California: Island Press. pp. 112–136.

Wyman, M. and Stein, T. (2010). Examining the linkages between community benefits, place-based meanings, and conservation program involvement: A study within the Community Baboon Sanctuary, Belize. *Society & Natural Resources*, 23 (6): 542–556.

Xiang, Z. F., Yu, Y., Yang, M. *et al.* (2011). Does flagship species tourism benefit conservation? A case study of the golden snub-nosed monkey in Shennongjia National Nature Reserve. *Chinese Science Bulletin*, 56 (24): 2553–2558. doi: 10.1007/s11434–011–4613-x.

Zhao, Q. K. (2005). Tibetan macaques, visitors, and local people at Mt. Emei: Problems and countermeasures. In: J. Paterson and J. Wallis (eds.), *Commensalism and Conflict: The Human-primate Interface*. Norman, OK: American Society of Primatologists, pp. 376–399.

Zhao, Q. K. and Deng, Z. (1992). Dramatic consequences of food handouts to *Macaca thibetana* at Mount Emei, China. *Folia Primatologica*, 58: 24–31.

Index